# JavaScript Three–Tier Architectures in AWS with React, Node and MongoDB

## Eric Bush

**Design, code, test, deploy, and manage in Amazon AWS**

*"Give me six hours to chop down a tree and I will spend the first four sharpening the axe."*
*- Abraham Lincoln*

BLUE SKY PRODUCTIONS INC.

Note: This book is a re-write of a previously published book titled – "Node.js, MongoDB, React, React Native Full-Stack Fundamentals and Beyond". Enough changes were made to the book that it was felt a title change would be warranted.

Front cover photo by Oleksandr Hruts on iStock purchased license.
Rear cover photo by Joe Straker on Unsplash.

Please send any comments to the author at jsdevstack@outlook.com.

_____

Paperback ISBN-10: 0-9971966-9-6
Paperback ISBN-13: 978-0-9971966-9-6

# Contents

# Acknowledgements

This book covers some phenomenal technology frameworks. One of those is Node.js. I am in awe of Ryan Dahl for his having conceived of it in the first place and for his having built it on top of some great work by others. Thank you, Ryan! I want to thank all those who have collaborated on the modules available through npmjs.com, they really make Node such a capable platform. This is open source at its best.

I also want to thank TJ Holowaychuk for his contribution to Node through his work in creating the Express web framework. It really does a great job organizing and simplifying the code needed to implement a Rest Web API.

I want to thank the brilliant people behind all the Amazon Web Services infrastructure. From those who envisioned it in the first place, to those who build it and those who run it 24 hours a day. Every year it adds amazing new capabilities for all to benefit from. This liberates application developers to code their applications and provide customer value. There is huge benefit from the many underlying services that do a lot of the heavy lifting for you.

Another great piece of technology is that of MongoDB. I want to thank those who developed and maintain MongoDB and thank the Atlas team for making it easy to host MongoDB in a PaaS environment.

I acknowledge all the ongoing efforts that to build, maintain, and evolve React and React Native. The React framework from Facebook is a perfect fit in the JavaScript stack.

I express my love to my sweetheart Loradel. Thank you for finding me. I will cherish you forever. ILYTM!

JavaScript Three-Tier Architectures in AWS with React, Node and MongoDB

# Preface

This book is not a copy of the API documentation of each framework but offers fresh content and real insight into the frameworks. Half the battle of learning any new technology, is learning the nuances and tools to get it all working. This is where this book will come in handy. Rather than just dumping a lot of code on you, I take the time to explain the development process and what tools you can use.

Many people will turn to the first few pages of a book to determine whether it will meet their needs. If you are just now deciding about whether you want to invest the time and money necessary to build up your skills in JavaScript full-stack development, I will state that you are doing well to have found this book. Let me give you a brief marketing pitch about why you need to pursue investing in full-stack JavaScript skills.

Technology trends come and go, but what this book covers are here to stay. The JavaScript technologies covered in this book have each been on a steady climb of adoption. What is exciting is that the UI, services, and database can come together in a unified stack, all using the JavaScript programming language. This book pulls together a lot of information that you might have a hard time finding on your own. With this book, you will not only learn about these technologies, but will also gain a longer lasting foundational understanding of a three-tier architectural design.

There is a huge development community behind this tech stack, so you will be in good company. Many top companies are finding success doing full-stack JavaScript development today. You will have to study hard, but it will be worth it.

**Level of skill required**
To get the most out of this book, you will need to have a basic understanding of JavaScript and HTML. If you don't already know the basic concepts of those, you can learn from online written content, video courses or another good book on the topics.

You will learn the basic concepts of a three-tier architecture. I always lead off with some material that covers the basic concepts that are involved in the architecture. When you get to the actual code implementation of the sample app, you will have this insight and not be lost.

If you are already familiar with the basics of the MERN stack, this book can get you to the next stage and lead you toward professional enterprise-level development. I make sure to dive deep into each topic and show some real implementation and deployment details. This book may save you time trying to figure out ways to test and secure your applications.

I have put myself in your shoes to give you the information that you need to gain traction. One of my goals is to give you knowledge that will help you get and keep a job.

**AWS and other knowledge**
This book utilizes AWS for its implementation. You get to learn about configuring and deploying to that environment. As mentioned, this book also contains substantial foundational instruction, such as details on security measures and information on how to perform testing.

**Choices**
This is certainly a great time to be a software developer. One of the things that makes it so great is the myriad of open-source projects available to leverage. There are now frameworks that exist to help you with every layer of your architecture from backend services to front-end GUI. There are literally hundreds of technology options for you to use in your development of a full-stack JavaScript application.

I will help you get started by narrowing things down to some of the essentials that are needed to get you up and running. In doing so, I will have more room to cover advanced topics to help you build quality enterprise applications.

I had to make some tough decisions on which frameworks and tools to focus on in this book and have done so based on the following criteria:

1. **It must be simple and quick to obtain value.**
2. **It must have wide adoption, with committed support for growth and future relevance.**
3. **It must have broad platform reach.**

After mastering what is covered in this book, you can turn around and evangelize full-stack JavaScript development as a reality today. It is great already and it will only get better.

Happy coding,
Eric Bush

*Note: Based on my personal preferences, I made the choice of React over Angular. For me, it is about how connected the code is to the actual UI rendering markup. React also is something that lends itself to being easily testable.*

# About the Author

Eric Bush began his professional career programming in assembly language for embedded hardware microcontrollers in airplane flight systems. He then worked developing CAD software on what were known as workstation computers. When Microsoft Windows came along, Eric started developing some of the first desktop applications available. After that, he transitioned back to client-server applications for used on Microsoft Windows Server. Finally, he made the move to enterprise cloud development in 2009 and has focused on full-stack JavaScript technologies, using MongoDB, Node.js, and React since 2014.

He has worked for some well-known companies including Walmart Labs, Microsoft, Tektronix, Mentor Graphics, Nike, Intel, and Boeing.

While working for the retail giant Walmart, Eric worked as a full-stack developer. He worked on backend scalable JavaScript Node.js cloud infrastructure and on front-end React code for the Walmart.com website store details pages. The backend work he developed serves up high-traffic product searches and cart capabilities used by millions of people each day. The work included many of the HTTP Rest endpoints currently in use. He was also involved in the DevOps work, following enterprise practices of scalability, quality, security, reliability, availability, and performance.

As a consultant at DocuSign in downtown Seattle Eric developed an internal enterprise application suite of tools to improve employee efficiency. This involved full-stack JavaScript with HTML/Angular that talked to a backend Node.js server which was integrated with a MongoDB database.

The current product development work Eric wrote is a platform to carry out performance testing of HTTP JSON Web API endpoints. Check it out at https://apizapi.com and sign up for free or paid usage.

As a consultant with Waypoint (https://waypoint.co/), Eric puts his considerable breadth and depth of knowledge into helping companies evaluate modern PaaS cloud architectures. Waypoint provides services for technical due diligence for mergers, acquisitions, and investments.

You can reach Eric via email at jsdevstack@outlook.com.

# Introduction

This book presents the technologies and best practices that span all the layers of a software development architecture. The core of this book focuses on specific JavaScript frameworks used to implement each architectural layer. Everything is based on MongoDB, Express, React, and Node. I will walk you through the design and development of a sample application to teach you the best practices in architecting, coding, testing, securing, deploying, and managing a RESTful Web Service and a Single Page Application (SPA).

This first section defines several key terms and introduces you to the sample application. This is important context you will need before you dive deeper into the materials presented in the rest of the book.

## What is a development stack?

A development stack is the collection of computer languages, frameworks and technologies used to construct a software application. These are the technologies you piece together for the front-end, backend, and everything along the side. A "development stack" is not the same thing as a software architecture but is related. A "software architecture" is a pattern of components and layers for building software. In other words, how to make use of the coding language and frameworks and host it to run. This book utilizes a particularly popular set of technologies referred to as the MERN (MongoDB, Express.JS, React, and Node.js) development stack. The main overall architecture pattern followed is this book is referred to as a three-tier architecture.

MongoDB, Express.JS, React, and Node.js are each called platforms, or frameworks. Framework take code, or some type of markup/configuration and execute it at a higher level of abstraction above the operating system level. In essence, they save you time that you would otherwise need to spend writing capabilities. Just image if you had to write your own database!

The MERN stack specifies four technologies. However, many more technologies come into play. The Node.js framework really gets its capabilities from modular plug-ins that extend its basic capabilities. Frameworks are usually extensible, meaning they are built to be extended in functionality. You will learn about many of the important extensions that can be utilized to extend the capabilities of Node.js, Express, and React.

*Note: I have chosen to develop with MongoDB as my database of choice. This is because it has a rich set of capabilities and is popular. MongoDB can be hosted in AWS as a PaaS service, which is what we will do.*

# The three-tier architecture

It is a good idea to modularize your code and build in layers so that you can more easily assemble your code. If you were building a house, you would not build it as one entangled mess. Instead, you would first lay a foundation and construct in layers. Each of the needed modules would fit together - flooring, walls, heating, plumbing, electrical, etc. Software construction can be thought of in a similar way and can be independently built and assembled.

Architecting software in layers and writing code as modules (also referred to as components) helps with testing, debugging, and makes it much easier to incorporate future enhancements. The latest buzzword in software construction is "microservices." You may also have heard of the concept of a service-oriented architecture (SOA). These are patterns that encourage separation of code into discrete pieces of functionality.

One proven approach to developing a software application is to divide it up into three distinct layers. The three layers (or tiers) are named data, service, and presentation. You can even utilize patterns such as model view controller (MVC) within a three-tier architecture.

Definitions of the three tiers are listed below. An assessment is provided on how each MERN technology fits into those layers. While this list gives you the briefest of introductions to each of these, the three parts of this book go into the details of each layer.

- **Data Layer (MongoDB):**
  This layer is where you persist your data in a database management system (DBMS). Data can be stored, retrieved, updated, or deleted. MongoDB is categorized as a "document-based" database because of the form in which it stores data. This book will cover a JavaScript API used to interact with MongoDB. A management portal will be used to work with MongoDB.

- **Service Layer (Node.js+Express.js):**
  This layer is where a web service API is exposed and contains business logic. It is the doorway to the data stored in the data layer. The service layer performs workflows that require more complicated computations and sequencing of operations. This is also where you might have HTTP endpoints or code that is scheduled to run periodically (e.g., AWS Lambda functions). Node.js was specifically designed for writing scalable server-side web applications. Express.js is used as a Node add-on module to simplify the building of a web service layer. Express gives support for HTTP request routing and sending back of responses.

- **Presentation Layer (React/HTML/CSS):**
  This layer provides the capability to display data and handle interactions. This could be on a computer or on a mobile device of some type. React provides a way to code up components that render a UI by formulating the UI controls that drive the display.

*Note: In my opinion, the middle service layer choice of Node.js is something foundational to a JavaScript-based development stack and should not be swapped out. The other choices of MongoDB and React could be swapped out and replaced with other similar technologies. For example, since MongoDB is accessed through a Node module, you could easily find another solution using whatever database you prefer. React is the highest layer and could be replaced with any number of frameworks that allow you to make HTTP web requests to the Node service layer and do the data binding (e.g., Vue.js or Angular).*

# Using JavaScript everywhere

You can understand from the book title that 100% of the development done in this book is in JavaScript. There are considerable benefits with having one single language used across all three architecture layers. For one thing, the code will all look the same and be similarly refactored and maintained. You can also benefit from using the same testing framework across layers. Some of the libraries you download can even be used in both React and Node.js code, or at least they have equivalent counterparts.

JavaScript has huge momentum and has a lot of online learning resources as well as tons of available open-source code to pull in. You will need to become adept at searching online before you start coding anything. Always search first, as there just might be a module out there that will do exactly what you need.

*Note: You can decide to use TypeScript with your full-stack coding. The code for the usage of Node.js can still be the same as found in this book as well as the usage of React and MongoDB. There may be minor differences, but those can be researched on the internet.*

*Note: JavaScript Object Notation (JSON) will be used as the data-interchange format. This is an excellent fit with a JavaScript application.*

*Note: There are several ways to write asynchronous code in JavaScript. You could use what are called Promises, or you could use callbacks, or you could use the async/await keywords. This book will stick with callbacks, but any code could easily be transitioned.*

# Using public cloud infrastructure

Every piece of code and all data storage should be hosted on public cloud infrastructure. This gives you characteristics like fast deployment and elastic scalability. This book shows you how to construct an application that hosts everything in AWS.

As far as cloud hosting strategies are concerned, they can be split into three categories. These are IaaS, PaaS, and SaaS. The "aaS" part of each acronym stands for "as a Service."

In **IaaS**, the "I" is for "Infrastructure" and means you can utilize the lowest level virtual machines (VMs) and have complete control of what operating systems and software you deploy. This means you are responsible for all update patching etc.

In **PaaS**, the "P" is for "Platform." This variation makes your life a bit easier as a developer. It still gives you the flexibility you need over machine deployment and scaling, but it frees you from the day-to-day maintenance of machines. You remove most of the work of IaaS needed to install and manage your own infrastructure. PaaS can be much simpler to setup and operate. After all, you want to spend your time on application development, not on infrastructure provisioning and maintenance.

In **SaaS**, the "S" is for "Software" and indicates a complete offering such as Salesforce.com, Azure SharePoint, or Amazon WorkMail. Think of this as an application that formerly would have been individually installed on your server at work or home computer, that can now be run in the cloud and be accessed from anywhere.

*Note: Related to PaaS there is also a hosting model called "Serverless". This means you don't even realize that there are machines at all. You just give code and then everything is sent out to be run. Examples of this are AWS Lambda and AWS Fargate.*

*Note: You will learn more about how deployments work in PaaS. AWS handles OS and hardware upgrades for you in a rolling manner that keeps your service up while that happens. And your deployments can use the same technique.*

# Microsoft Visual Studio Code

If you are out in the real world trying to make a living by writing software, the reality is that you must be able to combine many technologies together at once and target many different platforms. Central to this is the code editor that you use. In my opinion, Visual Studio Code (VS Code) is a great tool to use for editing your code. This is the tool I used to code and debug the sample application used in this book. You can certainly choose whatever code editor you like and are not required to use VS Code.

VS Code offers a rich editing environment as well as a set of integrated features like source code control, debugging. VS Code also has the capability to launch tasks using tools like Gulp, without needing to jump to a command line prompt. These tools can be used to automate the build and testing steps that you run frequently. There is a place you can go to download extensions that you might find useful.

Using VS Code, you can create a project, run it locally on your machine, and have access to IntelliSense, debugging, git commands, and app publishing. VS Code can be installed on Mac OS X and Linux as well as on Windows.

If you are starting from scratch and don't have any development tools installed, go ahead and install the following:

✓ **Visual Studio Code**. Did I mention how cool this is?
✓ **GitHub Desktop/Git**. This is used for source code control, but it can also be the means of deploying to AWS infrastructure through GitHub Actions. Git will be installed for you with GitHub Desktop.
✓ **Node.js** from https://nodejs.org/. Get a stable version that is labeled LTS. This will also install NPM (a package manager for JavaScript projects) at the same time.

# The NewsWatcher sample application

The coding concepts in this book are illustrated using a sample application I called NewsWatcher. This sample application will help you understand how everything comes together across all components of the application development stack and layers of the architecture. I will now walk you through the capabilities and architecture of NewsWatcher.

The basic capability of NewsWatcher allows a user to set up keywords to filter through news stories that are provided by a feed from the New York Times. The user can view, share, and comment on the news stories. It will be developed as a website to run on a desktop or mobile device browser. The book will not cover this, but you could take the React code and modify it a bit to create a React Native application to run natively on an Android or iPhone.

It is important to spend a short amount of time to talk about the vision of what NewsWatcher was meant to be. I needed a vision statement and some sketches of what it should look like. A rough plan for the development was laid out from there, including a prioritized backlog of features. I happen to prefer following the Kanban process to proceed through the work in a backlog. Here is the vision statement that I put together for NewsWatcher:

**Vision and Value Proposition Statement**
*Users of NewsWatcher will get news personally served up to them. NewsWatcher will be their trusted advisor to alert them and save important information, without their needing to constantly scan news outlet feeds. Users can do things like adjust filters to get the news they are interested in, see what their friends are looking at, and share comments about news stories.*

This vision statement is the longer-term goal of what to shoot for and would take many iterations to realize. From this vision statement, I was able to distill the vision down into a list of features to iterate on. The list that follows is what I came up with for NewsWatcher, and contains some requirements (I may have snuck in a bit of implementation detail also).

# Introduction

**Prioritized feature list that fulfills the vision:**
1. I can set up multiple news filters to hold news stories.
2. I can give my filter a title to identify it.
3. I can set up keywords for each filter with Boolean operations of AND and OR. For example, "Treatments AND insomnia", "Dogs OR Cats".
4. The filters are periodically scanned by the NewsWatcher backend service to search for news stories that match from a central pool of collected stories from a news service provider. Client-side processing will be freed up. Finding of stories proceeds even while devices are turned off or unable to connect to the internet. The server-side runs a periodic pull of stories from a news feed service.
5. News stories in folders can be scrolled through and clicked on to take me to the story content. The 20 most recent stories are shown for each filter.
6. Upon opening NewsWatcher, I see the home page of news stories.
7. I can click on and open a news folder to see the stories in it.
8. Stories in filter folders can be saved to an archive folder.
9. I can delete individual saved stories I select.
10. I can delete the filter itself with all its stories.
11. My account settings are stored server-side but cached on the device client-side also.
12. I can delete my NewsWatcher account.
13. There are globally shared news stories in which I can share a story to be seen by all NewsWatcher users. All other NewsWatcher users can see and comment on these stories. Stories will be kept for a week before being deleted. A maximum of thirty shared stories can exist; new ones will bump older ones off before the week is up. Any given user is limited to sharing five stories a week. There can be thirty comments per story kept. Offensive language in comments is blanked out.
14. I can view my news on whatever device I am logged in with.
15. The login expires and requires me to log in again periodically.
16. Two-factor authentication requires a text message code to be entered.
17. I can set actions on news folders to alert or email me when new entries are found.
18. I can set limits on when news alerts are sent to me. For example, wait at least ten minutes between alerts.
19. I can set times during the day when I don't want news alerts sent to me.
20. I can set NewsWatcher to download stories to my device when connected to Wi-Fi.
21. I can use a search entry box to do a search of all loaded stories.
22. I can designate the order I want to see the filters in.
23. I can email stories to others or share them via Twitter and Facebook.
24. iOS and Android phones have a Native app available in the app stores.

This book will only implement the core functionality of NewsWatcher to produce a Minimal Viable Product. Not everything will make it into the initial iteration, but it is nice to have a backlog ready to draw from.

## Wireframe Prototype

The following wireframe images were created in PowerPoint. PowerPoint has a great set of storyboard templates you can use to piece together a UI image and make realistic looking prototypes. There are also many nice websites you can use to sketch out a UI with.

### To create a PowerPoint prototype image:

1. Open PowerPoint and create a blank slide.
2. Click the **Storyboarding** tab, then click **Storyboard Shapes** on the ribbon to open the **Storyboard Shapes** library.
3. Start creating your prototypes by dragging shapes from the **Storyboard Shapes** library to your slide.

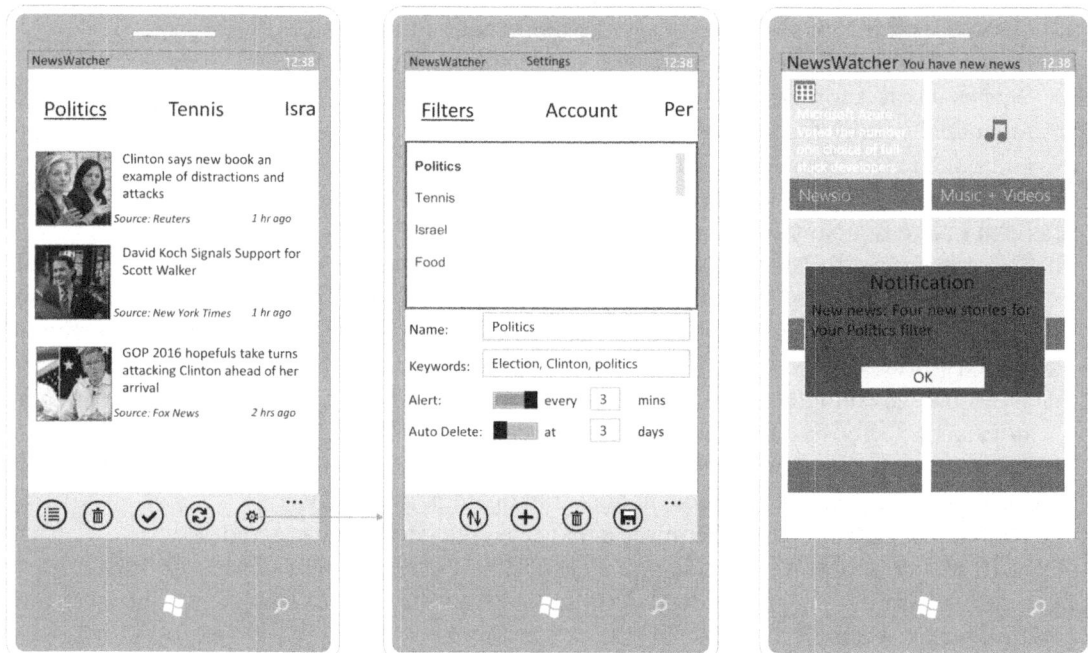

*Figure 1 - NewsWatcher wireframes*

**Note:** *If you are using PowerPoint 2013, you need to either have installed Visual Studio 2013 or later, or Team Foundation Server Standalone Office Integration 2015 on the same machine to create and modify storyboards. PowerPoint 2016 is installed with everything you need.*

### A peek at the result

Here are a few screenshots of what the actual Web UI ended up looking like on a mobile device. As any good design will evolve, it doesn't look quite like the wireframes.

# Introduction

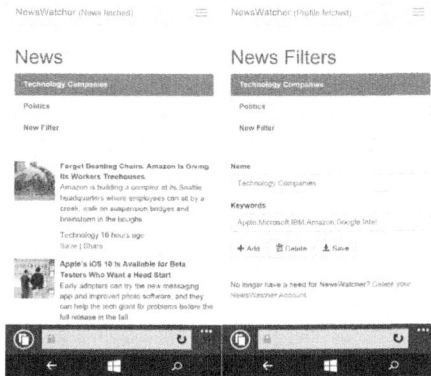

*Figure 2 - NewsWatcher concept representing UI on a mobile phone*

You can try the completed sample browser app on your mobile device or desktop by going to https://www.newswatcher2rweb.com. You can even try it out on your mobile device.

All the code for the sample application can be found on GitHub. You can download a ZIP file from https://github.com/eljamaki01/NewsWatcher2RWeb.

## A peek at the architecture and deployment topology

The three-tier architecture of NewsWatcher is depicted in the following diagram. The arrows show what parts of the architecture call into other parts.

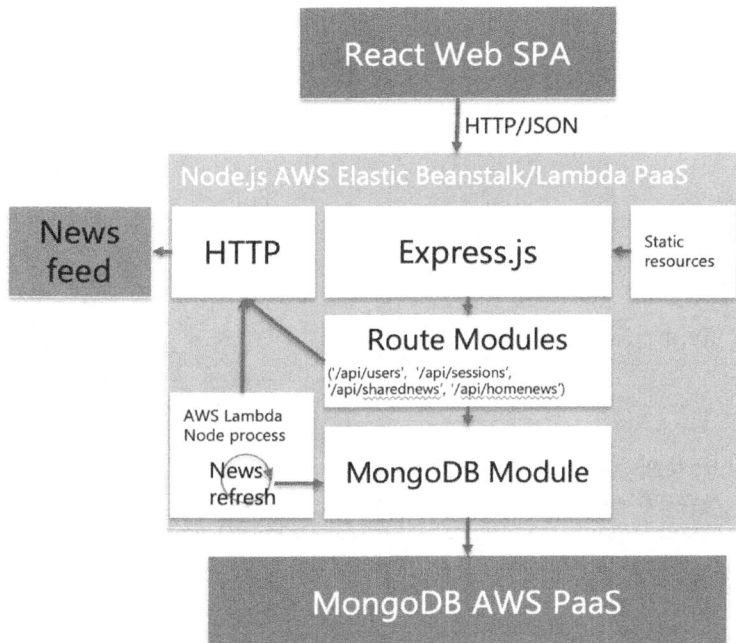

*Figure 3 - NewsWatcher architecture diagram*

# JavaScript Three-Tier Architectures in AWS with React, Node and MongoDB

The machine topology required to implement a three-tier application can vary widely and may even change over time. This is because the topology is really a function of the scaling that your application needs to achieve. You certainly would not start off with a topology that could scale to millions of users as it would be unnecessarily complicated to provision and maintain. You only pay for what you need; however, some things would be in place that will still cost you money.

The following diagram gives you a look at a starting point for an architecture topology. This topology gives you a fair amount of scaling to handle many users. Several parts here scale very well, due to horizontal usage of AWS virtual machines – Elastic Beanstalk, Lambda and the MongoDB instance.

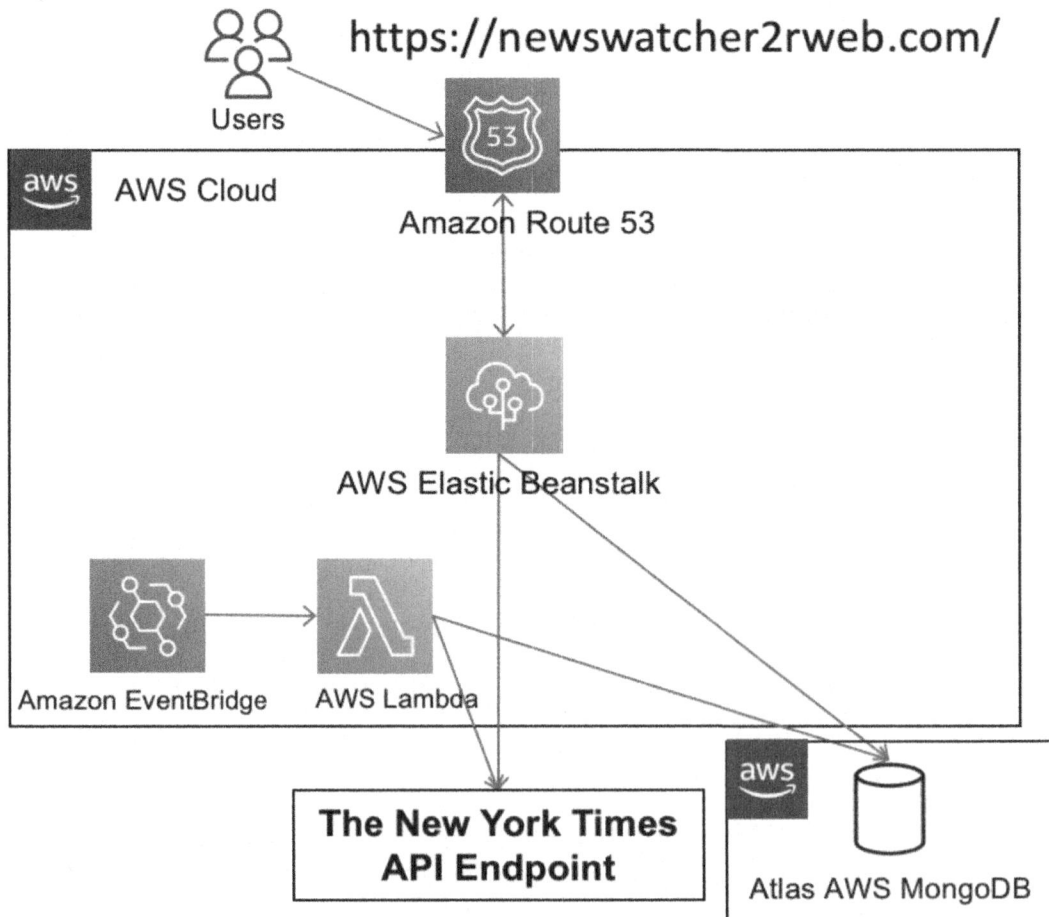

*Figure 4 - NewsWatcher service hosting topology*

# PART I: The Data Layer (MongoDB)

The first part of this book teaches you what a data layer is and how to implement one using MongoDB. This will be hosted using DBaaS/PaaS in AWS using the MongoDB, Inc., Atlas offering.

The first step in creating a data layer involves the modeling of the types of data you want to store. After that, you will learn what about MongoDB and how it can be used to implement a data model.

To finish out the first part of the book, you will construct a data layer that will support the needs of the NewsWatcher sample application. You will end up with a fully capable data layer that can be utilized by the service layer of the application architecture being built in the second part of this book.

To give full coverage to the topic, I will also cover what it means to manage the day-to-day operations of a MongoDB deployment.

# Chapter 1: Fundamentals

This chapter presents the concepts of a backend data storage system. I will show you what one is composed of and what capabilities are essential. After looking at backend storage systems in general, I will delve into the specifics of MongoDB and how it is integrated into the NewsWatcher architecture to fulfill the requirements for the backend data storage.

*Note: I will show you how to get MongoDB set up in a Platform as a Service (PaaS) environment with some basic development and monitoring. However, topics like database administration are out of scope for this book. Please see MongoDB and Atlas help resources for more on administrative topics.*

# 1.1 Definition of the Data Layer

The data layer of a software application architecture provides for the persistent storage of information. Anything of importance can be stored there, such as customer account information, inventory, orders, audit logs, tax computation tables, and anything else that can conceivably be stored in electronic form.

The term database management system (DBMS) is used to describe the commercial offerings that implement a data layer at the lowest level. Each available DBMS has capabilities that differentiate it from the others. MongoDB from MongoDB, Inc., is one DBMS system that is developed as an open-source project as well as being a commercial offering. See https://www.mongodb.com/ for their various offerings and capabilities. MongoDB Atlas (https://www.mongodb.com/cloud/atlas) is a commercial PaaS offering from MongoDB, Inc. You may find other companies that also host versions of MongoDB in a PaaS environment, such as AWS.

## DBMS capabilities

DBMS implementations can be placed into several possible categorizations. One possible way to categorize them is as follows: relational, key-value, hierarchical, graph, object-oriented, or document-based. Some DBMSs can also be said to be of the "NoSQL" type. Each category exists for a specific reason, and you will want to look at your specific needs and choose the DBMS technology that fulfills your needs the best.

A DBMS will provide the physical storage medium where the data resides. It should be robust to withstand a power failure. In a cloud-hosted DBMS, data eventually makes its way to being stored on non-volatile cloud storage drives in a secure data center. Data storage might even be geo-replicated between distant data centers for redundancy and high availability.

# PART I: The Data Layer (MongoDB)

A DBMS will provide for the storage and access features. These include the creation, retrieval, updating, and deletion of data. The acronym "CRUD" is commonly used to refer to these operations. A DBMS will typically provide the following features: Data storage, Transactions, Attribution, Auditing, Authorization, Data access, Indexing, Encryption, Notification, Programmability, Schematization, Security, Transformation and Validation.

Even though the DBMS itself provides some type of low-level programmatic access, there is often an additional layer referred to as a Data Access Layer (DAL). The low-level access is typically some wire protocol, such as TCP/IP. The DAL is a library for each of the supported programming languages (e.g., PHP, Java, Go, JavaScript etc.) on top of the wire protocol.

There are community written DALs as well as those that are commercially available. You can even write your own. The DAL will create an abstraction layer that hides the complexities of the data storage technology and may even allow you to switch to a different backend DBMS without affecting the upper layers of your application.

A DAL can greatly simplify your access code by allowing an interaction object structure that might not exist in the actual data storage system itself. For example, some DBMSs don't provide schematization of data. If you need that, you can possibly get it through a DAL. The following figure shows all the sub-layers within the data layer.

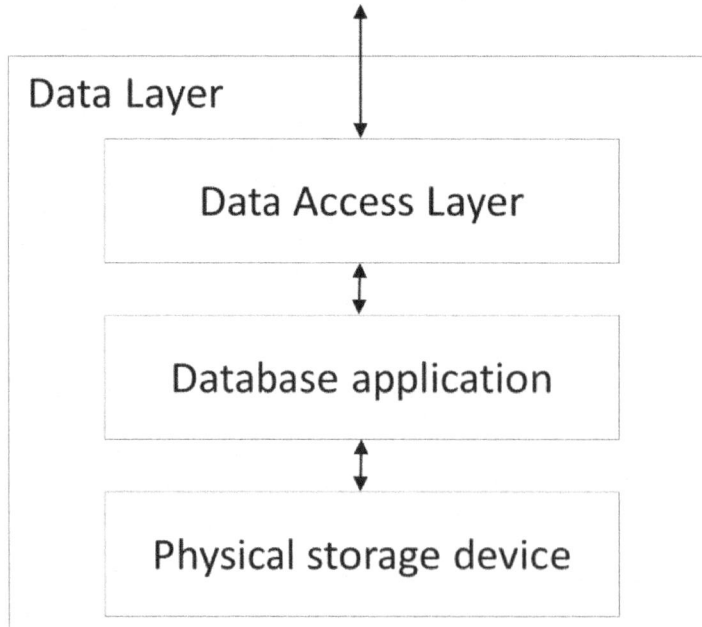

*Figure 5 - Data Layer with sub-layers*

# 1.2 Data Layer Design Process

Before storing any data in your data layer, you need to create models for what that data should look like. Knowing what you would like to store is the first step to take. With that information determined, you can model what the structural form of the data would be. This will require some thoughtful design to be able to organize your data in an efficient way. You will need to diagram a model for your data to aid in understanding all the nuances of the record types with their properties and relationships.

Often the data in a data layer outlives the applications that were written to expose it. User interfaces come and go, but backend data systems seem to live on. It can sometimes be hard to go back and modify your data model and actual storage after you have started using it, so take your time and design it for future growth.

*Note: Even when the lifetime of the DBMS technology platform is finally reached, the data may remain valuable. If you decide to migrate to a newer DBMS technology, the data can be exported from your old DBMS and then be reimported into your new DBMS. Most systems provide capabilities to export and import data in bulk. Some systems even provide an ongoing syncing capability for the transfer and transformation of the data.*

## Data layer planning

Make sure to utilize experts along the way before rolling anything out into a production environment. The following questions are crucial as you work through the initial data layer design. The answers to these questions should be carefully determined.

Data layer planning questionnaire:
- Is the data shared? If so, how is it shared between customers, applications, and processes?
- Will you have single-tenant or multi-tenant storage? What isolation of data is necessary?
- What are your data security and privacy requirements?
- Will you be storing data that is classified as high business impact (HBI) or storing any personally identifiable information (PII)? If the data is compromised, what are the legal ramifications? What compliance standards do you need to follow – SOC 2 Type 2, ISO 27001, PCI, HIPAA/HITRUST, FIRPA, GDPR/CCPA?
- Do you need data access roles to control access to the data?
- Are there any periodic processing jobs that run that will access the data?
- Will parallel access to a single record cause concurrency issue? Is there some type of locking or serialization of access required? Would request queuing or optimistic concurrency control (OCC) suffice?

# PART I: The Data Layer (MongoDB)

- Do you require transactional capabilities?
- Can writes be asynchronous or do you need immediate synchronous acknowledgment on a write?
- Is any lag time between a write and availability of that data for a read operation acceptable? In other words, do you need confirmation returned on all write operations that the data is truly written and available for read operations?
- What is the Service Level Agreement (SLA) requirement for each CRUD operation?
- What are the access volume and rates per minute? Will there be bursts of activity or is the activity evenly distributed across each day?
- What is the size of the data? How many records and how large will they become?
- How many users will access the data? What will be the needed data capacity per customer?
- Do you anticipate that the structure of your data will change?
- Do you need to keep a record of each data access, such as keeping an audit trail of accesses and changes?
- Will you be running data mining and business intelligence analysis over the data?
- Is there a strict data schema that needs to be validated against?
- What are the record types, contents, and relationships?
- What system dependencies are involved? How is the data transferred and how is it combined and verified?
- Are you storing media data that is in binary file form such as photos, movies, etc.?

# 1.3 Introducing MongoDB

MongoDB is classified as a NoSQL database. This means that it is non-relational and not schematized. No central schema catalog or data record structure definition is required, although one can exist if desired. Database records can basically be free form. Don't get concerned at this point about the non-relational nature of MongoDB records as you will learn how to set up a data model and how to validate data before it is stored in your data layer.

MongoDB is also classified as being document-based. It is a document-based database because it conceptually stores JavaScript Object Notation (JSON) documents. I say conceptually because it does not actually store JSON documents directly, but instead stores an internal binary representation. Documents are stored in Binary JSON (BSON) form.

Document-based storage with MongoDB really hits a sweet spot for modern application needs. It gives you the best of scaling and performance. MongoDB can be purchased as a PaaS-hosted service that runs on Amazon Web Services (AWS) infrastructure, in an easy to manage environment. You can, of course, download, install and run it yourself.

*Note: If you look at the database offerings of AWS, you will find several choices for data storage in the cloud, including DynamoDB, RDS/Aurora, Redshift, and S3. Each offering exists to fulfill different requirements, and each has its own advantages. MongoDB is offered through MongoDB, Inc. as well as in a "compatible" version with DynamoDB on AWS.*

The Atlas product from MongoDB, Inc., can be installed as a service and run as cloud-hosted services such as AWS. Through the Atlas management portal, you sign up to use MongoDB and then your database cluster (replica set or sharded cluster) is deployed to cloud infrastructure on your behalf.

*Note: MongoDB is itself built on top of another open-source component called WiredTiger that is a data storage engine. However, MongoDB is all that you will interact with directly.*

## Benefits of MongoDB with Atlas PaaS

The figure in the prior section showed an illustration of the data layer. This included three different sub-layers. The great news is that you get all three with MongoDB and its API access. Here are some compelling features to consider when evaluating the benefits of MongoDB, especially through the Atlas offering that uses cloud infrastructure:

- **Elastic storage capacity:** You simply dial up and down your storage capacity by moving to a higher performance plan or by adding more capacity. You can delete infrastructure that is no longer needed and then not be charged for it anymore. You can configure your resources through code or through a management portal. There is practically unlimited growth for "pay-as-you-grow" storage capacity.

- **Elastic performance:** You simply move your performance up or down by selecting a different plan. To achieve higher throughput, you can pay for the highest performance tier and get a sharded cluster and the highest input/output operations per second (IOPS).
- **PaaS:** MongoDB on AWS through Atlas is a PaaS offering. All the machine management of software and hardware upgrades are handled for you.
- **APM:** Capabilities for monitoring, alerting, scale management, and backups keep you in control.
- **Features:** Besides creating, reading, updating, and deleting documents, there are many features to use under the right circumstances. Features include views, indexing, replication, change streams, sharding (also known as partitioning), aggregation, capped collections, TTL indexes, access control, encryption, ACID transactions, and many more.
- **Auto-replication:** Data is automatically stored in redundant copies of the database. The replications are there for your safety to ensure availability through failover. You can also set up automatic backups.
- **Security:** MongoDB has an audit log to track all database operations. For added security, you can use an SSL connection to your database. MongoDB also has an encrypted storage engine available to protect data.

The Atlas MongoDB offering on AWS is ready to meet all your business and customer needs. It has a great feature set and is increasing in capabilities all the time.

# Try it out

You could download MongoDB and run it on your local machine. You also could do the work to set up a cloud-hosted VM or use a Docker container with MongoDB. My preference is to use MongoDB as a cloud-hosted PaaS solution.

MongoDB, Inc., makes it easy to sign up for a plan that then hosts MongoDB on your choice of Amazon AWS, Microsoft Azure, or Google Cloud Services. This is the approach I use throughout this book.

It takes just a few clicks to and be up and running with MongoDB as a hosted service in AWS. You don't need to manage the daily details of software updates and machine hardware maintenance to keep it up and running. There is no need to waste any time dealing with hardware failures, such as the replacement of drives or network cards.

You enjoy the benefit of having machine replication, scaling, load-balancing, failover, and backup. With the scaling of MongoDB, you only pay for the storage and performance you desire. You pay for what you use and can easily scale back down when you no longer need as much data storage or performance.

You are free to concentrate on the aspects of your application that deliver value to your customer and increase ROI for your company. These are some of the many benefits with PaaS cloud infrastructure.

# Pay for performance

As previously mentioned, you only pay for what you need. As with any cloud infrastructure, if you have an immediate need, you can pay for higher performance machines to run MongoDB. Thankfully, there is a free option through Atlas for your initial investigations.

You can pay for your higher scaling capability by purchasing a plan that comes with a higher storage allotment and is also configured with more machines for high-availability and auto-failover. If you need to scale up to more storage, you can keep adding database clusters as needed and spread your data across more machines with sharding.

You also have the flexibility to create multiple databases, each on different payment tiers. You would then place data on the more expensive plan that needs the highest throughput, while other data can be on the less expensive plan of storage. You can change things like the instance size, replication factor and sharding scale as needed at any time.

In some cases, a configuration change will still require downtime, such as a change in the instance size. This downtime would happen while the primary server is being migrated, but usually should be under a minute of time. See Atlas documentation for more details.

# MongoDB structure

With your Atlas account, you can create one or more database clusters. These clusters have databases that serve as the containers for what are called Collections. Databases hold Collections and they also hold your Indexes and User Accounts. Collections, however, are what contain your data in the form of BSON documents.

Each database can have a set of users with specific permissions. You can control the users of your database and give individuals read access and write access. This does not mean that everyone that connects to MongoDB through the NewsWatcher app needs their own user account. In a backend software platform, programmatic access happens through your service layer. Specific data access is controlled through a login mechanism via an API/SDK. That information will be covered later in this book.

*Note: An account is created when you first create the cluster for admin privileges. You can create code account access that works across all databases created on that cluster.*

PART I: The Data Layer (MongoDB)

The following illustration is a visual representation that shows the different resources that are part of the MongoDB managed platform service and how they relate to each other. The important thing to understand is that you can have multiple databases, with each database having multiple collections in it. Collections then have multiple documents.

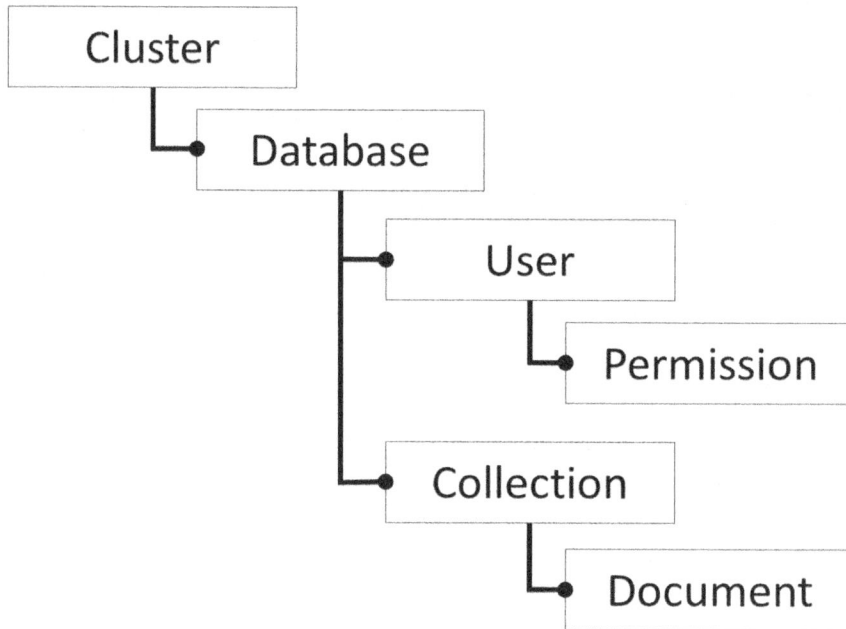

*Figure 6 - MongoDB resources*

# 1.4 The MongoDB Collection

A Collection is a container for storing documents and is the main resource with which interactions happen programmatically. You can write JavaScript code in your Node.js service layer for operating on the data in the collection.

Collections are a convenient way of separating out data in a MongoDB database. They collectively share whatever limit of total storage is set for the plan you are paying for.

To write data as documents (JSON/BSON) into a collection, you will want to use one of the APIs available to do that. You can also write data using the MongoDB command line tool or the Atlas portal. Most likely, you will want to use one of the coding language APIs available.

# 1.5 The MongoDB Document

Every database ever invented is fundamentally designed to store data as a set of records. Each individual record contains information that is useful for later retrieval. For example, in a relational database, the record is in the form of a row in a table, and you might have a table of customers where each row represents a single customer with their name, age, and email as part of their record. In AWS S3, a record takes the form of data with its descriptive metadata.

In MongoDB, a single record is represented by a document in a collection. The following figure illustrates how you can have multiple collections, with each collection containing multiple documents. Each rectangle is a document below:

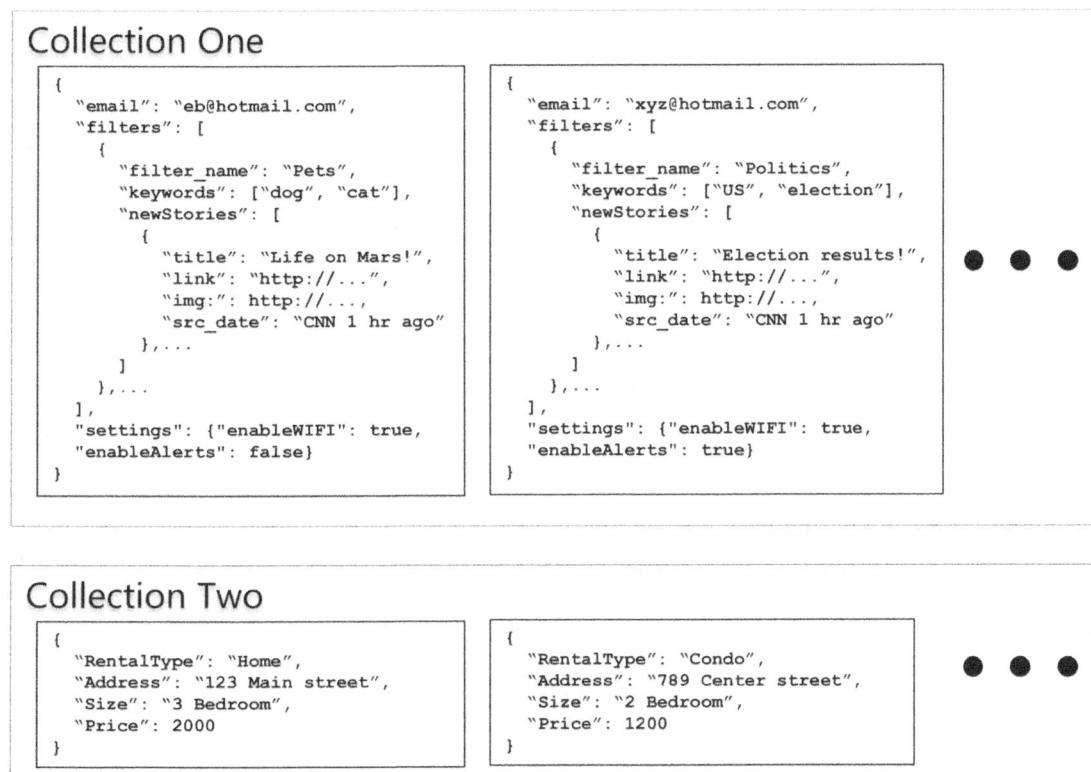

```
Collection One

{
  "email": "eb@hotmail.com",
  "filters": [
    {
      "filter_name": "Pets",
      "keywords": ["dog", "cat"],
      "newStories": [
        {
          "title": "Life on Mars!",
          "link": "http://...",
          "img:": http://...,
          "src_date": "CNN 1 hr ago"
        },...
      ]
    },...
  ],
  "settings": {"enableWIFI": true,
  "enableAlerts": false}
}

{
  "email": "xyz@hotmail.com",
  "filters": [
    {
      "filter_name": "Politics",
      "keywords": ["US", "election"],
      "newStories": [
        {
          "title": "Election results!",
          "link": "http://...",
          "img:": http://...,
          "src_date": "CNN 1 hr ago"
        },...
      ]
    },...
  ],
  "settings": {"enableWIFI": true,
  "enableAlerts": true}
}
● ● ●
```

```
Collection Two

{
  "RentalType": "Home",
  "Address": "123 Main street",
  "Size": "3 Bedroom",
  "Price": 2000
}

{
  "RentalType": "Condo",
  "Address": "789 Center street",
  "Size": "2 Bedroom",
  "Price": 1200
}
● ● ●
```

*Figure 7 - Example of MongoDB collections with documents*

Documents in this example, are represented as JSON with top-level curly braces enclosing each document. This is how they are shown textually, even though they really do not appear that way in the database itself.

23

This concept of storing documents in a document-based database has many advantages. For example, JSON is easy to formulate and consume in code. Another benefit is that it can contain complex hierarchical data in embedded structures. It also supports arrays.

# Datatypes in a document

The primitive data types supported in a MongoDB JSON document are the same ones that are available in JSON, because JSON is what you pass to be inserted. These are:

- Array
- Boolean
- Null
- Number
- Object
- String

In addition, an extended syntax in JSON can take the augmented BSON datatypes and preserve them. Some of the added data types of BSON are Date, Binary data, 64-bit integer along with many others. You will eventually be accessing MongoDB through code and will be using JavaScript objects with their supported datatypes. I will now give you a few tips for storing media data and currency values, as each has some interesting issues and techniques to make them work.

One of the issues you might come across is with the storage of binary data, especially media data such as photos and movies. Media data cannot realistically be inserted into a JSON file. You could use the BSON binary data type, but the main problem you would encounter is that the maximum BSON document size is 16 megabytes. A movie simply would not fit into one single document. It is best to store binary data in a separate storage system, such as AWS S3, and reference it from a MongoDB document.

Another issue is the representation of U.S. currency values, such as dollars and cents as decimal numbers in code. The problem comes when you attempt to perform floating point precision storage and arithmetic operations, such as might be required for investment calculations. You are not even safe in using the MongoDB 128-bit decimal datatype because in JavaScript you are still restricted to 64 bits for its Number data type. If Node.js were to calculate and print $0.1 + 0.02$, the result would be $0.12000000000000001$. This is due to the rounding errors inherent in CPUs because of how they represent floating point numbers.

A great solution is to multiply all your monetary values by a constant scaling factor. For example, if you were going to store the value of $1.99, don't store 1.99, but instead multiply that by 1000 and store an integer value of 1990. Then you can perform math on those integer values and convert back for display when needed.

# The JSON Document

One distinguishing characteristic of a document-based database is that it can store complex hierarchical objects without any predefined schema. This means that every document in a collection could have a different structure. This can be considered both an advantage and a disadvantage.

There are many advantages to having a schema-less database. The obvious one is that you don't need to specify the JSON format structure in advance. This means that there is no need to specify datatypes or have restrictions on them. It is simply a matter of creating the JSON with the name/value pairs that you desire and then inserting that into a collection. Here is a simple JSON document:

```
{
   "_id": ObjectID("59612a3dc17c5416d0a33041"),
   "myNumber": 99,
   "myString": "Hi there",
   "myBool": true
}
```

*Note: I refer to the name/value pairs in the JSON document as "properties," since its syntax is so close to the syntax of the property in a JavaScript object literal. The MongoDB online documentation, however, uses the term field/value for the pairs.*

If you open of the MongoDB, Inc., browser portal and look at some documents, you can see each in the UI with all their properties in JSON form. In this chapter, documents will be presented in the textual JSON form since you can visualize them better that way and, if needed, import them into a collection in that form.

Once you get to the chapter on Node.js and are using the API (also called the driver) to programmatically interact with MongoDB, you will see JavaScript code with the object literal syntax used. As an example, the JavaScript object literal syntax equivalent of the prior JSON example would be as follows (omitting the id property created upon insertion):

```
var mydoc = {
   myNumber: 99,
   myString: "Hi there",
   myBool: true
};
```

*Note: There is one restriction on JSON documents regarding usage in MongoDB. The restriction is that the property names in a single document must not be duplicated. This makes sense because, if you were to query and ask for a given property and it existed twice, that would be a little confusing. The core MongoDB storage system that stores BSON does not make this restriction, but the Node.js driver that accesses it requires it.*

# An example

To help you better grasp the concept of MongoDB, let's pretend you are running a business that sells books over the internet. The following document represents what you might want to use in a collection that holds customers of your online bookstore. Each customer document would hold information associated with that customer. You would want to store information about what books each customer had purchased. The order data for the customer is embedded in their document. You might also want to store their personal contact information. All of this can be placed in one single document as follows:

```
// Customer Document
{
   "_id": "77",
   "name": "Joe Schmoe",
   "age": 27,
   "email": "js@live.com",
   "address": {
     "street": "21 Main Street",
     "city": "Emerald City",
     "state": "KS",
     "postalCode": "10021-3100"
   },
   "booksPurchased": [
     {
        "title": "Agile Project Management with Kanban",
        "ISBN10": "0735698953",
        "author": "Eric Brechner",
        "pages": 160,
        "publicationDate": "20150326",
        "category": "Software Engineering"
     },
     {
        "title": "The Merriam-Webster Dictionary",
        "ISBN10": "087779930X",
        "author": "Merriam-Webster",
        "pages": 939,
        "publicationDate": "20040701"
        "category": "English"
     }
   ]
}
```

It is important to remember that a single document should only contain the information for that one main unique record. This means that you would not want to enter two customers into a single document as that would get confusing.

As previously mentioned, documents do not need to be completely uniform. This means that one document can contain properties that another document might not ever have, and both can exist in the same collection. For example, if you had a collection that contained products from the fictitious bookstore, it would obviously have documents representing books. It might also have magazines, maps, and puzzle documents as well. Of course, the properties of these products would need to be somewhat different.

Documents in a collection might each have some common properties such as price, title, description, and weight. For each product type, there would also be unique properties. A puzzle, for example, might have a property that states what the recommended age is for that puzzle.

The following example shows a collection of documents that have both common and unique properties. Notice how the document for a book differs slightly from the document for a puzzle, yet both can exist in the same collection.

```
// Bookstore products collection
{
   "type": "BOOK",
   "title": "Agile Project Management with Kanban",
   "ISBN10": "0735698953",
   "author": "Eric Brechner",
   "pages": 160,
   "publicationDate": "20150326",
   "category": "Software Engineering"
},
{
   "type": "BOOK",
   "title": "The Merriam-Webster Dictionary",
   "ISBN10": "087779930X",
   "author": "Merriam-Webster",
   "pages": 939,
   "publicationDate": "20040701"
   "category": "English"
},
{
   "type": "PUZZLE",
   "title": "Balloons in sky",
   "company": "ZipZap Toys",
   "age": "3-5 years old"
}
```

Just because documents can contain diverse content does not mean that you always want to have collections set up that way. You may decide to have all your collections contain uniformly structured documents. I will later describe circumstances that will help you make these type of design decisions.

*Note: I would advise you to stay away from multidimensional properties (a property that is an array of arrays). Otherwise, you may find yourself looking through your data and not remembering how you had set it up.*

# A common property of all documents

One thing that every single document will have, is an `_id` property. When a document is created, it will always have this unique id associated with it so that it can be identified. This serves as the primary key. You can set the value yourself. If you don't, MongoDB will set a value for you of type ObjectId. If you set it yourself, it can be of any type other than an array. It must, however, be unique across all documents in that collection.

If you ever need to directly access a document, you can access a document using the `_id` as the quickest way to query for it as there is an index created for this.

# Referencing external data

So far, you have seen that MongoDB allows for storage of JSON documents. As you know, the basic data types available with JSON can be used to represent most information you wish to store.

As was mentioned, the one thing that JSON datatypes are not well suited for is the representation of large binary data. For example, you would not really be able to store large media content such as photos, music, or video in a document.

To overcome this limitation, you can create a property that is a reference to where the actual media content is externally stored. You would use a Universal Resource Indicator (URI) to indicate its location, and create a parallel property that describes the type of data it consists of text, image, binary data, etc. That way, your code can interpret it correctly.

The externally referenced data can be any data you would like. It is up to you to set the type and then treat it as such when you retrieve it. Be sure to handle any needed deletion of your external data if it is supposed to be cleaned up when documents referencing it are deleted.

*Note: This chapter stated that MongoDB, as a document-based database, is not required to enforce any schema. MongoDB does have a capability where you can specify what properties should exist and what restrictions should be on them for documents. We will not utilize this capability in this book. I will later mention that you can use a Node.js npm module such as Mongoose to schematize your data and/or use joi to validate your schema. The validation MongoDB performs is also not as strict a definition as you have with a relational database. For example, relationship key specifications are available with a relational database that can also enforce what is referred to as referential integrity.*

# Chapter 2: Data Modeling

A data model for any DBMS defines the structure of records that are to be stored in a database. This includes information on how the different types of records relate to each other.

Because a document-based database is not a relational database does not mean that it doesn't have structure or relationships between document types. Specifying a data model in MongoDB consists of specifying what the JSON documents are, which collections they exist in, and how they relate to each other. I will show you how to use validation code to make sure any data going in conforms to expectations of your data model.

Sketching out your data model will be an important step in the creation of your data layer. It is helpful to design your document structures in advance so that you can look at all aspects of your data. You can match your data storage needs against the characteristics that a MongoDB database offers and do your data modeling accordingly.

To visualize your data model, you can use whatever diagraming tool you like. You can even scribble it out on a piece of paper, although you might want ot keep it in electronic form for easier editing. Ultimately with MongoDB, the JSON format is what you need to specify. Let's now cover some of the bigger design decisions that need to be made with a document-based DBMS.

# 2.1 Referencing or Embedding Data

Records in a relational database each contain keys to identify them and to act as references to each other. A relational DBMS has mechanisms to specify and enforce the integrity of these references.

You may have heard the term "normalization" used in the context of relational database designs. Normalization requires the separating out of data into different record types and relating them to each other. For example, you might have two types of records in your database, one that represents people and another that represents the pets that are owned by the people. There would be a key to relate an owner to one or more of their pets. The following illustration shows this normalized structure:

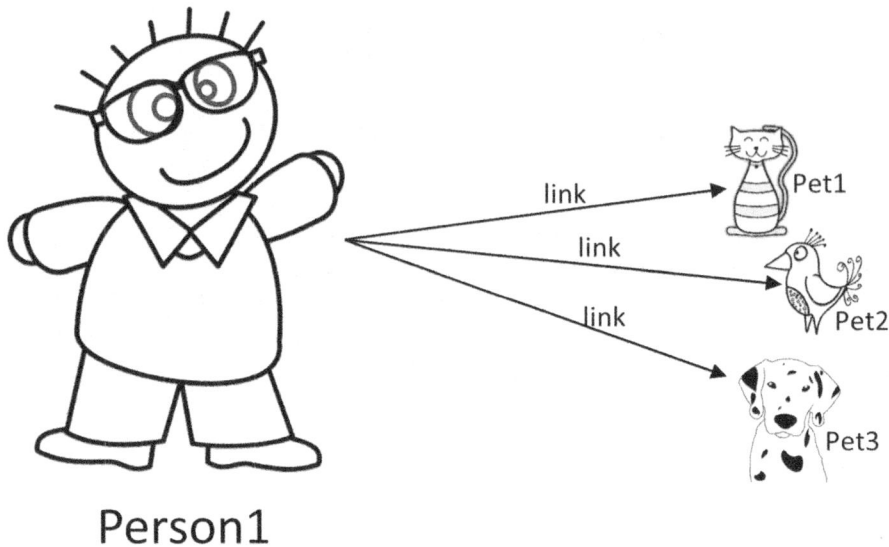

**Person1**

*Figure 8 - Normalized structure*

In a relational DBMS, you set up the keys to relate the record types to one another. The relational DBMS provides functionality to be able to query and join the records together. This means you can query the database to return a person and all the pets they possess. This is done through a single query operation.

Let's now bring this conversation into the MongoDB world. Following a normalized model with MongoDB, you would create separate documents for each person and each pet all in the same collection. Keep in mind that there is no concept of a schema for a collection required, and no concept of relational keys or cross-document join queries. You need to design your own properties on each document to relate pets to their owners. With people and their pets separated out across MongoDB documents, you have achieved a normalized database model.

A normalized model, however, is not necessarily the best way design for storing data in a document-based database. You really want to think more about how to denormalize your data. Denormalized data means that everything is all bundled together.

To denormalize, you bundle pets together with a person. You do this by simply placing pets as an array property that is embedded in each person's document. Imagine stuffing all the pets in the pocket of the person. This way, people and their pets are always found together with no relational links required. You also do not need a join operation if you denormalize your data. In fact, the classic relational join operation is not fully supported in MongoDB (see $lookup). The following visualization shows how the person and pets travel together:

*Figure 9 - Denormalized structure*

As was mentioned, document-based databases like MongoDB do encourage joins of records such as cross-table joins done with a relational database. This is because cross-document joins do not make sense in document-based databases. This means that you need to get comfortable with keeping your database designs denormalized. Denormalization is more efficient and works well in a document-based database. What you do is make use of the array and embedded object properties in your JSON.

In the previous chapter, you saw a JSON document representing a customer of an online bookstore company. What was shown was a denormalized data pattern. Each customer document contained information about the customer, but it also contained information about each book they had purchased. You end up with larger and more complex documents when you denormalize your data.

Even if you end up with large, complex documents in MongoDB, the querying capability supports retrieval of just the portions of the documents that you need. For example, if you kept pets embedded in a person document, you can do a query to just return a subset of properties of the person and the pets. Perhaps you just want to know the names of all the pets and do not want to retrieve anything else about the pet or the person. That is easy to do.

In summary, it can be stated that database normalization has you separate out your data into different distinct record types and has you set up keys as reference links between them. On the other hand, denormalization, with document-based databases, has you bundle as much data together as possible by embedding data that is related.

# 2.2 When to Use Referencing

It is obviously not going to work well if you always embed all your data and end up with huge, complicated documents. You need to make decisions as to what properties will be embedded in a document and what makes more sense to pull out into separate document types to reference.

You can keep all your different document types in one collection. There are, however, reasons why different document types should be kept in separate collections or even separate databases.

## Performance implications

Your data model design will have an impact on the performance of all database operations. For example, the performance of reads and writes could improve if you spread data out into separate documents. Doing this results in decreased data transfer amounts as well as more efficient document updates.

The downside of referencing data from one document to another is that it will be slower if you need to combine data and present it together at any point in time. Imagine the code required to piece together a person with their pets if pet documents are kept separate.

Let's go back to the online bookstore example. Imagine that you have two document types: customers and books. With these documents, you keep track of all the book purchases of individual customers. If you keep only the book IDs in the customer document, certain queries will run slower, such as listing a customer with the book titles they ordered. For each customer, you must look through the `booksPurchased` array property and then do individual fetches of data for each book ID listed. Here is how this normalized model looks:

```
// Customer documents
{
   "_id": "77",
   "type": "CUSTOMER_TYPE",
   "name": "Joe Schmoe",
   "age": 27,
   "email": "js@gmail.com",
   "address": {
      "street": "21 Main Street",
      "city": "Emerald City",
      "state": "KS",
      "postalCode": "10021-3100"
   },
   "booksPurchased": ["0735698953", "087779930X"]
}
```

Chapter 2: Data Modeling

```
// Book Documents
{
    "_id": "7865",
    "type": "BOOK_TYPE",
    "ISBN10": "0735698953",
    "title": "Agile Project Management with Kanban",
    "author": "Eric Brechner",
    "pages": 160,
    "bookReviews": [
        {
            "reviewer": "Joe Schmoe",
            "comments": "Wish I had this years ago!",
            "rating": 4
        },
        {
            "reviewer": "Jane Doe",
            "comments": "Forever a classic.",
            "rating": 5
        }
    ]
}

{
    "_id": "7866",
    "type": "BOOK_TYPE",
    "ISBN10": "087779930X",
    "title": "The Merriam-Webster Dictionary",
    "author": "Merriam-Webster",
    "pages": 939,
    "bookReviews": []
}
```

Imagine that you want to query and find all books that are over 500 pages in length and that were purchased by customer Joe Schmoe. To do this, the query must first search the collection to find the Joe Schmoe document and then, for every book id listed in the Joe Schmoe document, look up that book document and see if the pages property value is over 500, and then finally return those documents. This involves more than one document.

In relational databases, this can be done in a single query. In MongoDB, it must be done in separate queries and involves code to piece everything together as needed.

This is not necessarily a bad thing and is just part of what you need to do in a document-based database when you separate out data into related documents. But again, remember that denormalization relieves you from doing these join operations across documents. You will later see that there is a compromise between the two options.

# When to reference

Let's cover the basic scenarios that would cause you to split data across documents with a normalized design that uses references. The following are some of the common reasons to do this:

- **The data is rarely used in queries:**
  If you find that you rarely reference certain properties in a document, then you might see this as a sign that those properties belong in a separate document or even in a separate collection that can be referenced as needed. For example, the address property in the customer document could be taken out and placed into a separate document if you decide that it is rarely needed. Just remember that you are gaining the faster interaction with the primary data at the expense of slower data retrieval processing to piece it together later.

- **The data is common across documents:**
  The book detail is certainly data that will be duplicated across customers. If you had thousands of customers who each had purchased a certain book, it would not be necessary to keep duplicating all the metadata for that book. One big problem with duplicated data is that if you need to alter one of the properties of a book, it requires altering it across all the duplicates instead of in just one central document. In this case, it would be wise to keep properties in a separate book document if they are shared and updated frequently and make a reference to them.

- **The data has a mutual or cross-reference relationship:**
  The information in the `bookReviews` property originates from customers who submit reviews, but each review is also specific to a single book. The question arises — should the book reviews exist with the book being reviewed or with the customer giving them? You might not want to update the individual book document and keep adding to it every time a new customer adds a review. Nor do you want to fetch all the book reviews every time you fetch the document for a book. This is where you can make the case for the book reviews for each book to be in their own separate document with an array property or place each review in individual documents and then reference by both the customer and the book documents.

- **The data will grow very large:**
  An array property might grow to have many entries, yet an array property of a document cannot grow unbounded. Remember that there is a 16-megabyte limit on the size of an individual document in MongoDB. You would therefore set up several related documents where each has an array that comprises sections of the total set.

# 2.3 Reference Relationship Patterns

At one point I showed you a customer document that had a list that contained the IDs of the books a person had purchased. This design keeps the documents for the actual book details separated out. This is one of several patterns you can find useful for referencing data.

I will stress again that this is totally under your control to implement. MongoDB does not provide any built-in recognition of relationships. It is important to realize that for our purposes in this book, MongoDB will not verify the referential integrity of the data as a relational DBMS might do. It is up to you to manage that. If a given book is deleted from a collection, you also need to delete all the references to it. For example, if a customer document had a `booksPurchased` array property with an entry in it with an ID of 5777, there is nothing in MongoDB that verifies that a book document exists with an ID of 5777.

Here are some common patterns that can be used for referencing data across documents in MongoDB:

- **One-to-One:** This is where you separate out a piece of infrequently accessed data. For example, you could pull out the address information of a customer and put it into its own document. This diagram illustrates a one-to-one pattern:

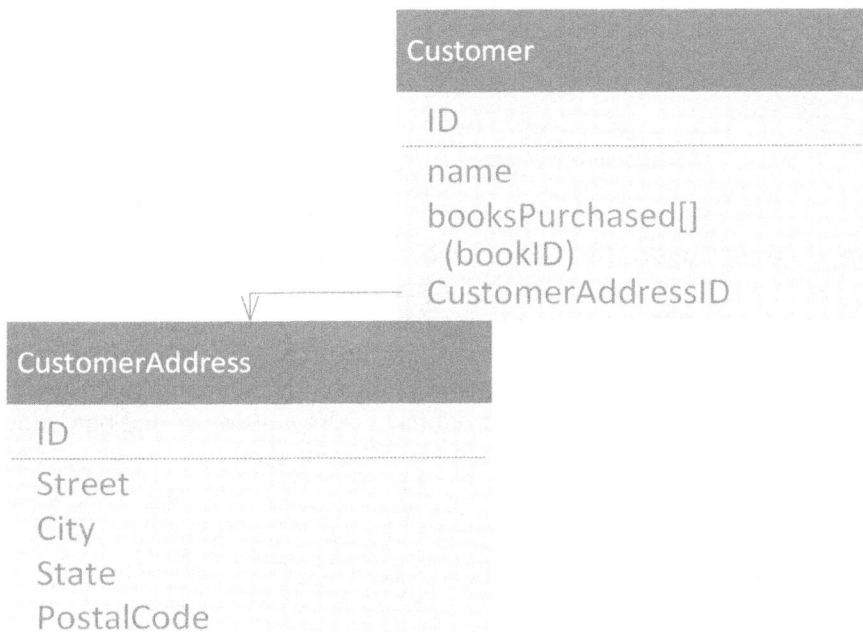

*Figure 10 - One-to-one relationship pattern*

- **Many-to-One:** This is where one document may be referenced by many other documents. You have already seen this example where a single book can be referenced by multiple customers. Each entry in the `booksPurchased` array has an ISBN10 id to reference the book document with. This diagram illustrates a many-to-one pattern:

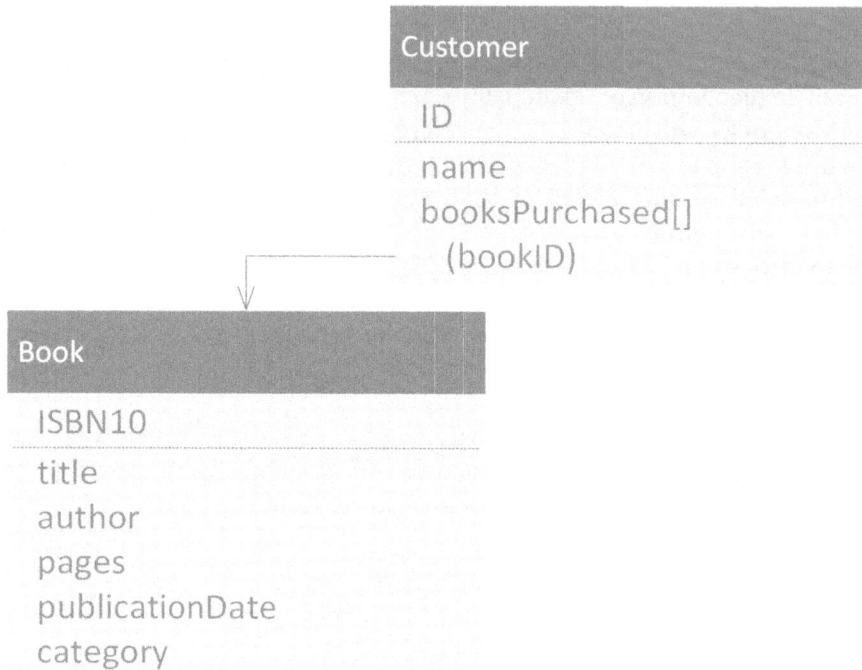

**Customer**

ID

name
booksPurchased[]
    (bookID)

**Book**

ISBN10

title
author
pages
publicationDate
category

*Figure 11 - Many-to-one relationship pattern*

- **Many-to-Many Association:** This is where you have two separate document types where neither references the other, but instead have a third document type in common that ties the two together through an association. This third document can also contain information relevant to that association. For example, you might have a document for purchases that records the details of the transaction of a book purchase. Thus, you have many purchase documents that reference many customer and book documents. This diagram illustrates a many-to-many association pattern:

**Purchase**

salesID
customerID
bookID
date
amount
discountCode

**Customer**

ID

name
age
email
address
> street
> city
> state
> postalCode

**Book**

ISBN10

title
author
pages
bookReviews[]
 (reviewer
 date
 comments
 rating)

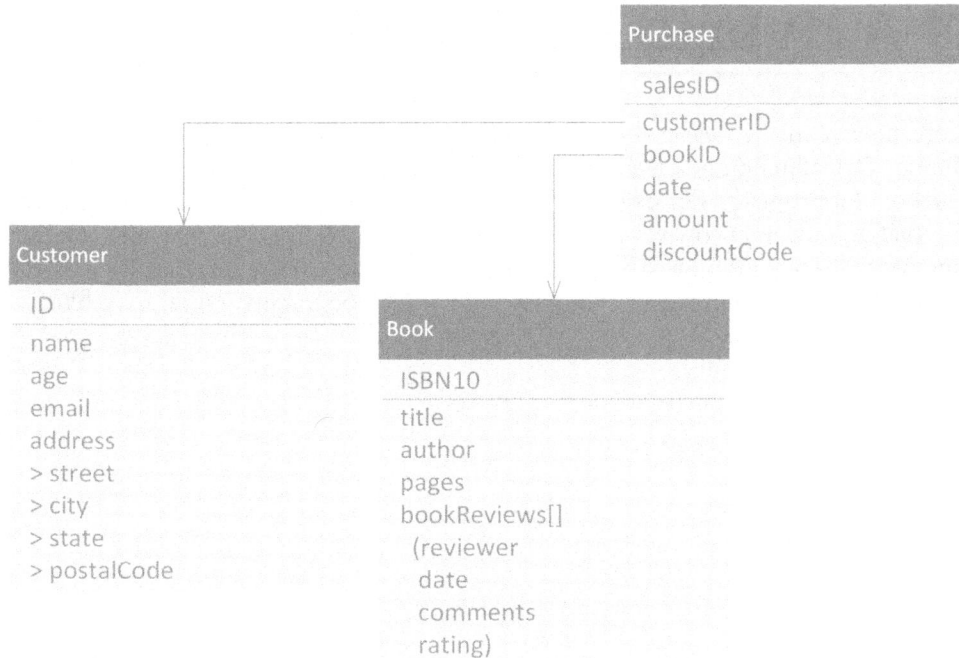

*Figure 12 - Many-to-many association pattern*

Many-to-many relationships can have documents referencing each other in a circular manner. Try not to allow it to get that complex, as it presents many difficult data integrity issues and complicates the service layer implementation.

***Note:*** *It is useful to create a visual diagram of your document types as you go through your data model design process. You are then able to see the structure and relationships more clearly. Several diagram standards have been created over the years to visually represent objects in object-oriented languages and for representing records in databases. All you really need is a simplified visual representation of JSON objects in your data model. You can draw a rectangle shape for each document type and list all the properties inside as I have shown in the previous illustrations. I chose to have the id property exist above a line. Those properties below the line are the rest of the document properties. I have chosen to use a greater than sign to show sub-object properties. An array of objects is shown in parentheses, and it is understood there would be zero or more of these in an actual instance of the document. You might also want to list the data type of each property out to the right of the name.*

# 2.4 A Hybrid Approach

You are now ready to learn about a hybrid approach that can give you the best of both techniques for referencing as well as embedding data. Why not combine the two techniques in a sort of compromise? For example, if you find that you often need to list the titles of the books that a customer owns, you can duplicate a portion of that information across documents. This means you store the book ID and the title, even though it is duplicated information. The following is an example of what the documents would look like:

```
// Customer Document
{
   "_id": "77",
   "type": "CUSTOMER_TYPE",
   "name": "Joe Schmoe",
   ...
   "booksPurchased": [
      {
         "ISBN10": "0735698953",
         "title": "Agile Project Management with Kanban"
      },
      {
         "ISBN10": "087779930X",
         "title": "The Merriam-Webster Dictionary"
      }
   ],
   ...
},

// Book Document
{
   "_id": "77",
   "type": "BOOK_TYPE",
   "ISBN10": "0735698953",
   "title": "Agile Project Management with Kanban",
   "author": "Eric Brechner",
   "pages": 160,
   "publicationDate": "20150326",
   "category": "Software Engineering"
},
```

You can see that the compromise was to keep the title duplicated across document types as the title is frequently needed. The rest of the properties are separated out. Be aware that you would want to synchronize changes to any duplicated properties that existed. For example, if you need to make a change to the `title` property in the book document for a given book, you will want to have some background process that would query all occurrences of that book in the customer documents and update the title in those.

# 2.5 Differentiating Document Types

You may have several different document types existing in a single collection. Storing multiple document types in one single collection allows you to use a single API connection in your code. API connections are tied to a database and are for a single collection.

With all document types in the same collection, you need a way of differentiating them from each other so that queries can find each specific type. One way to solve this is to include a type property in each document.

Let me use the previous example of documents for persons and for pets. Person documents would be declared with: "type": "PERSON_TYPE" and pet documents would be declared with "type": "PET_TYPE". For example:

```
// Person and Pet types in same Collection
{
   "type": "PERSON_TYPE",
   "_id": 1,
   "name": "Ian",
   "Pet club membership": true,
   "pets": [12, 24],
},
{
   "type": "PET_TYPE",
   "_id": 12,
   "name": "Kirby",
   "breed": "Cavalier dog",
},
{
   "type": "PET_TYPE",
   "_id": 24,
   "name": "Kaitlyn",
   "breed": "Siamese cat",
}
```

This allows a query to narrow down results to just returning the document type you want and then you can add in whatever further criteria you are looking for.

# 2.6 Running Out of Space in a Database

There are ways to deal with cases where the number or size of documents becomes a problem. Think about what would happen if you kept adding documents to a collection that existed on a single Solid-State Disk (SSD)? Obviously, at some point, you are going to reach the storage space limit that is set for a single database you are paying for. This does not necessarily mean that you need to separate out and reference data across document types in separate collections. You can keep data embedded and grow the number of documents indefinitely.

The method for achieving storage capacity scaling is through what is called sharding (also known as partitioning). This means your collection is more of a logical concept and is spread across multiple storage locations. The unit of scaling for MongoDB is called a replica set. A given document, however, must only be found in one replica set (shard) of the sharded cluster. The nice thing is that MongoDB hides that from you, and your query or update does not even realize what is going on.

For customer documents, you could have things set up to distribute customer documents out across different replica set storage. In a later chapter, I will discuss this type of data partitioning. Don't worry if you do not fully understand this concept just yet. You may never need to implement sharding unless you run into really large amounts of data and must maintain fast read and write times.

# 2.7 Access Control

There are several ways to interact with MongoDB documents. One is through the Atlas management web portal, another is through a mongo shell command line interface, and another is through API access, such as in Node.js JavaScript driver code. While you also have the mongo shell, that is not something you will need for the purposes of what this book is teaching. You will learn just the bare minimum to use the mongo shell, such as for importing documents in bulk. If you want to learn more about the mongo shell, please refer to MongoDB documentation.

From the Atlas management portal, a user account can be added to provide authentication for that user. You can give a user read-only privileges if you need that restriction in place. Account administration through the portal is not central to the topic of this book, so please refer to the Atlas management portal documentation if you desire more information on that.

As mentioned, the primary type of access discussed in this book is done programmatically through a MongoDB API, such as one that is provided for Node.js developers. The details of this will be explained in the Node.js section later in this book.

You will later see how the NewsWatcher sample application has a UI that goes through the middle-tier service layer where interaction with the database takes place. The middle tier will then be able to authenticate on behalf of the user, access the database on their behalf, and restrict access to just documents that a particular user is allowed to see.

# Chapter 3: Querying for Documents

You may have heard of Structured Query Language (SQL) as the language used to query for records in relational databases. When using SQL, you submit queries to your DBMS and receive resulting records that match. MongoDB supports querying, but it does not use SQL as it has its own unique query syntax.

The lowest level interface to MongoDB has a means of interaction to accomplish operations such as queries. Everything at the lowest level is done through a TCP/IP connection that has a well-defined wire protocol for the operations that it supports. There are around nine operations you can make with this protocol. It is not necessary to fully understand all of this yet; what you need to know at this point is that other people have reduced the complexities of using this protocol by creating specific language drivers you can use. This book will be concerned with using the Node.js JavaScript driver. The following diagram shows the overall access layers:

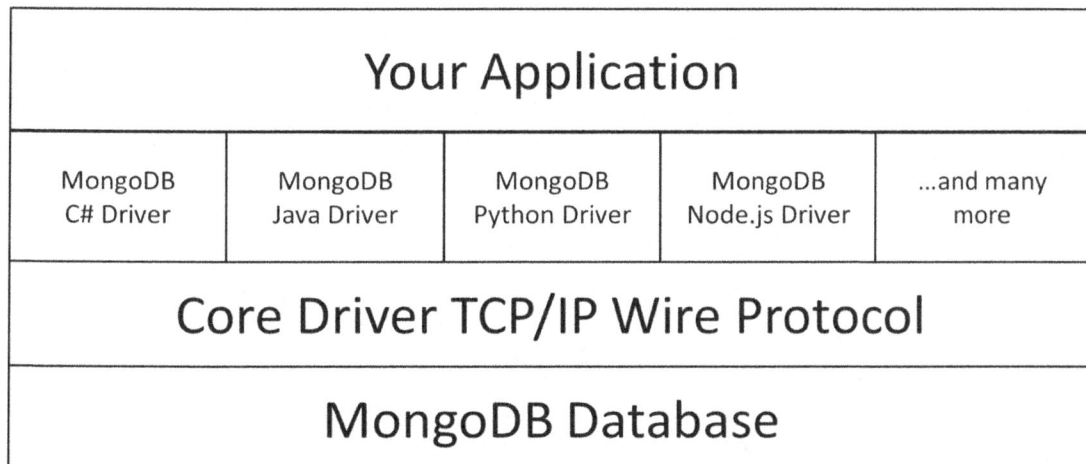

| Your Application | | | | |
|---|---|---|---|---|
| MongoDB C# Driver | MongoDB Java Driver | MongoDB Python Driver | MongoDB Node.js Driver | ...and many more |
| Core Driver TCP/IP Wire Protocol | | | | |
| MongoDB Database | | | | |

*Figure 13 - MongoDB access abstraction through APIs*

In the service layer part of this book, I will show you how to use the MongoDB Node.js driver for operations on the data such as create, read, update, and delete (CRUD). This current chapter will only cover the specifics of syntax for query operations in general.

Regardless of which of the CRUD operations you perform, you need to specify your query criteria as part of that request. You can explore this topic now because you don't need to write any code to try out your queries. You can use the portal application to try them out and experiment with the syntax as you like.

# Example documents

Carefully review the example JSON documents shown below. You can imagine these documents being used by an online bookstore, although there obviously would be a lot more data available than what is in this example. This JSON will be used to help illustrate how to construct your queries.

```
// Customer Documents
{
   "_id": "77",
   "type": "CUSTOMER_TYPE",
   "name": "Joe Schmoe",
   "age": 27,
   "email": "js@gmail.com",
   "address": {
      "street": "21 Main Street",
      "city": "Emerald City",
      "state": "KS",
      "postalCode": "10021-3100"
   },
   "rewardsPoints": 99,
   "booksPurchased": [
      {
         "id": "1098",
         "title": "Agile Project Management with Kanban"
      },
      {
         "id": "1099",
         "title": "The Merriam-Webster Dictionary"
      }
   ]
}, {
   "_id": "78",
   "type": "CUSTOMER_TYPE",
   "name": "Jane Doe",
   "age": 37,
   "email": "jd@gmail.com",
   "address": {
      "street": "100 S Bridger Blvd",
      "city": "Paradise",
      "state": "UT",
      "postalCode": "84328"
   },
   "rewardsPoints": 0
}
```

```
// Book Documents
{
   "_id": "1098",
   "type": "BOOK_TYPE",
   "title": "Agile Project Management with Kanban",
   "ISBN10": "0735698953",
   "author": "Eric Brechner",
   "pages": 160,
   "format": "Paperback",
   "price": 27.66,
   "publicationDate": "20150326",
   "category": "Software Engineering",
   "bookReviews": [
      {
         "reviewer": "Joe Schmoe",
         "date": "20140321",
         "comments": "Wish I had this years ago!",
         "rating": 4
      }, {
         "reviewer": "Jane Doe",
         "date": "20150923",
         "comments": "Forever a classic.",
         "rating": 5
      }]
}, {
   "_id": "1099",
   "type": "BOOK_TYPE",
   "title": "The Merriam-Webster Dictionary",
   "ISBN10": "087779930X",
   "author": "Merriam-Webster",
   "pages": 939,
   "format": "Paperback",
   "price": 11.26,
   "publicationDate": "20040701",
   "category": "English",
   "bookReviews": [
      {
         "reviewer": "Joe Schmoe",
         "date": "20100101",
         "comments": "A terrific volume to keep handy.",
         "rating": 4
      }, {
         "reviewer": "Jane Doe",
         "date": "20120817",
         "comments": "Wish it came in an audio book format.",
         "rating": 2
      }]
}
```

I will now introduce you to the basics of the query syntax. The syntax is robust and much of it is beyond the scope of this book and something you may not ever need to use. For example, I will only briefly mention how to use the MongoDB aggregation features and its syntax. For more information on that topic and other supported syntax intricacies, refer to MongoDB documentation.

# Syntax overview

In a later chapter, you will be querying a collection using a MongoDB NPM module to write code in JavaScript that runs in Node.js. That module provides functions such as `find()` or `findOneAndDelete()`. The first parameter of those functions is the query criteria that specifies the matching to take place across all the documents in a collection.

If you are familiar with SQL, this is similar to what a WHERE clause does. Here is an example query using the `find()` function with a greater than operator in the query criteria. This query below will return all the documents in the collection whose age property contains a value greater than 35. This is showing you how you would call it in code, and this is similar to how you would format it for use in if you were to use the mongo shell.

```
db.collection.find({age: {$gt: 35}});
```

For the function call above, there is the possibility that the query criteria won't match anything. This is okay, and no documents will be returned. On the other hand, if there are a lot of documents returned, then you need to use the Node.js driver capability to fetch results in batches. You will see how to actually access the results of the `find()` function in code.

Besides the query criteria I just showed you, there are also criteria you can provide for what is called the projection criteria. The projection criteria determine the properties that will be returned from each document. Here is another example using the same `find()` function with the optional second parameter that specifies the projection.

```
db.collection.find({age: {$gt: 35}}, {name: 1, age: 1});
```

This query will find all documents in the collection whose age property contains a value greater than 35. With the projection specified, only the name, age and _id properties will be returned. _id is always returned unless you specify otherwise.

As mentioned, the second parameter in this query is the projection criteria. This is where you list the properties you want to be returned. The number one following the colon determines if the property is included or if it is excluded (1 for include and 0 for exclude). If you do not specify projection criteria, then the complete document is returned.

I will keep using the example of the bookstore customer documents, but for now, assume they only have four properties each (_id, name, age, and email). The following diagram shows the two criteria (query and projection) in the function call and how they determine the output:

## Customer Documents

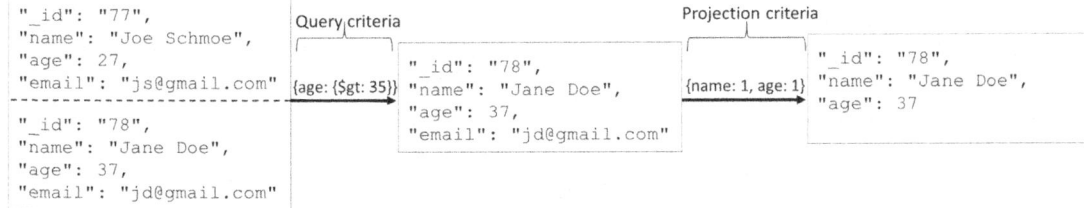

*Figure 14 - Criteria flow*

You can see that the query criteria determine what documents pass through to the result set. The projection criteria select what properties you want for each document in the result set.

You can connect to your MongoDB hosted cluster through the portal application, add a database, add a collection and some documents, and then try out some queries on your own.

*Note: You cannot just create a query and assume that it will end up being efficient. The execution time of a query can vary greatly. To address performance issues, you either must create indexes that can speed up your queries or think about a more efficient way of modeling your data. The topic of index creation is covered later.*

# 3.1 Query Criteria

Query criteria are actually optional on a function such as the `find()` function. It is, however, something you will almost always be using. If you call `find()` without any query criteria parameter, every single document in the collection will be returned.

The query criteria are tests applied to the collection to see what documents are to be included as part of the result set. If you want to, you can narrow down the result to return a single document. For example, you can query by the `_id` property with an equality test. The `_id` property is unique, so each document can be uniquely identified with it. Here is an example of code to query by `_id`. The result it returns is also shown:

```
// Query
db.collection.find({_id: {$eq: "77"}}, {address.state: 1});

// Results
{
   "_id": "77",
   "address": {
      "state": "KS"
   }
}
```

# PART I: The Data Layer (MongoDB)

You don't need to write any code to try this out, but you can use the portal application to try out queries. If you wish, you can go to Chapter 6 to learn how to create your PaaS hosted MongoDB cluster, database, and collection. You can use the Atlas management portal to import documents to use as you experiment with queries. Refer to the Example Documents section for the documents you can create in order to try your own experimentations.

Once you have your test database, test collection, and documents imported, you can open the management portal and go to your collection to enter some test queries in the Documents tab. Here is what the UI looks like if you are trying out a query:

## newswatcherdb.newswatcher

STORAGE SIZE: **176KB**　　TOTAL DOCUMENTS: **22**　　INDEXES TOTAL SIZE: **116KB**

Find　　　Indexes　　　Schema Anti-Patterns ⓪　　　Aggregation　　　Search Indexes ●

FILTER  `{"address.state": {"$eq":"UT"}}`

QUERY RESULTS **1-1 OF 1**

```
_id: "78"
type: "CUSTOMER_TYPE"
name: "Jane Doe"
age: 37
email: "jd@gmail.com"
> address: Object
rewardsPoints: 0
```

*Figure 15 - Application querying capability*

Each query criteria can utilize one or more operators. The example above uses the $eq operator. The real power of the query is in the use of the criteria operators. I'll now go over what those are and show you some examples.

# Criteria operators

You have seen operators such as `$gt` and `$eq` used in the examples in this chapter. Those operators stand for greater than and equal to. There are many more operators that you can use in your query criteria to filter documents. The following operators are currently supported:

- **Comparison**
  - `$eq`
  - `$gt`
  - `$gte`
  - `$in`
  - `$lt`
  - `$lte`
  - `$ne`
  - `$nin`
- **Array**
  - `$all`
  - `$elemMatch`
  - `$size`
- **Bitwise**
  - `$bitsAllClear`
  - `$bitsAllSet`
  - `$bitsAnyClear`
  - `$bitsAnySet`
- **Element**
  - `$exists`
  - `$type`
- **Evaluation**
  - `$expr`
  - `$jsonSchema`
  - `$mod`
  - `$regex`
  - `$text`
  - `$where`
- **Geospatial**
  - `$geoIntersects`
  - `$geoWithin`
  - `$near`
  - `$nearSphere`
- **Logical**
  - `$and`
  - `$or`
  - `$nor`
  - `$not`

PART I: The Data Layer (MongoDB)

Each of the operators uses its own unique syntax. The $eq operator uses the following syntax:

```
{<name>: {$eq: <value>}}
```

The name is what you want to test against. It can be a top-level property, or it can be a property within the hierarchy of the JSON. The value is a string, number, or other value that matches the data type of the property. Here is an example that tests a second-level property, which is referred to as an embedded document field in MongoDB documentation:

```
{"address.state": {"$eq": "UT"}}
```

You can even specify a property of an element found in an array. If you look at the example document, you see that booksPurchased is an array and id is a property of each of the elements of that array. In the example below, the returned document is the complete document, as the query criteria are only used to find a document match and not specify properties to return.

```
{"booksPurchased.id": {"$eq": "1098"}}
```

Here are a few operators from some of the categories to give you an idea of how they work. For more detailed information on each of the operators, see the MongoDB documentation.

## Comparison
With the comparison operators, you need to first specify the property name you are interested in testing against. This is followed by the comparison operator and finally the value you want for that comparison test. The only exception to this is with the $in and $nin operators, which use an array and not a single value.

I will use the same example shown previously that does a query for a single document by its _id value:

```
{"_id": {"$eq": "77"}}
```

To just test the equality of a property, you can shorten the syntax to the following:

```
{"_id": "77"}
```

You can use more than one comparison operator at a time, such as you need to do to test ranges of values. All operator tests need to pass their test for a given document to be included in the results set.

Here is an example that queries for books that have between 100 and 200 pages:

```
// Query
{
   "pages": {
     "$gt": 100,
     "$lt": 200
   }
}
```

**Logical**

Logical operators let you string together several tests in a row to perform the desired logical testing. The syntax for the logical operator $and is:

```
{$and: [ { <expression1> }, { <expression2> } , ... , { <expressionN> } ]}
```

Logical operator syntax starts with a Boolean operator such as $and. It then contains an array of expressions that can be made up of individual comparison operators, as we have seen previously. The following query looks for books that are less than 200 pages in length and which are also in the Software Engineering category.

```
// Query.
{"$and": [{"pages": {"$lt": 200}},
          {"category": {"$eq": "Software Engineering"}}]}

// Results. Assuming you also have the projection criteria of {"_id": 1}
{
   "_id": "1098"
}
```

If you are only carrying out this one level of Boolean operation, then you don't really need the $and operator. Instead, you can just list the conditions one after another. Here is the same example query without the $and operator:

```
// Query.
{"pages": {"$lt": 200}, "category": {"$eq": "Software Engineering"}}

// Results. Assuming you also have the projection criteria of {"_id": 1}
{
   _id": "1098"
}
```

# PART I: The Data Layer (MongoDB)

Here is a query that uses both the `$and` and the `$or` operators. This example queries for books that have less than 200 pages and are in either category specified.

```
// Query.
{
  "$and": [
    { "pages": { "$lt": 200 } },
    {
      "$or": [
        { "category": { "$eq": "Software Engineering" } },
        { "category": { "$eq": "Science Fiction" } }
      ]
    }
  ]
}

// Results. Assuming you also have the projection criteria of {"_id": 1}
{
    "_id": "1098"
}
```

If you find your query has a lot of `$or` operations to match on many different values for the same property, then you can use the `$in` operator. The value to match can even be a regular expression. The example below shows how easy it is to use this to list all possible matches:

```
// Query using IN
{
    "category": {
      "$in": [
        "Software Engineering",
        "Science Fiction"
      ]
    }
}
```

## Element

The element operators `$exists` and `$type` are for selections based on whether a property exists and if it is of a certain datatype. The following example shows the syntax for `$exists`:

```
{ name: { $exists: <boolean> } }
```

A typical use for `$exists` would be to use this operator inside another operator. In this example, you want to make sure the document has the `publicationDate` property. It may be that a book has a price set, but has not been published yet, so the `publicationDate` property is not there yet.

```
{"$and": [{"price": {"$lt": 30}}, {"publicationDate": {"$exists": true}}]}
```

50

**Evaluation**

If you struggle to get exactly what you want in your query, you may find the evaluation operators are just the help you need. Here is an example that shows the use of a regular expression with the option for a case-insensitive test. This example will find all books that have a title that starts with the word "agile" no matter the letter casing:

```
{
  "title": {
    "$regex": "^agile",
    "$options": "i"
  }
}
```

You may want to search if specific words or phrases exist for a document with large amounts of text. You can use the $text operator in this case.

```
{"$text": {"$search": "MongoDB"}}
```

*Note: To do so, you need to first create an index of type text on the properties you want to use it on. For example, you could create the index on the title property and then search for books that contain certain words in their title.*

For those rare occasions where you just cannot get what you want with any of the available operators, you can resort to writing JavaScript using the $where operator. Here is a test that checks to see if a person has purchased more than a single book. This requires JavaScript because the length property of the booksPurchased array is only accessible through the API returned object.

```
{"$where": "this.booksPurchased.length>1"}
```

*Note: If your hosting is set up to use the Atlas free tier, you cannot presently try this query out from the portal. The same is true for your code.*

# Object and array properties

In the examples shown, testing properties have been single values, such as a string or a numeric data type. But what if you have a document that has a property that is an array of strings? What if you have an object property? Even better, what if you have a property that is an array of objects?

The address property in the bookstore customer example document ends up as an embedded document in MongoDB BSON. You can do a search for an exact match on an embedded document and specify individual names to match as shown in a previous example.

For arrays, you can do an exact match on the full contents of the array or just on specific values existing somewhere within the array. If the array holds objects, you can search for the element entry and sub-property off that. Here is a previous example query that is doing this:

```
{"booksPurchased.id": {"$eq": "1098"}}
```

Just as you can test multiple single property values, you can also do that for properties that are arrays objects. Let's say you wanted to search for books with book reviews by Joe where he gave a four-star rating. Here an example of how that would look:

```
{
    "bookReviews.reviewer": "Joe Schmoe",
    "bookReviews.rating": 4
}
```

For a simple array of strings, you could match for that exact array. To search for documents where one string entry in an array exists, you could do an equality test. Here is an example document with a property that is an array of strings:

```
{
    "favoriteColors": ["green", "red", "blue"]
}
```

To include that document shown above, here are the query criteria you can use:

```
{"favoriteColors": "green"}
```

Array searches and projection capabilities in MongoDB are powerful. If you take the time, you can learn how to match on things like an element in a specific index or do something like return the first numeric element that is larger than some value. You will have to learn this yourself as you go, as it is tricky to explain all possible nuances. See the MongoDB documentation. For example, look at the `$elemMatch` operator documentation.

# Data type mismatch problem

Equality comparisons can end up producing an undefined outcome if the property data type specified does not match up with the test value data type. For example, you cannot test for a number value on a property that is a string.

The test statement syntax of MongoDB does not work the same as it does in the JavaScript language. The following query will not work because of the data type mismatch:

```
// The selection will not work, as the _id property is a string
// in the document, and you are comparing it with a number
{
   "_id": {"$eq": 77}
}
```

Data type coercion happens in JavaScript. With the following JavaScript code sample, you can see that data type coercion happens. A Boolean test between a string and a number will work in JavaScript.

```
// JavaScript uses "==" for equality testing.
// Coercion rules apply and both equality tests evaluate to true
var v = "77";
v == "77"; // true result
v == 77;   // true result as coercion happens
```

# 3.2 Projection Criteria

Just because you have your query criteria returning the proper result set of documents, does not mean that you are done. You may also want to set up projection criteria to return the set of properties that you really need. You already have seen this demonstrated, but now you can look at this in more detail.

There may be cases where documents with all their properties are what you want returned. This may be the case with a sparse document, making it reasonable to return the whole document every time. With larger, more complex documents, you can benefit from restricting the properties being returned. Limiting what properties are returned saves on the amount of transferred data.

## Inclusion and exclusion

You have already seen the projection criteria in use, so you know most of what you need to know. Just to review, if you don't provide projection criteria, then the complete document is returned. If you do provide projection criteria, then you can specify the inclusion or exclusion of whichever properties you would like. Exclusion means to return all properties except the ones you list. The inclusion and exclusion syntax are as follows:

```
<name>: <1 or true or 0 or false>
```

True means to include and false means to exclude. You cannot mix both inclusion and exclusion in the same projection criteria. The only exception to this is if you are using inclusion criteria, you can also specify one single exclusion if it is to exclude the _id property.

Here are some examples of different projection criteria with a comment added to state whether they are valid or invalid:

```
{"address.state": 1}        // Valid
{"address.state": 0}        // Valid
{"name": 1, "age": 1}       // Valid
{"name": 1, "age": 0}       // Invalid
{"age": 0}                  // Valid
{"name": 1, "_id": 0}       // Valid
```

This is extremely handy. Let's say you want to create a list of people with their addresses. You could use the following projection criteria:

```
{"name":1,"address":1,"_id":0}
```

The following is the result set returned:

```
{
    "name": "Joe Schmoe",
    "address": {
        "street": "21 Main Street",
        "city": "Emerald City",
        "state": "KS",
        "postalCode": "10021-3100"
    }
}
{
    "name": "Jane Doe",
    "address": {
        "street": "100 S Bridger Blvd",
        "city": "Paradise",
        "state": "UT",
        "postalCode": "84328"
    }
}
```

# Missing properties

Since MongoDB can be schema-less, it is possible that any number of documents in a collection that you are querying might not even contain the given property that you have specified in your selection criteria. For example, it is possible that booksPurchased is a missing property in some of your documents by your own design. This is important to consider when you are constructing your query criteria and projection criteria.

The following example query will return two documents, but the second document will not have the booksPurchased property. This is because the second customer has not bought any books yet.

```
// Projection criteria
{ "name":1,"booksPurchased":1}

// Results
{
   "_id": "77",
   "name": "Joe Schmoe",
   "booksPurchased": [
      {
         "id": "1098",
         "title": "Agile Project Management with Kanban"
      },

      {
         "id": "1099",
         "title": "The Merriam-Webster Dictionary"
      }
   ]
}
{
   "_id": "78",
   "name": "Jane Doe"
}
```

If you want this second document left out completely because the property does not exist, use a query selector as shown here to only get those with a non-null value:

```
// Query criteria
{
   "booksPurchased": {
      "$ne": null
   }
}
```

Of course, the property could still exist and just be a zero-length array and it would be returned in that case.

# Arrays

There is a special operator named `$slice` that allows you to return just specific portions of array properties. Examine the following document that has a property containing an array of colors:

```
{
   "favoriteColors": ["green", "red", "blue"]
}
```

Here are a few examples of the use of different operators like $slice, $, and $elemMatch to pull out different elements from the array property shown above:

```
// Return first two elements
{ favoriteColors: { $slice: 2 } }

// Return first element
{ favoriteColors.$: 1 }

// Return first element that matches
{ favoriteColors: { $elemMatch: { $eq:  "red"} } }
```

For more information on these operators, see the MongoDB documentation.

# 3.3 Querying Polymorphic Documents in a Single Collection

In the section on data modeling, I mentioned that you may decide to normalize some of your data. You might like to store some of the data in completely different document types in the same collection. To do this, your query must always include a way to pick out just the documents of a particular type that you want returned.

Using the bookstore example, if you had the customer and the book documents in the same collection, you could include a type property in each.

The following is an abbreviated example showing this approach. Examine the following documents:

```
// Customer documents
{
   "_id": "77",
   "type": "CUSTOMER_TYPE",
   "name": "Joe Schmoe",
   ...
}
{
   "_id": "78",
   "type": "CUSTOMER_TYPE",
   "name": "Jane Doe",
   ...
}
```

```
// Book Documents
{
   "_id": "1098",
   "type": "BOOK_TYPE",
   "title": "Agile Project Management with Kanban",
   ...
}
{
   "_id": "1099",
   "type": "BOOK_TYPE",
   "title": "The Merriam-Webster Dictionary",
   ...
}
```

In every query, you need to include query criteria that are specific to the type of document you want. Here is an example of that query criteria:

```
// Query criteria for retrieving all books
{"type": "BOOK_TYPE"}
```

# Chapter 4: Updating Documents

The previous chapter covered the topic of querying or the 'R' for Read operations in the CRUD acronym. This chapter will cover the 'U' for the Update operation. This chapter will give you an overview of how the update operators work.

When updating a document, you can provide the complete document for uploading. As an enhancement, MongoDB allows you to do things like specifying that a single property be updated. There are many update operators you can choose from, and these can be combined in a single atomic update submission for a given document.

To use the update capability of MongoDB, you first need to provide the query criteria to identify the document or documents to be updated. This criteria uses the same syntax as already covered as the query criteria. What is new here is that update criteria are added as another parameter. Here is the syntax for how the update criteria parameter is structured:

```
{
    <operator1>: { <name1>: <value1>, ... },
    <operator2>: { <name2>: <value2>, ... },
    ...
}
```

*Note:* *Create and Delete operations will not be covered. A deletion operation uses the same query criteria syntax that a read does. All MongoDB CRUD operations will be covered once again in the service layer discussion where code is presented.*

# 4.1 Update Operators

The following are the currently supported update operations used for single value properties:

- $currentDate
- $inc
- $max
- $min
- $mul
- $rename
- $set
- $setOnInsert
- $unset

Here are a few examples to illustrate some of these update operators, using examples that only update a single property at a time. You can combine multiple operators on different properties in one update submission. These will all be committed at the same time and will result in an atomic operation at the document level.

While there are many operators you can use, I will only give examples of a few of them. For every update call, you need, as the first parameter, the query criteria to identify the document(s). The second parameter specifies the property to update. If the query criteria identify more than one document, the update happens on all those documents identified.

*Note: The Atlas management portal currently does not allow the use of the update syntax, so only query searches work. That is why I am showing the examples with code.*

# $set

The `$set` operator is the way you replace the value of a property. The following example sets a new value for the `rewardsPoints` property for the selected person:

```
db.collection.update({_id: "77"}, {$set: {rewardsPoints: 1000}});
```

If the property previously did not exist in the document, it is created. `$set` can be used to do a complete replacement of any value. It can also do a replacement of a complete array or an embedded document. It also works to replace a specific property of an embedded object.

`$rename` can be used to give a property a new name. `$unset` will delete a property.

# $inc

The `$inc` operator is used to update the integer value of a property by a specified amount. You can add or subtract from any value. Using the bookstore example, you could add rewards points to a customer. Here is an example of what that would look like:

```
db.collection.update({_id: "77"}, {$inc: {rewardsPoints: 10}});
```

# $min and $max

The `$min` and `$max` operators are used to test a given value and only replace it if the value is less than or greater than the test value. Here is an example:

```
db.collection.update({_id: "77"}, {$max: {rewardsPoints: 2000}});
```

In this example, if the `rewardsPoints` property had a value of 1000 to start with, it has a value of 2000 after this update.

# 4.2 Array Update Operators

The following operators are for use with array properties to perform updates:
- $
- $[]
- $[<identifier>]
- $addToSet
- $pop
- $pull
- $pullAll
- $push

## $push

The $push operator is used to add another element to an array property. Here is an example that adds a new book review to a book document:

```
db.collection.update({_id: "1098"},
  { $push: { bookReviews: {
    reviewer: "Skylar",
    date: "20150923",
    comments: "It was really profound!",
    rating: 5
  }}});
```

The $addToSet operator is similar to $push except it first checks to see if an identical entry exists and only adds the new element if it is not already present in the array. You can use the additional operator of $each to add multiple elements at once. The $sort operator can be combined with the $push and $each operator to keep the array sorted. Combine the $position operator with $push to specify the point of insertion.

## $

The $ operator is used to specify that the update is to happen for only the first element of an array that is found to match the query criteria. The part of the $set that uses bookReviews.$.rating uses the .$ to signify that the replacement should take place on just the first element match that is found. This example does a search for any document that has a bookReviews element with a rating of 4 and then updates the first matched element:

```
db.collection.update({_id: "1098", bookReviews.rating: 4},
  { $set: { "bookReviews.$.rating" : 5 } }
)
```

60

## $pop and $pull

The $pop operator is used to remove elements from an array property. This example removes the first book review:

```
db.collection.update({_id: "1098"}, {$pop: {bookReviews: -1}});
```

You can use $pull to remove all entries from an array that match what you specify. You can use $pullAll to specify more than one match for the removal criteria.

# 4.3 Transactions

A single write operation might modify multiple documents if the selection query matched more than one document. The modification of each individual document is atomic, but it is not atomic across all the modified documents. It is possible to isolate a single write operation across multiple documents using the $isolated operator, but only if there is no sharding involved.

You may need to modify two documents simultaneously or even modify different properties on each. If you have a requirement to change two documents simultaneously, then you need to work this out on your own. For example, if you want to take rewards points from one document property and add them to a different document, this requires what is termed a multi-document atomic transaction.

This is a new capability in MongoDB 4.x. Previously, you built that capability yourself. Search "two-phase commit" in MongoDB documentation.

You can imagine how important this is to get right in an application that manages transactions across financial accounts. MongoDB now supports multi-document ACID transactions. ACID stands for Atomicity, Consistency, Isolation, and Durability. A document-based database may not need this in most circumstances, and you should only turn this on if you really need it. In the code, you would call an API to start a session that surrounds several operations that you want transacted. Then you would either abort it or commit it. It would look as follows:

```
Const s = client.startSession()
s.startTransaction()
…code to do multiple updates, inserts etc. on different documents.
---> on an error exception you would call await s.abortTransaction()
---> on success you would call await s.commitTransaction()
s.endSession()
```

# Chapter 5: Managing Availability and Performance

In an ideal world, you can store an infinite amount of data, access it from anywhere in near-zero time, and never have any data loss or corruption. Reality is that it takes a lot of work to approach these ideals. As you design your data model and subsequently try it out, you need to tune your DBMS for consistency, availability, and performance. You can now consider what mechanisms are at your disposal to approach these ideals.

It takes a fair amount of time to fine-tune each aspect of the management of your MongoDB database. Many times, you will be faced with tradeoffs. This chapter will look at some of the aspects that can be "fine-tuned" for specific access scenarios.

In a PaaS environment, some of this should be less work than has been traditionally required in the past with a DBMS. MongoDB has done a great job of automating some tasks and operations that would have require a lot of manual configurations.

# 5.1 Indexing

Imagine that you have a problem finding personal belongings such as your car keys, the TV remote, your favorite pair of socks, your wallet or purse, etc. Perhaps when you try to head out the door, you find yourself frantically searching for your car keys every morning.

One approach to solving this would be to keep a whiteboard right next to your front door that has two columns. One column lists the item you need, and the other column lists the actual location of that item. The whiteboard might look like this:

| Car keys | Left side pocket of the jacket hanging in the entryway closet. |
| TV remote | In the pile of toys in the room of baby Jane. |
| Purple socks | Laundry room floor under the pile of towels. |
| Wallet | Under the couch cushion in the TV room. |

Imagine the huge time savings this could provide. I saw one study that stated that, on average, a person spends a whole year of accumulated time looking for lost items over their lifetime.

A database index uses the same concept as the whiteboard lookup table, with the goal of serving database records faster. A database index works by creating a separate lookup list that allows for faster querying. This alleviates the need to search through all documents to find the one(s) you are looking for. For example, let's say you had a lastName property in every document of a collection. If you created a query that was looking for a particular person with the last name of "Smith," how would a query find it quickly? The slowest way to search would be to start looking at all the documents one by one until a document was found with "Smith" in the lastName property. That type of search has no choice but to search each and every document in an unsorted storage system. In a huge collection, this would be a major performance problem.

Indexing can speed up your search by creating a separate sorted list of last names to search against. Each entry would point to the corresponding complete document. A query for the last name of Smith quickly finds those entries using something like a binary search.

Here is a representation of random documents, with one row per document that exists in a MongoDB collection. This is only an abstract representation of how it is stored. To search for "Tuttle," you start a sequential search from document to document until you find a match on the lastName. In this case, it would be the last one found.

## Customer Documents

| lastName | zipCode | rewards | age |
|----------|---------|---------|-----|
| Smith | 27896 | 77 | 32 |
| Williams | 43890 | 3 | 18 |
| Adams | 99054 | 654 | 55 |
| Tuttle | 12345 | 567 | 21 |

*Figure 16 - Customer documents unsorted and with no index*

If you add a second list that contains the lastName property along with the name, you could have a link to the customer document. This second list can be sorted, and your searching will be much faster. A quick binary search of the index list will find the lastName to match and then use the link to get the document.

Implementing a good index can be a critical part of your work to maximize the efficiency of your queries. It is well worth your time to measure and analyze the performance of your database and to fine-tune it with indexes. There are reports in the paid Atlas tier subscriptions that you can bring up that help you look at index performance.

The following example illustrates the concept of linking through an index.

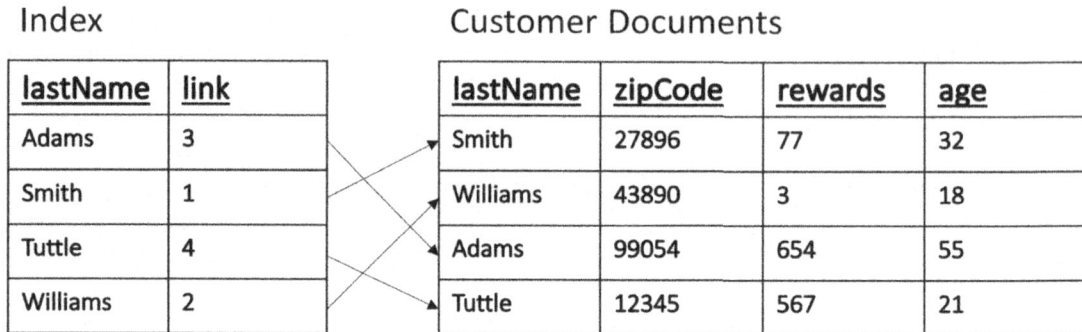

Index

Customer Documents

| lastName | link |
|----------|------|
| Adams | 3 |
| Smith | 1 |
| Tuttle | 4 |
| Williams | 2 |

| lastName | zipCode | rewards | age |
|----------|---------|---------|-----|
| Smith | 27896 | 77 | 32 |
| Williams | 43890 | 3 | 18 |
| Adams | 99054 | 654 | 55 |
| Tuttle | 12345 | 567 | 21 |

*Figure 17 – Last name index for Customer Document querying*

You can use the management portal for creating all your indexes. There are ways to do this programmatically, but for areas of the application that only need to be set up once, I always prefer to do this in the portal.

***Note:*** *Index configuration can get rather complex, and I can only cover the basic common scenarios pertaining to the sample application. You will have to go to the online documentation to get more comprehensive details on what is possible.*

# Single-property index

The simplest way to learn about indexes is to learn how to set up an index on a single property. This section goes over how this works for a single value or object property. The section topic that follows explains the subtleties of what happens if the property is an array data type.

Let's go back to the example of the online bookstore. If you look at the requirements for your querying, you can see that you need to be able to search for customers by name. If you had hundreds of thousands of customers, it is certainly going to improve the performance of this query if you create an index on the name property.

The syntax used to create an index is like the syntax used to set up your other criteria. This example shows how to specify an index on `name`:

```
{"name": 1}
```

That is how simple it is. You can also add an index on a property of an embedded document. For example, what if you wanted to look up all customers who reside in a certain postal code? To make this query run faster, place an index on the sub-property as follows:

```
{"address.postalCode": 1}
```

You can place an index on the address property as a whole, but then you have to put in a complete address for the query criteria with properties in the same order for the match to succeed.

*Note: The _id property that exists on every document is automatically indexed, so you never need to add an index for that.*

# Array property index

If a document property is an array, you set up an index on it in the same way as for a single value property, with just a few restrictions. One restriction is that you cannot have more than one array property index at a time. For example, if your documents have multiple array properties, you can only set up an index on one of the array properties.

The index then is used to match on anything found in the array. The index can only be used for individual element matching and not for matching the array as a whole. As an example, let's say you had the following documents:

```
{"name": "Kiara", "favoriteColors": ["blue", "yellow", "cyan"]}
{"name": "Tristan", "favoriteColors": ["green", "red", "blue"]}
{"name": "Halli", "favoriteColors": ["juju", "nana", "mango"]}
```

You can set up an index similar to the previous example:

```
{"favoriteColors": 1}
```

Now you can set up a query with query criteria to match a color that may be found in the array. If you search for "red," then the document for Tristan will be included in the results set. What is happening internally is that the index contains all array element entries across all documents. That results in three entries in the index for Kiara, alphabetically sorted by the color name. Each of those three would point back to one document. Similarly, there will be three entries for Tristan and Halli.

You can also set up an index if the array contains elements that are objects, as with the following example:

```
{"name": "Skylar", "clothes": [
   {"type": "school dress", "color": "tan", "size": 5},
   {"type": "school pants", "color": "tan", "size": 6},
   {"type": "church shoes", "color": "white", "size": 3}]
{"name": "Korver", "clothes": [
   {"type": "pajamas", "color": "green", "size": 3},
   {"type": "tennis shoes", "color": "brown", "size": 2},
   {"type": "winter coat", "color": "blue", "size": 3}]
```

The following example will set up an index just for one property on the object in the array:

```
{"clothes.color": 1}
```

# Multiple-property index

Many times, you will have queries that specify a match for more than one property. This is used when you want to narrow down your search even further, such as when you want to run bookstore specials to encourage people to use rewards points.

For example, what if you were looking for all people less than 20 years old who had no rewards points so you could give them some points as a free bonus offer to try them out. The following example creates an index on both `age` and `rewardsPoints` so you can search for customers that way:

```
{"age": 1, "rewardsPoints": 1}
```

This is called a compound index. Be sure to understand how this really works. The underlying index sorts all entries by `age` and then sub-sorts entries by `rewardsPoints`. This means that you cannot use this index for a simple query for rewards points only. For example, querying for people with over 1000 points won't use the index. You can, however, use this index to query just by age. This underscores that the order of properties listed for the index is vital.

You could create two separate indexes to query separately by age and rewards points as well as for combined usage in either order. MongoDB will first try something called an index intersection for you to increase performance. For example, you could create the following two indexes:

```
{"age": 1}
{"rewardsPoints": 1}
```

With this configuration, you can still query by the combination of age and rewards points. But you can also query just by age or just rewards points. You can also reverse the combination and query by rewards points and then age, in that order. The only downside is that each index you set up means more storage overhead.

# Index sort order

Up to this point, I have been using the numeric value of 1 in all the index examples, which instructs MongoDB to create the index in ascending sort order. You can alternatively specify −1 and get a descending sort order if you create the following index:

```
{"age": -1}
```

This only matters if you want to return multiple documents of more than one property through a corresponding multi-property index and you want the result returned in a specific sort order. For example, querying for all people with the last name of "Smith" and returning the documents in descending order by age.

Alternatively, you can add a command to sort the result that is returned. With the API usage, there is a `sort()` function that can follow the `find()` function that will process the sort order for you, but it will be done outside the index and will be slower.

# Other types of indexes

The index examples used so far are used for exact value matches and range type queries. So that you are aware, you can investigate and utilize other types of indexes that may suit your needs. Some of these other index types supported by MongoDB are geospatial, hashed, and text indexes.

Let's say you are keeping a database of restaurants along with their menus, reviews, and location coordinates. With a geospatial index, you can then perform queries to take a person's current location and find all restaurants within a certain radius of their position.

If you have large amounts of text in a property, you can use what is called a text index. For example, if you are storing news stories and you want to have an index on the story content, you may find a text index highly performant, which means it provides greater efficiency/performance.

An index called a hash index can be used on properties that hold a single value (such as string, number) or that have an embedded document, but cannot be used with array properties. The hash index is populated with hashed values. A hash index by itself does not work for range-based queries, so you may want another separate single-property index for that.

For embedded documents, the index gives a hash of the complete embedded document. This then is a potentially faster lookup since the search is comparing hash values to find a document. A hash index might make more sense when combined with sharding. This will be covered in an upcoming chapter.

# Index creation options

One of the options you can use with the creation of an index is to specify that the values for a given property are unique. For example, in the bookstore example, each customer has an `email` property. You could make the index require that all emails must be unique across all documents. Here is an example that creates an index with the `unique` option:

```
{"email": 1}, {unique: true}
```
You should create an index like this before any data is populated. Index creation would fail if multiple documents existed that have the same value for their email property. Once an index is created with the unique option, any document creation will fail unless it has a unique value on the indexed field with this option set.

The `sparse` option can be used to create an index that only has entries for documents that contain that property, as shown in the following example. If a document does not have the `email` property, then it won't be included in the index. Thus, any subsequent query for values to match on using the `email` field would ignore those documents, so be certain that's your intent. If an index exists for a property, then MongoDB will use it, so that is why the rest of the documents would not be searched that are missing that property.

```
{"email": 1}, {sparse: true}
```

## Tuning

This about wraps up the discussion on indexes. As I stated at the start of this chapter, there are no perfect solutions in the world of databases. This is true with indexing. What you need to do is to completely understand what your query needs are first to understand how best to create your indexes.

Under certain circumstances, you may even consider not having any index. For example, if your database has 95% of the operations being writes and 5% reads, you might not want to create an index. This is because an index will slow down your write operations. There are, however, configuration settings you can still make to speed up writes.

If you have the opposite situation, and read performance needs to be fast, and writes are a small percentage of the load, then definitely create indexes. You can always measure your performance before and after to make sure that your indexing decisions are valid.

*Note: Once your database is up and running and you are running code queries against it, you can get a diagnostic report on an individual database. This will tell you how well your index is performing. This is done by either using the* `explain` *option or the* `explain()` *method, depending on the API call you are making.*

# 5.2 Availability Through Replication

A single point of failure is never good, no matter what service you are using. For example, in the days of the telegraph, there may have been only a single telegraph line connecting two cities. That configuration creates a service that has a single point of failure. Cut the single line and communication is severed. Having two telegraph lines gives you redundancy. It also gives you greater throughput if you put both into usage at the same time.

The same redundancy is necessary with data stored on hard drives. Perhaps all your family photos are on a single hard drive. What if that hard drive fails? Doing backups to make copies is necessary. Application data must *never* be at risk of being lost or being unreachable. Therefore, some form of data redundancy is necessary with MongoDB.

# MongoDB replica set

If you have a single machine for your MongoDB database, you can configure a MongoDB single-node. When that goes down, then you cannot access your database until it comes up again. This is fine for occasional usage scenarios or for development experimentation.

MongoDB has the ability to configure what is called a replica set. This gives you multiple parallel, redundant copies of all data. This ensures that your data will be safe and available. This is what you always get by default through the Atlas portal if you utilize that PaaS.

In a replica set, you have multiple duplicated databases. Only one machine is designated as the primary at any given time. If the primary database server goes offline, a secondary server will take over. The following diagram gives you a general idea of what this looks like:

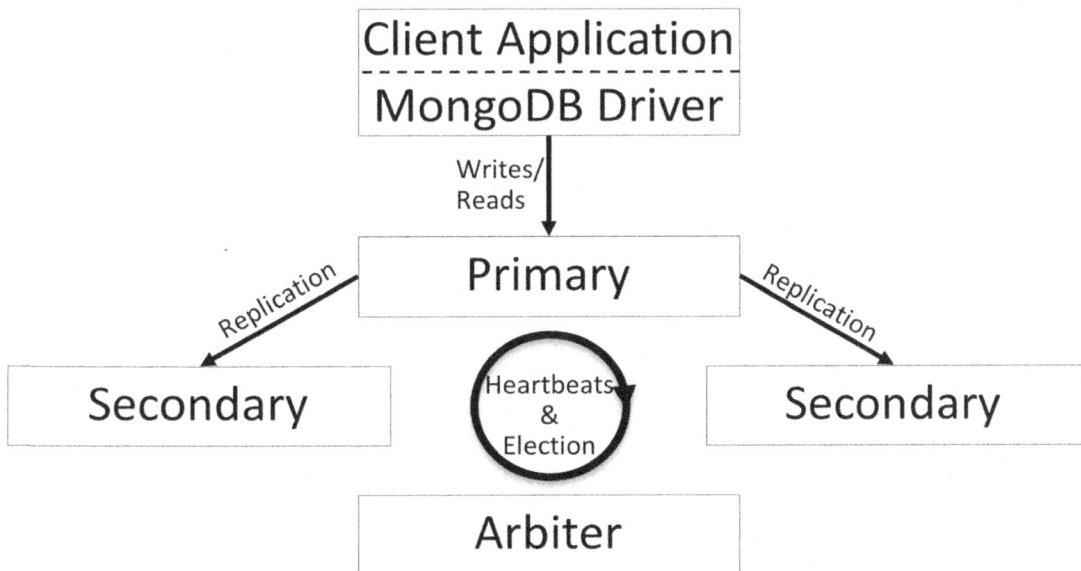

*Figure 18 - MongoDB replica set*

The configuration will look slightly different based on which plan you select from Atlas. The basic idea is that the primary database server receives and fulfills all read and write requests.

69

All the while, the secondary database servers are kept up to date with all changes. Each database server can be on its own dedicated AWS EC2 virtual machine in different availability zones.

Each server is constantly being checked with a heartbeat signal. If the primary database goes down, the two secondary servers and the arbiter detect that, and one of the secondary servers is switched over to become the primary database server. The arbiter is really just there to break any tie votes if needed.

A replica set allows for faster reading of data because multiple copies exist, and data can be fetched in parallel from each replica copy if that is what you want. You must designate reads to be fulfilled by secondary servers if you determine that is justified. Be aware that you can get stale data that has not yet been updated by a replication process. You can keep adding secondary machines to achieve better read performance.

*Note: With a PaaS solution, you generally do not have control over the replica set configuration unless you work with the provider to get something customized. There are pre-determined configurations you select when you purchase a plan. Nothing is preventing you from implementing an IaaS solution and setting up your own virtual machines and replica set configuration if that works better for you.*

## Secondary consistency

There is a complication in having replications available. Any write to the primary storage collection eventually must make it to all the copies. Therefore, you must make a choice as to how that replication is accomplished.

MongoDB has a setting called "write concern" that allows you to specify if you want a majority of replicas to report that the write has taken place before it is acknowledged or failed. You don't have to require this because all writes eventually make their way asynchronously to all replica database servers, yet you may choose to require that acknowledgment.

# 5.3 Sharding

The replication previously discussed stores the same data on multiple machines to provide emergency backup to ensure availability. Sharding is another technique that also spreads data out across machines. With sharding, a given document appears in only one replica set of a sharded cluster and will be in the primary and secondary machines of that shard.

The purpose of sharding is to allow you to grow the amount of data you can store and increase the performance of operations. Both concepts of replication and sharding can be applied at the same time in an architecture. Sharding is just the increasing of the number of replica sets that you designated as individual units.

Sharding spreads the data across multiple replica sets. The multiple replica sets in a sharded cluster act as if they are one single collection. The sharding technology knows where to go for any given read or update to make it easy for you to use.

Sharding helps when you have large datasets and want to maintain high throughput. For example, you may have a lot of data constantly being accessed. This can become a bottleneck with a single SSD. If you distribute the load across multiple SSDs, then the CRUD operations don't conflict as much.

MongoDB can be set up to take care of everything for you for replication sharding issues. You also can select a plan from Atlas that has it set up for you that best fits your budget.

Atlas has various plans you can choose from, depending on how much you are budgeting to spend. With the free-tier plan, there is a hard limit with one single SSD block storage for your database, so you can only go up to a certain size and then you can't grow beyond that. Atlas currently only offers replica set and sharded plans (multiple replica sets), and you can't just have a single MongoDB service on its own.

Cluster plans go up to a certain amount of storage. However, working with Atlas support people, you can keep increasing the horizontal scaling of the sharding by adding more storage. Additional replica sets can, in theory, be added to accommodate your largest data storage needs.

# Reasons for sharding

The concept of data sharding (also called partitioning) was invented to help approach the ideal of being able to store an "infinite" amount of data and retrieve any part of it in a minimal amount of time. Let's dig a little deeper into the scenarios that will cause you to implement a strategy for sharding. Here are some reasons to implement data sharding:

- **Running out of room:** With a limit to storage for a single database SSD, you might simply outgrow that capacity.

- **Machine performance:** You may reach utilization limits for RAM, CPU, and SSD access.

*Note: With Atlas PaaS databases, you can choose a configuration with sharding already configured for you. In the case of Atlas, you can pick a preconfigured machine architecture and then set up how your sharding will act. If you want to go the IaaS route, then you must configure this yourself.*

# How sharding works

Here is how sharding works. Imagine that you start out with a single MongoDB database server and on that server, you have a single collection. Each document you create can have a property that has a random capital letter chosen from A through Z. You may also set up an index on the letter property. A JSON document that you may want to insert can look like this:

```
{
    "letter": "G"
}
```

At this point, no matter what the letter property value is, all documents will be created in the same database collection. The box below represents a single replica set (primary and two secondary machines). This example shows what this would look like:

A-Z

Collection

*Figure 19 - Collection in a single replica set*

Later, at some point, you realize you need to add a whole lot more documents and want to achieve a higher level of throughput on your read and write access. The above single-node configuration can then be made into what is called a multi-node sharded cluster.

MongoDB will start balancing documents between the available shards (replica sets) in the cluster to create a more evenly distributed storage allocation. It actually does this in chunks. With ongoing additions and deletions of documents, MongoDB keeps it all balanced based on the sharding key. You can choose either a hash or a range strategy for your sharding.

Your documents end up being distributed over the three shards in the cluster. See the following figure for a visualization of the distribution using a range sharding strategy:

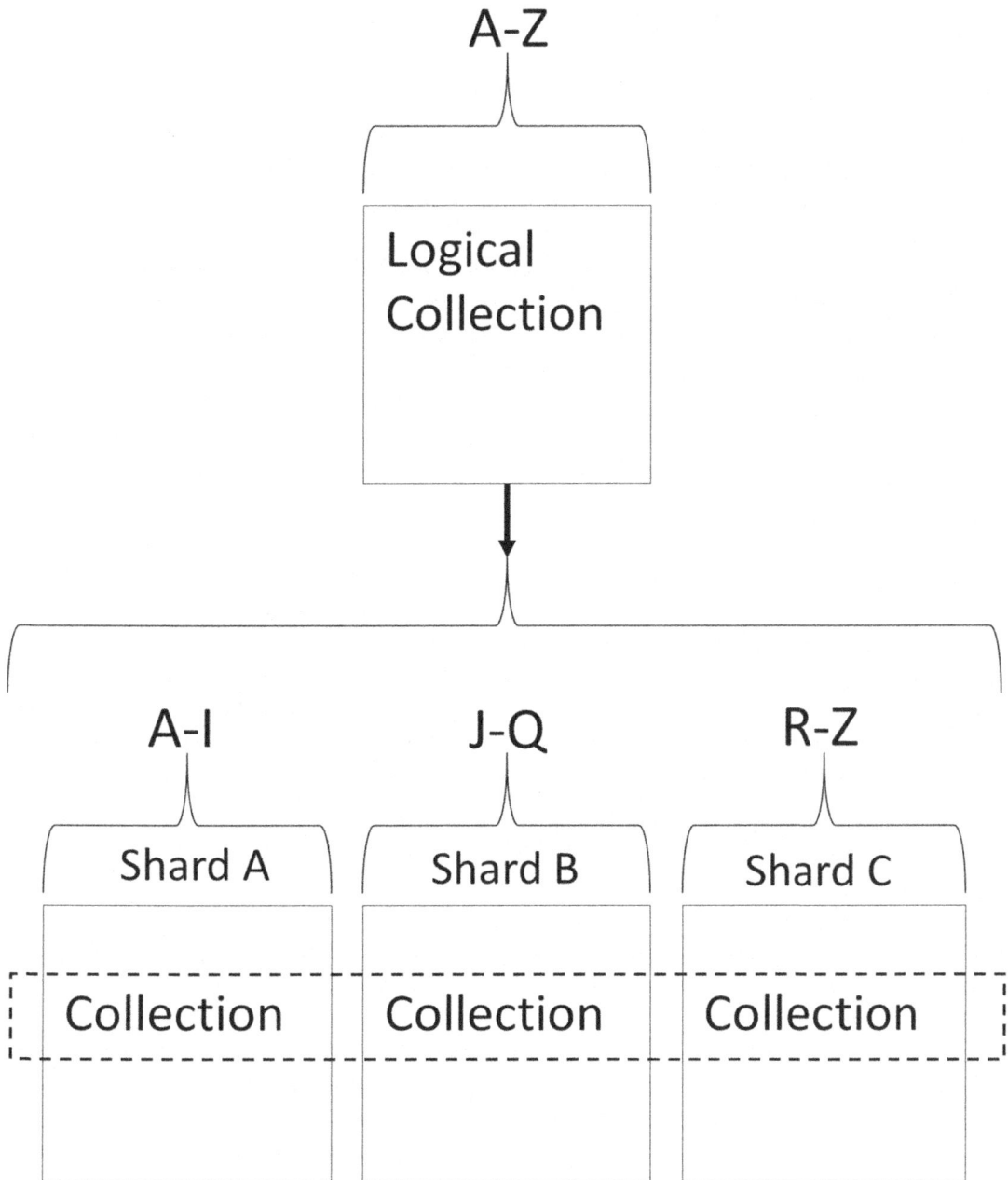

*Figure 20 - Collection distributed over three shards*

As it turns out, each shard is a replica set. When a database request comes into the cluster, MongoDB does all the work to route the request to the proper replica set shard. Your code is shielded from the fact that this is going on. A single logical collection does all the work for you to coordinate across the actual shards that have the real collections.

I won't go into the architectural diagram showing the components to set this up, but you can look it up online if you really want to implement an IaaS configuration on your own instead of using the PaaS solution. When you use the Atlas PaaS solution, you most likely will enlist a support engineer to help you if you want to customize your sharded cluster.

*Note: MongoDB has a way to take shards out of your cluster with a mechanism that lets MongoDB know that this is your intention. Once you do so, MongoDB begins migrating data off the soon-to-be decommissioned shard. Once that is done, that shard can be freed up.*

# Sharding key

Your shard must be set up with what is called a sharding key. Setting up a sharding key is similar to how you set up an index. With an index, you specify a property that you want to use for a speedy lookup using some determined algorithm, such as a range or a hash search.

Look at the previous figure and you will see three shards. The sharding key, in this case, is the property that MongoDB will use to determine what shard each document exists in. For this example, it would have been the letter property.

Each document can only exist in one single shard. A sharding key is thus used as a sorting property. If I create a document with the letter property set to 'M', it can be stored in the middle shard because of the designated range strategy that places it there.

A fair amount of thought is needed to select the proper sharding strategy and to select a property to key off. Just remember that you must know what your queries are going to look like. Don't forget that you may have queries that cross shards, like those using range criteria. Imagine if you want documents from the previous example that had a letter greater than D and less than L. The shard service knows it needs to send the query to both the first and second shards and then your code will process the result set of documents you want.

Indexes still exist on each shard. The query lookup will first go to a shard and then the shard replica set uses any applicable indexes to find the document(s). You can have multiple indexes, but only one sharding key. The sharding key must be the same as one of the indexes. In our example, we had an index for the letter property, and that was also used for the sharding key.

If your documents represent customers, you can look up individual customers by their last name. You can then use a hashed sharding key. That way, queries can narrow the location to one single shard and then quickly retrieve documents from that shard using the index. A hashed sharding key most likely can give you a more uniform distribution of documents across shards for a fairly even retrieval cost. This is great for locating documents using a specific query that can zero in on the document.

Range queries might not be as effective with sharding. If you do know you have a good distribution of range values, then perhaps a range strategy is best. Range sharding is efficient if you have queries where reads target documents within a contiguous range of values.

You also must consider what your queries will look like and what your document composition will look like. You will certainly have a performance problem with hashed sharding strategy if you try and do a range type of query that causes all the shards to be searched.

If your query does not actually utilize the sharding key property, then the service has no choice but to send the query to all shards in the cluster and then collect all the results. But at least, you should have considered what the index should be first, for making that effective.

Like an index, a sharding key can consist of multiple property names. You could thus use a compound key such as last name, first name, and city. Sharding keys cannot be created for a property that is an array.

You may be thinking that if you keep adding documents with the same or similar key value that they all can go to the same shard and then the disk for that shard would run out of memory. This is not the case. There is actually a process called the Balancer going on in the background regardless of the sharding type (range or hash) that moves data around between shards in chunks. You don't set the shard boundaries yourself for the Balancer as MongoDB figures that out for you. The previous figure is just a fictitious illustration of a possible balancing.

You are ultimately limited by the available disk space and that is why you should keep adding shards as you reach the limit of document storage. As a new shard is added, the Balancer does the work to spread documents evenly out across all available shards.

*Note: Let me make sure you have all the terminology down. The Atlas portal lets you create a **cluster**. This is the set of machines that hold your databases. In their terminology, **a cluster** can be either a **replica set** (primary and secondary machines), or a **sharded cluster** (multiple replica sets that are each called a **shard**).*

# Chapter 6: NewsWatcher App Development

This chapter takes some of the concepts that you have learned and applies them to a project using the Atlas PaaS offering, hosted in AWS, to create a data layer. You will learn how to get the data layer up and running and learn some best practices along the way.

In this chapter, you will go to the Atlas portal and create the MongoDB database and collection resources for the NewsWatcher sample application. To get started, you must first have an active Atlas account.

*Note: MongoDB is an open-source project and you can download it for free and run it on any machine you like. However, this is not the approach taken in this book. You can investigate that option if it better meets your needs. There are other MongoDB PaaS hosting options out there besides Atlas, so do your research.*

You will be setting up the following resources:

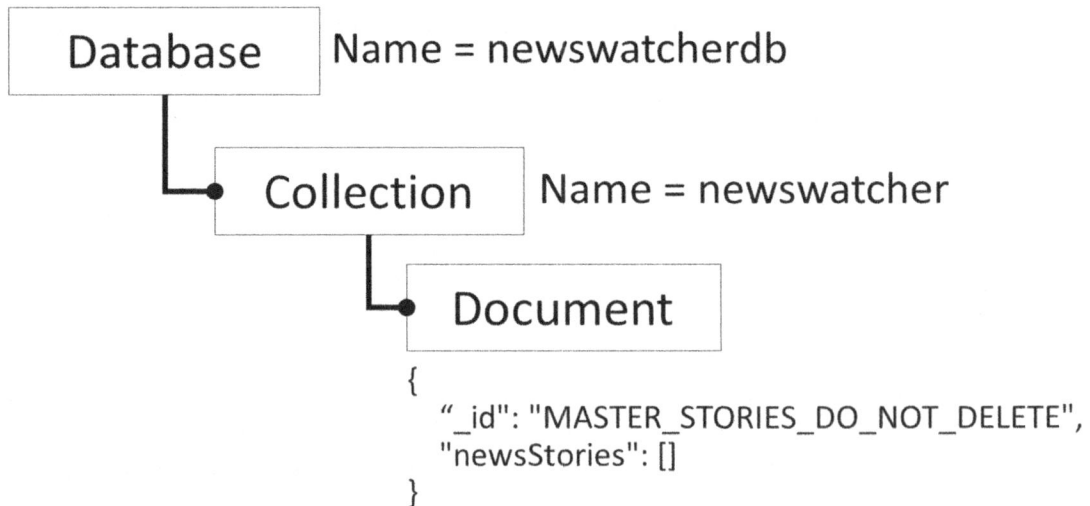

*Figure 21 - NewsWatcher MongoDB resources*

The first document you will add to the collection right now is one you will add manually. It is required for the functionality of the NewsWatcher app. You also need to manually add a few other documents for testing purposes to try out a few queries. Later, you will see how these documents will be added through JavaScript code in your Node.js process.

# 6.1 Create the Database and Collection

The first task will be to create the database. For the sample application, you can select the option that gives you free hosting. This will be sufficient for your development and testing purposes until such time that you need to scale for greater storage and performance.

You also can study the other configuration offerings available through MongoDB, Inc. You can try them out for a few days since you are only charged for the time you have them available, and you can easily delete them when you no longer want the charge.

It is worth the cost to try out some of the other configurations that allow for other capabilities such as sharding. You might want to take some time to look through the plans and pricing pages on the MongoDB Atlas site to familiarize yourself with what is possible. For example, the amount of RAM and storage and IOPs changes per the different plans. Thus, it is important for you to understand your usage needs to be able to select a plan.

When you sign up, you can select the hosting provider such as AWS, the hosting location, and the specifics about the configuration that determine the charge.

To create a database cluster for the NewsWatcher app with the Atlas portal for free hosting, you can go to the MongoDB, Inc., website (https://www.mongodb.com/). It is fairly self-explanatory from there. First, create a project in your account. Then look for the **Build a Cluster** button. You also will provide a username and password. This sequence will look a bit different after you already have an account and are doing this for the second time. Please refer to the online documentation on how to use the Atlas portal. For example, you will need to create an organization and project before you create a cluster. The following figure shows you what the Atlas portal looks like when you use it to create a database cluster.

After a few moments, you will be set up and ready to start using your free MongoDB database from MongoDB Inc., hosted on AWS EC2 machines. If you look in the MongoDB Atlas portal, you also will see your cluster now shows up.

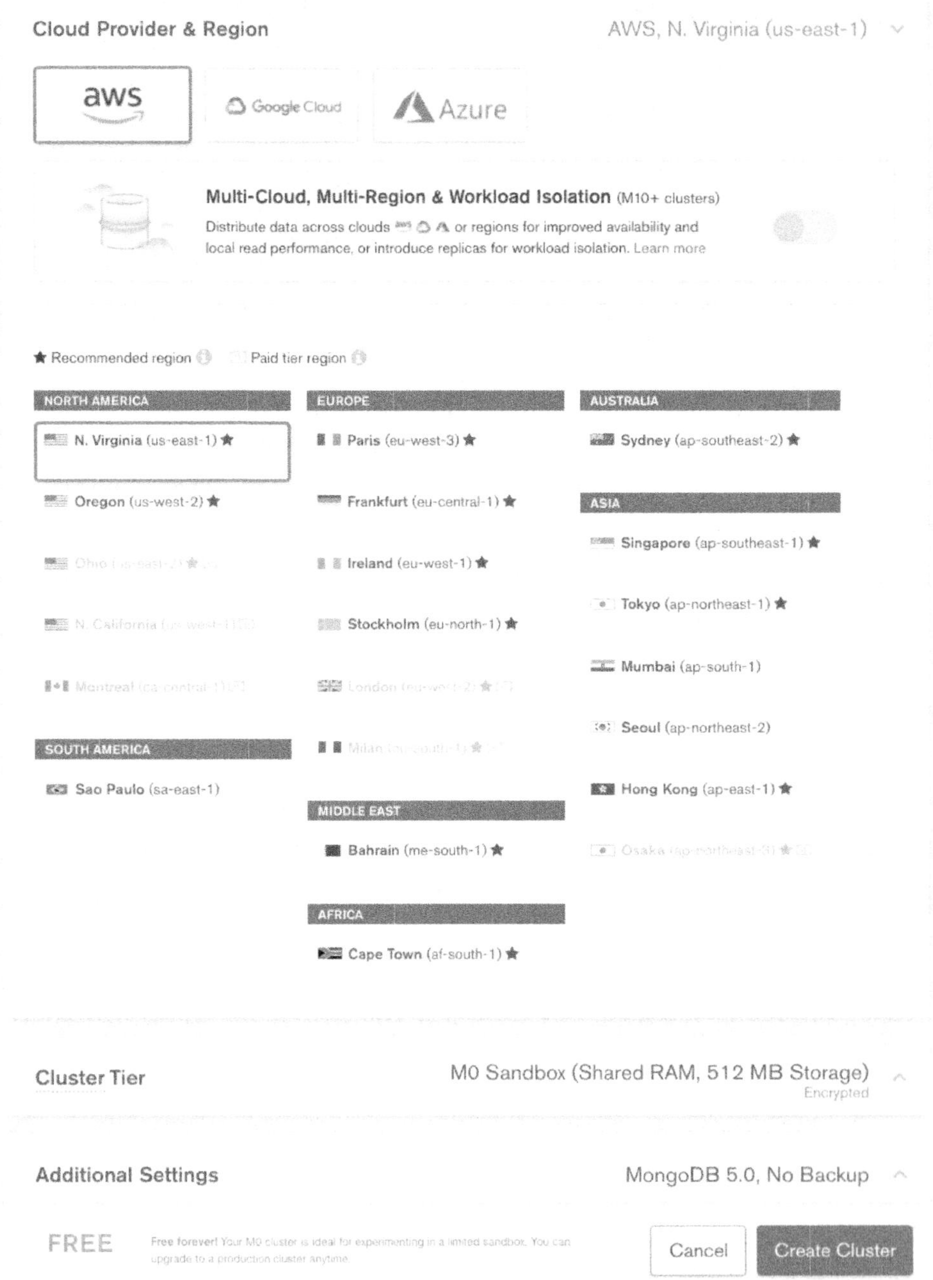

# PART I: The Data Layer (MongoDB)

**Cloud Provider & Region**                          AWS, N. Virginia (us-east-1)  ∨

aws          Google Cloud          Azure

**Multi-Cloud, Multi-Region & Workload Isolation** (M10+ clusters)
Distribute data across clouds aws ☁ ⚠ or regions for improved availability and
local read performance, or introduce replicas for workload isolation. Learn more

★ Recommended region ⓘ     Paid tier region ⓘ

| NORTH AMERICA | EUROPE | AUSTRALIA |
|---|---|---|

**NORTH AMERICA**
- N. Virginia (us-east-1) ★
- Oregon (us-west-2) ★
- Ohio (us-east-2) ★
- N. California (us-west-1)
- Montreal (ca-central-1)

**SOUTH AMERICA**
- Sao Paulo (sa-east-1)

**EUROPE**
- Paris (eu-west-3) ★
- Frankfurt (eu-central-1) ★
- Ireland (eu-west-1) ★
- Stockholm (eu-north-1) ★
- London (eu-west-2) ★
- Milan (eu-south-1) ★

**MIDDLE EAST**
- Bahrain (me-south-1) ★

**AFRICA**
- Cape Town (af-south-1) ★

**AUSTRALIA**
- Sydney (ap-southeast-2) ★

**ASIA**
- Singapore (ap-southeast-1) ★
- Tokyo (ap-northeast-1) ★
- Mumbai (ap-south-1)
- Seoul (ap-northeast-2)
- Hong Kong (ap-east-1) ★
- Osaka (ap-northeast-3) ★

**Cluster Tier**            M0 Sandbox (Shared RAM, 512 MB Storage)  ∧
                                                          Encrypted

**Additional Settings**                  MongoDB 5.0, No Backup  ∧

FREE    Free forever! Your M0 cluster is ideal for experimenting in a limited sandbox. You can
        upgrade to a production cluster anytime.          [Cancel]   [Create Cluster]

*Figure 22 - Create New Cluster page, MongoDB, Inc., portal*

# Add a database and collection through portal

You need to create a database with a collection inside of your cluster as follows:
1. Open the MongoDB Atlas management portal.
2. Click **CREATE DATABASE** (hover over the "+" plus sign at the bottom of the left pane) and type in the **Database Name**. I entered "newswatcherdb". Enter a name for **Collection Name**. I entered "NewsWatcher".
3. Click **CREATE DATABASE** in the form.

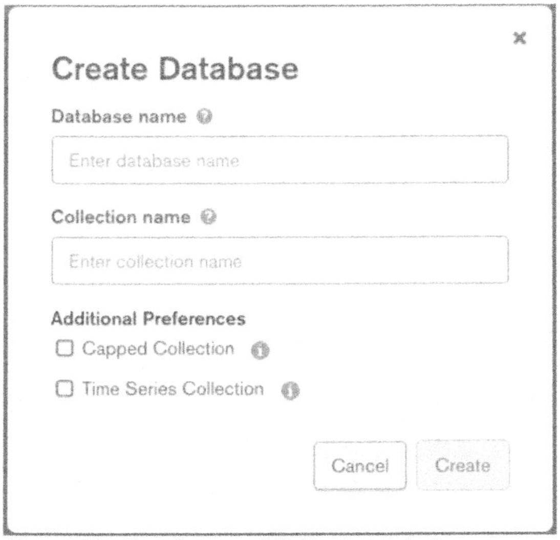

*Figure 23 - Database and Collection creation*

**Note:** *If the portal will not allow you to perform edits, it is probably because you are connected to a secondary machine in the cluster, and not the primary.*

You can now manually create the first required document. Later, you will see how this document fits into your data model. To create the document, do the following:

1. Click on the newswatcherdb database.
2. On the NewsWatcher collection, click **INSERT DOCUMENT**.
3. Paste in the document content as shown below. Make sure to select a data type of Array for the newsStory property. Click **INSERT**:

```
{
    "_id": "MASTER_STORIES_DO_NOT_DELETE",
    "newsStories": [],
    "homeNewsStories": []
}
```

```
Insert to Collection

VIEW  {}  ≣

1 ▾ {
2        "_id": "MASTER_STORIES_DO_NOT_DELETE",
3        "newsStories": [],
4        "homeNewsStories": []
5   }
6

                                    Cancel   Insert
```

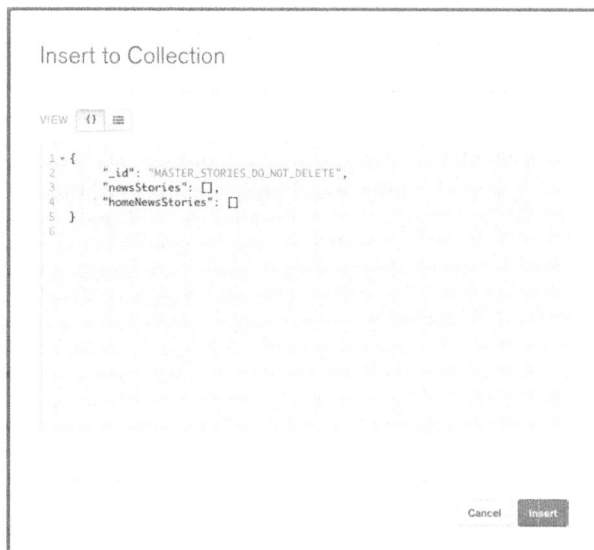

*Figure 24 - Create a document*

After clicking **INSERT**, you will see the document on the collection page. Isn't PaaS wonderful? There is no setup or maintenance or worrying about if you have the latest version of MongoDB. Once created, you can edit any property/field of a document. It is not the easiest editor to use, but it can work for simple changes. In edit mode, you can also change the Type with the dropdown you will find there on the right.

# 6.2 Data Model Document Design

We can now diagram the structure and relationships of the document types that you will need for the NewsWatcher application. This is definitely an iterative process where refinements are made over and over until it is correct. Even after you have implemented a data model, you may find that it does not give you the performance you expected, and you may end up tweaking the design.

Think about what the requirements are for the NewsWatcher application, and you will understand what is needed. First of all, NewsWatcher will have users that log in. Thus, you have identified that there is a need for a user document.

A second document type is for the news stories that users share and comment on. This model is completely denormalized, so there are no keys to link any documents together and there will not be a need for any type of join operations.

The third document you see holds the master list of news stories. There will be code that is run every few hours to collect news stories and store them in that document. You will have multiple User and SharedStory documents, but only one MajorStories document. The following diagram shows a partial list of the needed documents:

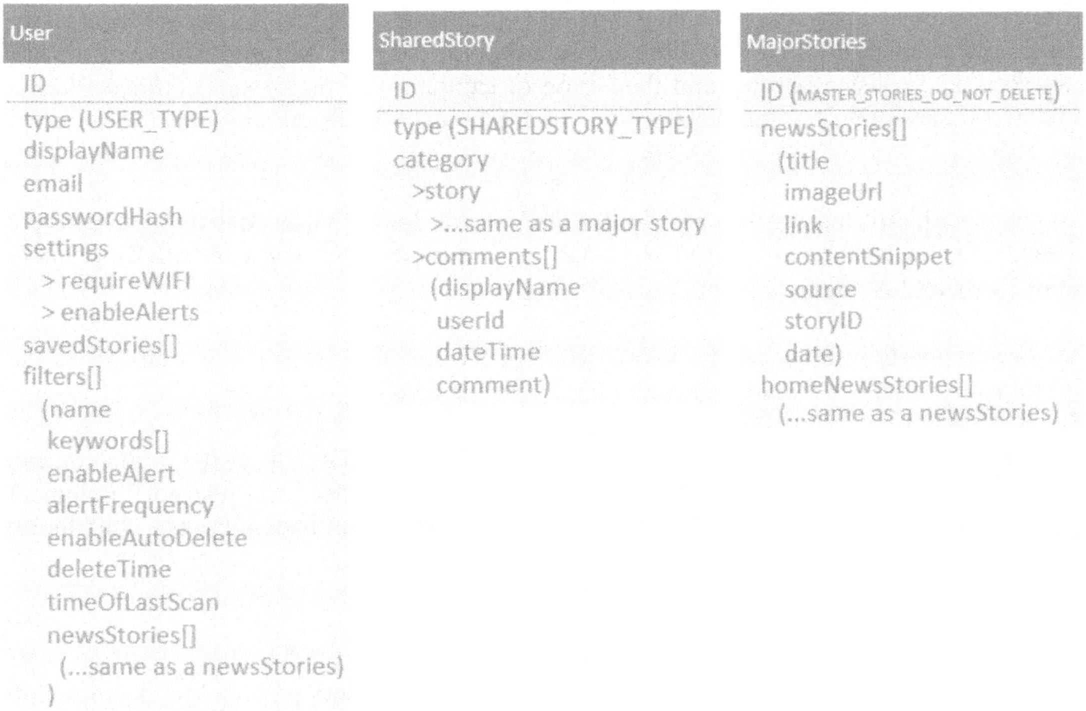

```
User                          SharedStory                   MajorStories

ID                            ID                            ID (MASTER_STORIES_DO_NOT_DELETE)
type (USER_TYPE)              type (SHAREDSTORY_TYPE)        newsStories[]
displayName                   category                        (title
email                           >story                        imageUrl
passwordHash                      >...same as a major story    link
settings                        >comments[]                   contentSnippet
  > requireWIFI                   (displayName                 source
  > enableAlerts                  userId                       storyID
savedStories[]                    dateTime                     date)
filters[]                         comment)                   homeNewsStories[]
  (name                                                        (...same as a newsStories)
  keywords[]
  enableAlert
  alertFrequency
  enableAutoDelete
  deleteTime
  timeOfLastScan
  newsStories[]
    (...same as a newsStories)
  }
```

*Figure 25 - NewsWatcher documents*

Let's look at what the User document contains. In there, you will want to include an email address for each user. This will uniquely identify your users and allows them to sign in. Users must also enter a password. You can safely store a hashed value of the password (you should never store a password in plain text). Then you can let users pick a display name that other users will see when a user comments on a shared story. You should never reveal their email to anyone else.

Next, there should be certain global values that can be used for user preference settings. You can put that in a sub-hierarchy called `settings`. For example, you might want to give users the option of not using any cell phone data and restrict the app to using Wi-Fi only.

You can assume that there will be some users who would like an alert feature to be immediately notified when news stories come in. You could create a Boolean value for that.

The compelling feature of NewsWatcher is the ability to have the application scan for news a user cares about. NewsWatcher users are not the type of people who want to go to some general overall curated news page but are interested in customizing their own specific filtered view of their news. This is done by filtering news stories with keywords.

Users can set up as many filters as they like, so you can conclude that the design requires an array of filters. Each filter will need to contain a title for the filter, keywords, time of the last news scan, and a list of stories and their time of capture. The list of stories for a filter is populated by scanning the master story document to see if there are any matches with the keywords.

NewsWatcher has the ability to save off interesting stories for each user so that they separately appear. This is what the savedStories property is used for. We won't actually implement that property at this point.

The other properties shown in the user document are for other features as outlined in the requirements, but those features won't be implemented here.

This will give you a good start at a Minimum Viable Product (MVP) to go out with. You can model all of this in your diagram with some thinking in advance, and just not implement everything yet. You can feel confident that your data model can accommodate your future needs.

# Entering some test data

At this point, you can open the page for the NewsWatcher collection and add a document to the collection for testing things out. You can insert documents with the Atlas portal as described previously. Here is some JSON you can copy into the insert UI.

```
{
  "type": "USER_TYPE",
  "displayName": "Bushman",
  "email": "nb@hotmail.com",
  "passwordHash": "XXXX",
  "date": 1449027434557,
  "settings": {
    "requireWIFI": true,
    "enableAlerts": false
  },
  "savedStories": [],

  "filters": [
    {
      "name": "Technology Companies",
      "keyWords": [
        "Apple",
```

```
            "Microsoft",
            "IBM",
            "Amazon",
            "Google",
            "Intel"
        ],
        "enableAlert": false,
        "alertFrequency": 0,
        "enableAutoDelete": false,
        "deleteTime": 0,
        "timeOfLastScan": 0,
        "newsStories": []
    }
  ]
}
```

You will see the document added. It has an automatically assigned _id created since we did not provide one. You can open the Atlas portal and view your created documents and delete or edit them as desired.

For now, you can go ahead and experiment by creating a few more User documents in this same collection. Later, documents will only be added through code. At this point, all you need to be interested in is being able to test out some queries before developing the next layer of the application. You can get a feel for how the portal UI is used and learn about how queries are constructed before you put those into code.

When you read through the MongoDB documentation, you will find that there are ways to do bulk importing or exporting of documents. For example, tools such as mongoimport and mongoexport allow you to run from the command line. For example, an export might look as follows:

```
mongoexport -d test -c records -q '{ date: { $lte: new ISODate("2017-09-
01") } }' --out exportdir/myRecords.json
```

# 6.3 Trying Out Some Queries

You may have entered a few documents by hand in a collection. Once that is accomplished, you can now try some queries against that data through the Atlas portal. At this point, you just want to get a feel for what the tool looks like and to be ready to learn about how queries are constructed before you put those into the service layer code.

You will use the same Atlas Portal UI shown earlier to run queries against your MongoDB collection. That is where you utilize the criteria syntax to query and specify what you desire to see in the output.

Try some queries like the following:

```
{"type": "USER_TYPE"}
{"type": "USER_TYPE", "email": "nb@hotmail.com"}
```

Now set the **PROJECT** (projection criteria) in the options area (expand the UI to see it) to be `{"displayName": 1}` and try the query again.

If your query syntax is incorrect, you will be notified of the error. However, if you mistype the name of a property you want to project or query for, you will not get an error but will get an empty result instead. For example, try the projection criteria property name as `{"blah": 1}`. If you do this, you will not get an error but will get an empty result set.

Keep in mind that MongoDB is a schema-less database, and it assumes that the "blah" property could be there in the future, but it just is not there right now. Properties can come and go in a schema-less document-based database.

# 6.4 Indexing Policy

You can write code that uses the MongoDB API that runs to create your needed indexes. My approach is to not put things in the code that are one-time configurations. I instead prefer to use the Atlas portal to create indexes. You also can use the mongo shell to run a command to create the index.

For the NewsWatcher application, now imagine that you have a query in the service layer that will look up a user by their email. You will want to add a specific index for that by doing the following:

1. Open the portal application, click the newswatchedb database.
2. Click the NewsWatcher collection.
3. Click the **Index** text then click the **CREATE INDEX** button.
4. Enter **Fields** as "{email: 1}"
5. Enter **Options** as "{email: { $exists: true}}"
6. Click and then **Confirm**.

You need to check **Create unique index** as you don't want email addresses duplicated across users. This is a way to uniquely identify an account for a person. You have also set up what can be called a "sparse index" using the **Partial Filter Expression**. This means that documents that don't have the "email" property will not be used in the index. This will give you lower storage requirements and offer better performance with the index maintenance.

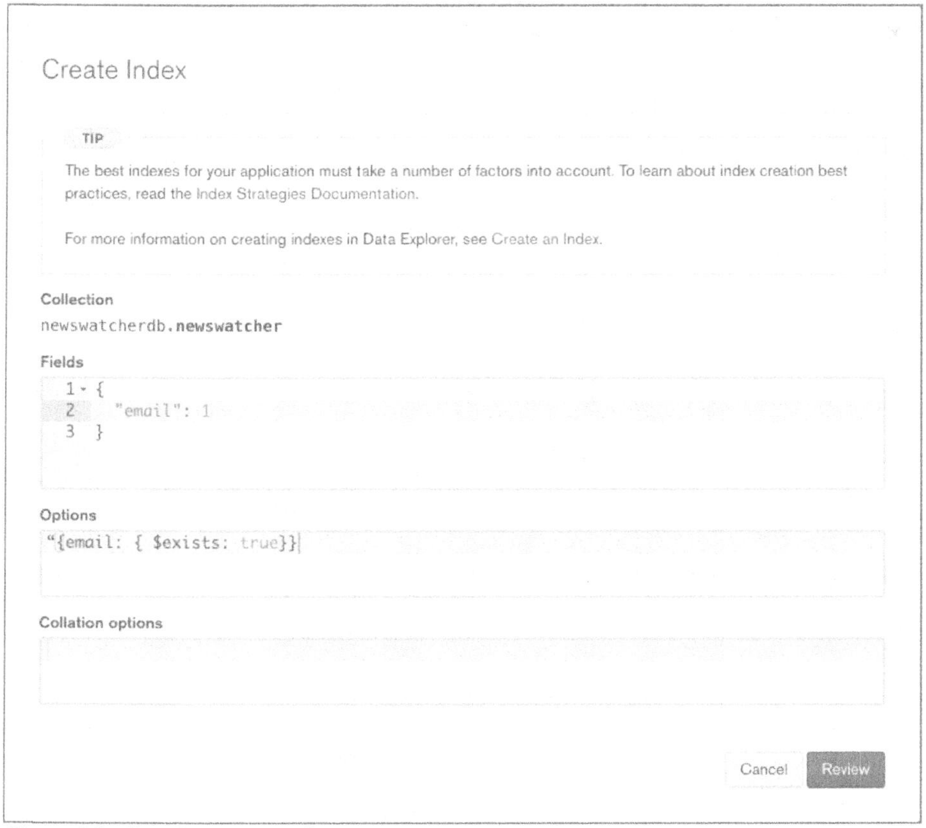

*Figure 26 - Creating a new index*

# 6.5 Moving On

This completes the work to get the data layer up and running. You can see that it made use of the Atlas portal for setting up your configuration. You did not even need to write any code yet. We are postponing the writing of any Node.js JavaScript server-side functionality until you work on the service layer.

## Testing

For the NewsWatcher application, the service layer will end up being the proving ground for the data layer. The service layer connects directly to the MongoDB collection and perform CRUD operations. There will be functional tests put in place to prove that the data model works. You will also be able to take care of the nuances that go along with the data layer, such as performance tuning and concurrency issues.

In reality, the best way to develop software is to work on it in terms of vertical slices of functionality. This means that for any features you have thought up, you will implement it in all three architectural layers at once.

# Chapter 7: DevOps for MongoDB

In this chapter, we will go over some of the operational responsibilities for managing a MongoDB database. For example, with NewsWatcher, you know that the data layer stores user accounts and news stories. You can think about what concerns you may have with that. Daily DevOps work will involve the monitoring of the database. You can take a look at what the Atlas management portal will let you monitor.

You will want to make sure that data access is secure and performant. You can set up replication, sharding, and indexes. Once those are set, you should leave them alone until some change comes along that causes you to tweak them for a specific reason.

# 7.1 Monitoring Through the Atlas Management Portal

One view you can look at is the view of machines in your database replica set. Here is a screenshot that shows all the machines in a replica set (primary and two secondaries):

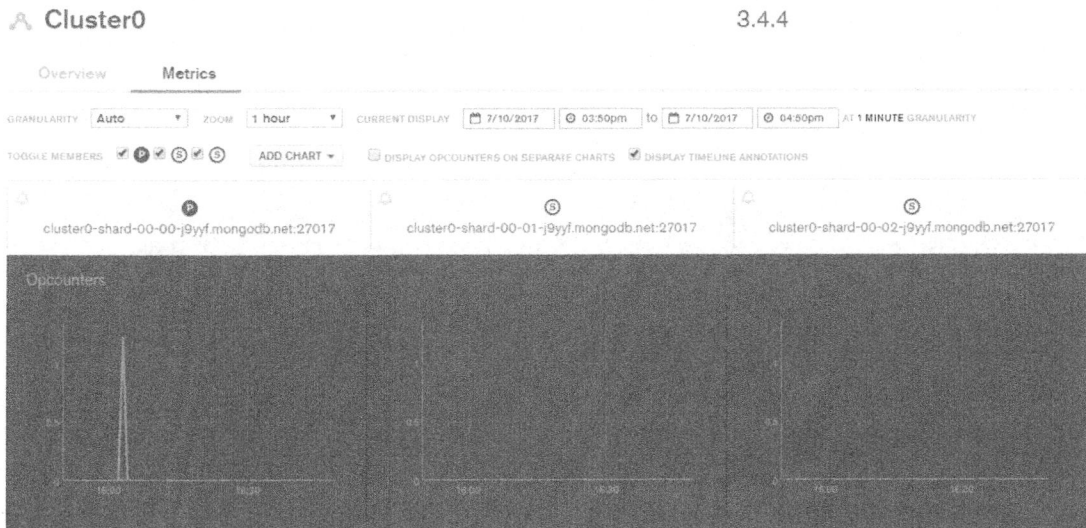

*Figure 27 - Atlas management portal server view*

*Note: The free tier offering is limited in what it offers in the Atlas portal. For example, you get the use of a Data Explorer with the paid tier. The Metrics page for a paid tier also has a lot more information available, –such as Sharded Cluster Metrics, Replica Set Metrics (more metrics), Real-Time Tab, Status Tab, Hardware Tab, DB Stats Tab, and Chart Controls.*

# Telemetry charts

You can drill down further into the performance metrics of each of these machines by clicking on them. There are charts provided in the Atlas management portal to show you the server utilization numbers. The real-time telemetry values will let you know things like how much storage you have used up.

Here is an image showing the Atlas management portal monitoring page with the telemetry that is shown by default. In a replica set, you will have more than one machine, so you must pick the primary or one of the secondary machines if you want to separately see their telemetry.

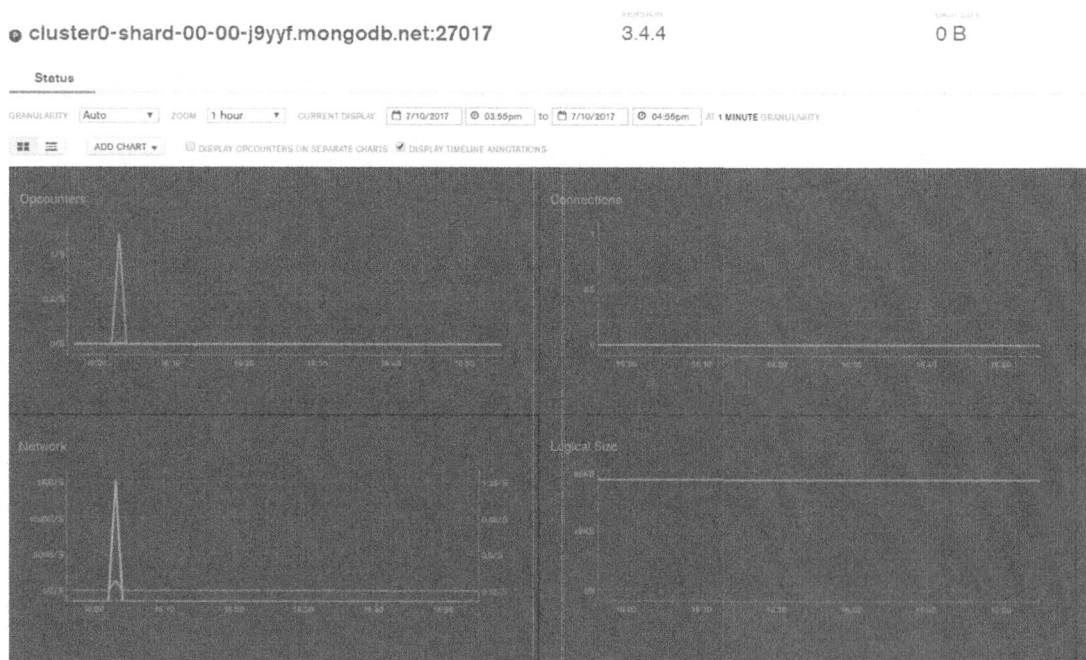

*Figure 28 - Atlas management portal metrics page*

# Telemetry alerting

You will want to be aware of how much storage is left for your database. You can set up an alert to notify you when you are reaching this limit. You might also be concerned about machine performance and set up some alerts around specific performance measurements. If you see your machine performance degrading, then you can shift to a more capable configuration with the PaaS offerings of MongoDB, Inc.

If you go into your Atlas portal, you can click the "bell" icon in the upper right. Then you can select the appropriate tab from there. For example, there is a tab to view and acknowledge alerts that have been triggered.

Click on the **Alert Settings** tab to see what alerts you have by default and add any you need. In the figure below, you can see that there is an alert that triggers when you have reached 90% of your storage capacity.

*Figure 29 - Atlas Alert Configuration page*

You can select from a wide choice of possibilities for how you get notified of an alert being triggered. Click on the **Add** button to see these. The selections include - Atlas User, Email, SMS, HipChat, Slack, Flowdock, PagerDuty, Datadog and much more. Click the "leaf" icon on the upper left and then **Alerts** from the menu on the left if you want to set an alert that is on your account, such as your billing cost going over a certain amount.

# 7.2 The Blame Game

Once your data is secure and has been tuned for the best performance, you don't need much in the way of day-to-day care. However, believe me when I say this – your potential troubles are not over by any means. From my experience, you will be spending your time caring for the integrity of the data as much as anything else. This is especially true if there are a lot of other systems integrating with yours that touch the data at some point.

Unfortunately, every time someone sees a data corruption problem, they come to blame whoever oversees the DBMS. You will hopefully have confidence that most of the time the accusations are unwarranted, and you will be able to track the problem down to some supporting system. For example, some external system that is feeding you data may suddenly have missed, intentionally altered, or corrupt data. It is a good idea to put data validation measures into place at all the points of integration.

You also would be wise to put some handy scripts in place to allow you to diagnose issues and fix them. For example, you may need to recover data from a backup snapshot or reimport data in bulk from a dependent system.

# 7.3 Backup and Recovery

The good news is that you don't have to worry about disaster recovery; the bad news is that you must worry about disaster recovery. It all depends on what your definition is of a "disaster."

With a replica set in place, MongoDB stores multiple copies of your data that are always in sync with the primary server. This means that, within a region, you have redundancy in case of network or drive failures. This is the case with the different AWS availability zones that each EC2 server is in for your replica set. You have this set up for you through the PaaS plan you selected and don't necessarily have to deal with it directly yourself.

Replication means that your data is safe from drive failures, machine reboots, power outages, network outages, and similar incidents. If the drive goes bad that contains the primary copy of your MongoDB database, you are covered. MongoDB and AWS will take care of rotating this drive out and moving you to a new primary drive and adding a new replacement backup.

# You still need backups

Data replication is not the same as performing a data backup. Just because you have replication does not mean you are protected from somehow losing or mangling your own data by mistake. It is a good idea to institute a backup process to periodically store a snapshot of all your data. That way you can recover from inadvertent corruption or loss of your data.

Backups are useful in many scenarios. For example, you might have some bug that was introduced in your code that causes all your data to get corrupted. You then need to roll back to the database copy you had before the data was corrupted.

You could, for example, write some code to copy data from one MongoDB region into another region and keep that as a backup.

You could also save a collection as a file for safekeeping on some local machine you have or place it in EBS or S3 storage in a compressed form. Then you can do a restore programmatically or use a tool to import everything from your snapshot. You can export a JSON file and store it yourself if you don't want to spend the money on database collections or other online storage being used for backups. Just remember that you should be encrypting all backups for proper security reasons.

The Atlas management portal has a Backup selection on the left-hand side that lets you create an immediate backup or schedule a time each day for automatic backup. There is a pre-determined retention policy for each specific time-related snapshot. The core MongoDB project has backup utilities you also can use to perform data backups.

# PART I: The Data Layer (MongoDB)

# PART II: The Service Layer (Node.js)

Part Two of this book will teach you what a service layer is. You will create an HTTP/REST API that interfaces to the data layer. This will set things up to continue to the development of the presentation layer of the platform. It is the UI that talks to the service layer. Node.js/Express.js and JavaScript are the technologies of choice for this service layer.

Many decisions go into creating a service layer. The first thing to design is what type of interface is needed over the data. This involves separating out the different types of data that your REST interface will expose. This requires you to think about the JSON payloads that get transferred back and forth for each request.

You will learn how to use a npm downloaded Node module to call into the MongoDB data layer developed in the first part of this book. At the end of part two, the application will be fully functional and ready to integrate with the presentation layer.

This book uses Amazon Web Services (AWS) to host the Node.js application.

The extremely important topic of testing will be covered, and you will learn how to use the Jest test framework to run your tests. You will learn about functional as well as performance load testing.

In order to give full coverage to the topic, I will also discuss what it means to set up all aspects of the day-to-day operations of Node.js for actual data center operations management. You will learn how to manage a PaaS environment and how to do debugging of issues. Security will be an important topic that also is covered.

*Note: Many people refer to Node.js simply as Node, and I often do the same herein.*

# Chapter 8: Fundamentals

This chapter presents the fundamental concepts of the middle-tier of the three-tier application architecture that is being outlined for you in this book. You will learn what the middle-tier is typically composed of. We can then get into the specifics and show you how Node.js can serve as a middle-tier service layer.

# 8.1 Definition of the Service Layer

The service layer provides the core capabilities of a three-tier architecture. The whole idea of a service layer is to build an abstraction layer over business logic and data access.

If you simplify down the concepts of a three-tier architecture, you can say this about the lower and upper layers – the lower data layer just stores data and the upper presentation layer just displays the User Interface. That leaves the middle-tier service layer to do all the rest of the work. In most applications, you will find more code in the service layer than in the other layers. The following diagram shows this simplified view:

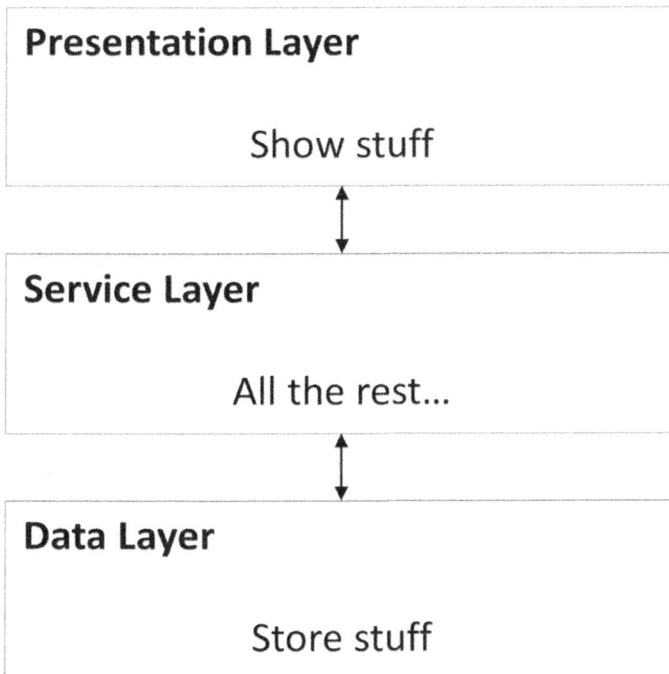

**Presentation Layer**

Show stuff

**Service Layer**

All the rest...

**Data Layer**

Store stuff

*Figure 30 - Simplified three-tier architecture diagram*

It would not be reasonable to have the presentation layer handle the business workflow logic. You also do not want to expose business workflows in the data layer. The data layer should be kept as simple as possible and should only handle the CRUD operations and perhaps more difficult data transaction logic. You will normally find the more complex business services code in the services layer.

*Note: Some architectures split the middle-tier out into a services and a business layer. These two concepts are combined in this book. You will find that all the needed functionality of this layer can be accomplished within the single technology framework of Node.js, along with the use of numerous npm packages. You will also see a separate web app tier, however React as a client side SPA consumes that functionality.*

# A contract of interaction

A service layer can be created with no particular UI in mind. For example, major companies expose their APIs so anyone can interface with their backend and write their own UI. Companies such as eBay, Twitter, and Facebook have been successful at this. For example, you interact with the eBay API to bid on items.

In the case of the NewsWatcher sample application, there is just one single UI that was written, yet anyone else could write a different UI on top of the exposed REST API.

Regardless of whether you are tied to one single front-end UI or if your service layer is open to allow many applications to connect, you need to think in terms of a strict contract of interaction. This means you must define the connection routes upfront and the JSON messages that are required.

I will not be using any specific connection standard like you see with Simple Object Access Protocol (SOAP). You can also explore things like Swagger on your own if you are interested in making your API generally available to people in an easy-to-consume and formal way. You can also explore an AWS offering known as the API Gateway. This can be used to surface a formal API contract that sits in front of your Node.js service layer.

# An abstraction layer

A service layer is built in a way that abstracts away the complexities that go on in the backend. This will shield the presentation layer client-side code from any tight coupling. The client is also protected from any backend rework. In many cases, the client doesn't even need to make any changes when backend code is rewritten.

You find architectures where a single call to the service layer results in a series of backend calls that are each processed in what can be called a workflow. This coordination falls squarely in the service layer to hide the complexity. The multiple backend services comprise your overall architecture and can be unified through a single Node.js entry point.

96

In large enterprise platforms, you should never build one single monolithic service that does everything. For example, you may have one service that does all the storage and retrieval of all user account information. Another service might contain billing account information, and yet another might deal with order information. Take the time to split up your platforms into discreet services. Each of these will serve a role, be self-contained, and operate independently. Research what is called a microservices architecture. NewsWatcher breaks out key functionality into separate code but is still deployed as mostly one monolithic application.

You will want your presentation layer to only have one single service layer entry point that coordinate calls to other independent microservices that each exists as autonomous services. You will be grateful you have done this, as you can make rather significant changes in the lower layers and minimize the code that must change in the client.

Perhaps your backend might already consist of "legacy" systems that were not written with Node. These backend systems might be written in different languages and be running on your own proprietary, on-premise platforms. In this case, you can decide to write a gateway service layer with Node and have that code interface to your backend systems. Node is great for routing and orchestration. It can work to process requests asynchronously with a high degree of concurrency. AWS offers an API Gateway that you can also investigate.

All or part of your systems can have their own Node.js interfaces. It is up to you to decide what is worth your time and investment to do. The following diagram illustrates the gateway layering that could exist in a services layer:

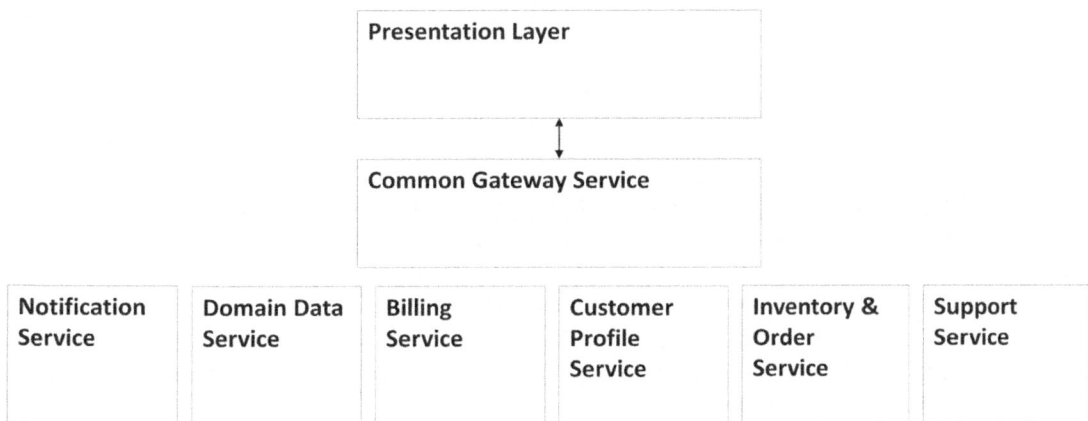

*Figure 31 - Service layer gateway concept*

If you decide you need to replace the billing system, the gateway service provides the abstraction and protects the client from any changes.

## Service layer planning

Knowing what operations and workflows are needed is the first step in creating a service layer. Use the following questions to help you determine the design of your service layer:

- What operations and workflows are needed?
- What are your security and privacy requirements?
- How are people authenticated and authorized?
- Is there a need for a pub/sub notification system to deliver push notifications?
- Is there any type of domain-specific configuration required?
- Is programmatic resource management required? Scripting or templates?
- Are all data interactions encrypted?
- Are multiple systems going to call this? Would a message contract schema be appropriate?
- What validation can be made at the API interface to not let anything invalid in?
- What meaningful errors need to be returned?
- Do you need user roles and access control?
- Do you need caching on top of your database layer?
- Is there any periodic processing of the data? Is it periodic in the background or real-time? In batches?
- Will you need to queue up work and have workers process work asynchronously?
- What are the SLA requirements for all operations?
- What are the access volume and the rates per time period?
- Will there be bursts of activity or is access evenly distributed?

The answers to these questions should be carefully considered. Consult experts along the way before you roll anything out into a full production environment.

# 8.2 Introducing Node.js

The simplest way to describe Node.js is to state that it is a runtime that executes JavaScript code that you provide. You might think I just described a browser environment for you. After all, this is what the Chrome browser can do. The Chrome browser has a JavaScript engine called V8 that is used for executing client-side JavaScript code.

Node.js, however, is running on the server-side. To accomplish this, a person named Ryan Dahl actually took the same V8 engine mentioned and made use of it in a server-side process, and then built additional functionality that makes sense to have running in server code.

You can think of Node as an abstraction layer on top of your operating system to make your code run as platform independent code and get many of the capabilities that an operating system provides (file system, processes, network, etc.). Node.js provides the overall runtime execution environment.

# Platform independence

Let me put it this way - you could go and write an application in C++ for Windows that implements a web service that listens to and responds to requests and interacts with the file system. But then what would you do to take that application and make it run on Linux? You would have to port it, which requires a rewrite of that C++ code to use operating system libraries found on Linux machines!

To get that to run on a Linux OS, you would port your code to use system calls that are available on Linux. This image shows that you would be writing your code over and over for each platform.

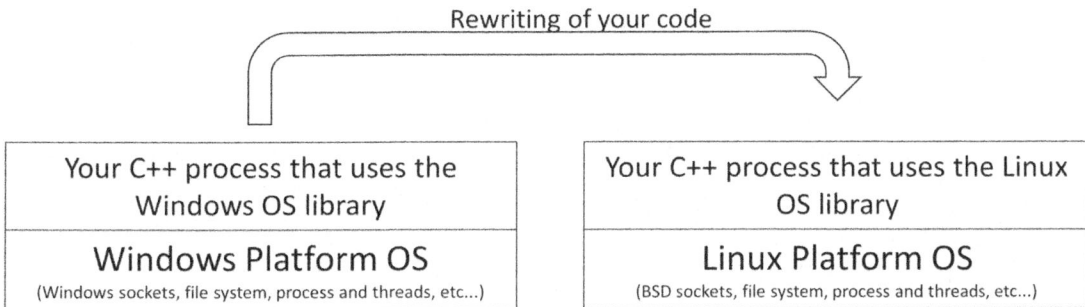

*Figure 32 - Code porting across from OS to OS*

Node lets you write your application code once, and then Node handles the lower-level porting of Operating System level calls for you. Node.js acts to abstract away the platform OS capabilities. Not only that, but Node allows you to write all your code in JavaScript!

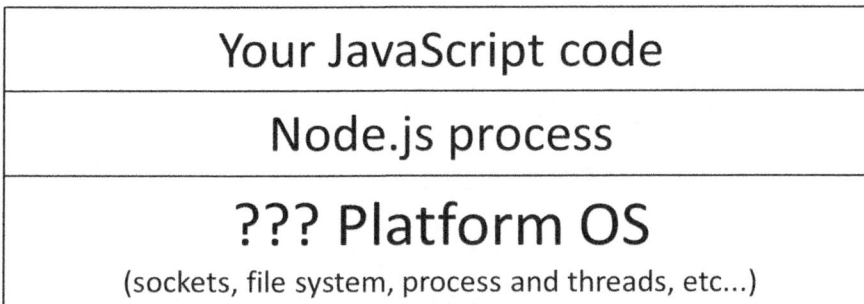

*Figure 33 - Writing code once*

Node.js is an open-source project and has been ported to run on different operating systems. Much of the core code of Node.js is written in C/C++ to enable native integration with underlying operating systems and achieve the fastest possible performance. It also utilizes the Google V8 engine to execute JavaScript. V8 actually compiles client JavaScript code to native machine code, such as for x86 machine architectures, for faster execution.

Node.js is well-suited for network-based I/O applications. Using Node, it is extremely simple to piece together a web server similar to IIS or Apache usage. You can easily set up a web service to expose an HTTP/REST API that works with JSON payloads. Node really fulfills a lot of purposes that allows it to satisfy all the requirements of a middle-tier service layer.

# Extensibility of Node

The real power of Node comes through its extensibility. Node was written to provide the core runtime of execution, scheduling, and notification capabilities. Its functionality is then greatly increased through the many extension modules written for it. For example, there are modules for functionality such as WebSockets, data caching, database accessing, asynchronous processing, authentication, and many others.

Node is widely adopted, and people are constantly improving it and creating new modules for it. There is a large community of developers that has generated many extremely useful modules to give rich functionality to your application. Since the patterns of using these modules are all similar, it is extremely easy to consume them without any learning curve.

# JavaScript bliss

Since Node.js executes JavaScript, there is a consistency throughout the application stack that you are building. JavaScript design patterns utilized in the construction of modules become the self-contained components that then are available to consume. You can download and integrate those offered by others as well as create your own to be shared.

There are also unit-testing frameworks that work with the JavaScript language that you will use to test your Node.js service layer.

JavaScript is utilized as more than just a client-side scripting language in a browser. It is now a formidable server-side language as implemented with Node.js.

*Note: Lest I get a slew of emails accusing me of living in a fairytale, I will make a brief comment on the sensibility of using JavaScript in enterprise applications. It is true that JavaScript can be a challenge with respect to delivering on quality. However, with proper design patterns and testing, you can today create large enterprises services of superb quality with Node.js and JavaScript. Many large companies have already done so. Perhaps as the JavaScript language evolves, more features will convince the skeptics that it is here to stay.*

# 8.3 Basic Concepts of Programming Node

A simple Node program that you can write is a program that outputs text to the terminal process window. The following is an example of what this looks like:

```
console.log("Hello World");
```

You can place this text into a file, save it to disk, and then have the Node process execute it. If you had Node installed, you could open a command prompt and type the following, substituting the name of your file for *<filename>*.

```
node <filename>.js
```

If you haven't already done so, you should go ahead and install Node on your machine at this time. Go to https://nodejs.org/. You will see a download link labeled "LTS" and another labeled "Current." You want the Long Term Support (LTS) version, as that is the stable one. The other one is from the latest code under development and has not been sufficiently proven.

Go ahead and create a file named server.js, place the `console.log("Hello World");` line in it, and run node as shown with this file as an argument. You need to be in the same directory as your file.

You will also notice that the process does not stay running. In this case, once your code in your file has been executed, the Node process exits. This does not have to be the case. I will soon explain what causes a Node application to keep running so it can continually perform server-side processing, such as listening for and processing web requests.

## The REPL

Node has several different options to control how it runs. If you were to leave off the JavaScript file, Node would default to what is called REPL mode. Read Evaluate Print Loop (REPL) is the mode where you get a prompt and can enter JavaScript to be executed as you type it in. REPL is a common thing for execution frameworks to provide. For example, MongoDB provides something similar, called the mongo shell.

In this book, you will not be using the REPL and only need to be concerned with the main means of invoking Node as already shown, passing your JavaScript file as an argument.

# Node executes JavaScript

Here is another file to try. This one illustrates a bit more code that Node can execute. Running this will result in "HI THERE 343" being displayed:

```
var x = 7;
var s = "Hi there";

function blah(num, str) {
    if (num == 0) {
        return "Can't do that";
    }
    return str.toUpperCase() + " " + Math.pow(num,3);
}

var result = blah(x, s);

console.log(result);
```

This illustrates some basic JavaScript language capabilities. If you look at the JavaScript specification, you can see more of what is possible. You have datatypes, operators, structured programming, logic control, built-in objects, and much more.

If you have previously programmed browser scripts, be careful about what you assume is available in JavaScript. For example, there is functionality available in Node that are not available in browser JavaScript and the other way around is also true. Of course, some functionality only makes sense in the browser context as well as other functionality only making sense on a server machine.

What you will find is that people have done the work to try and get comparable functionality across both and have ported back and forth some capabilities that both sides of development have needed.

As was mentioned, the Node process will run and exit upon execution of your lines of code. This is because Node.js will only run while it knows it has code to execute. Try running the following code:

```
while(true){
  console.log("Hello World");
}
```

To stop the Node.js process, you will have to press <Ctrl> C on your keyboard or close the window. Later, you will see some code that requires Node.js to run forever because it has been set up to respond to events that could perpetually happen.

# Using built-in modules

As previously explained, Node extends the basic capabilities of JavaScript by providing a set of built-in modules. There are quite a few of them, and you can review them if you go to https://nodejs.org/en/docs/. Click on an API version on the left to see the capabilities. You will see things on the list such as HTTP, Net, OS, Crypto, File System, Console, and Timers.

You can now see how to use the Node.js provided functions such as `console.log()` and `setTimeout()`. Here is some sample code that takes advantage of these added capabilities:

```
setTimeout(function () {
  console.log('World!');
}, 1000)

console.log('Hello');
```

If you are not familiar with how `setTimeout()` works, you have to be aware that this schedules a callback function to run some number of milliseconds in the future. Thus, the string `Hello` prints first and then one second later you see `World!`.

Let's now look at how easy it is to create an HTTP Web server with code that Node would execute. The following example is all the code needed if you use the built-in HTTP module:

```
var http = require('http');

var server = http.createServer(function (request, response) {
  response.writeHead(200, {"Content-Type": "text/plain"});
  response.end("Hello World\n");
});

server.listen(3000);
```

If you execute this with Node on your local machine, you then can open a browser and navigate to http://localhost:3000/ and see your `Hello World` message appear.

The only non-obvious line of code is the first one. This is a function defined in Node.js that you call to load the HTTP module that provides the web server capability. Those who write browser scripts in JavaScript use an 'import' statement instead (see Node documentation).

Node uses the concept of modules as its extensibility mechanism. The `require()` function simply returns an object with functions and properties placed on it that are specific to that module for you to use. In this case, you call `require('http')` and get an object, and then use the `createServer()` function from that object. Some things like the `setTimeout()` function are made available globally without you needing a `require()` statement.

# Using external modules

In some programming languages and runtimes, you mostly rely on what comes built in. For example, this is common with the combination of C# and .Net. In Node, however, you will constantly be looking for external modules to add to your application to give you many of your capabilities. As a matter of fact, since there are so many of these external modules provided online, you will need to become good at searching for something and then determine the best option.

There is a package manager site you can go to for searching and downloading NPM modules. Go to https://www.npmjs.com/ to take a look. Be aware that some of what is found in NPM are code modules you can use, while other downloads are actually tools you can run, such as the testing tool called Jest.

When the NPM modules are utilized, the layers of code now look as follows:

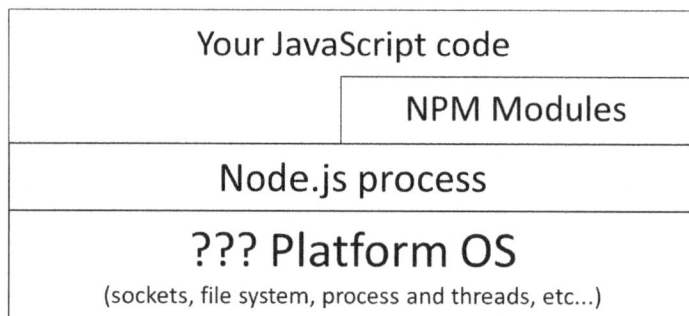

Figure 34 - Adding NPM modules to the architecture

This gives you the highest level view of the different blocks of code.

*Note: Regular Node modules are written in JavaScript and distributed for others to use. There is, however, a way to write a module in C++ that is referred to as an add-on. If you do that, you can get better performance (more sustained compute intensive code) and also access to operating system APIs that are native to a machine that may only be available in C++ libraries. N-API is a newer capability provided by Node for building native add-ons. Learning to create add-ons is not necessary for most projects and the topic will not be covered in this book.*

To use external modules from NPM, you need to have them installed alongside your own JavaScript file. There are two steps needed to use an NPM code module:

1. Create a file named package.json.
2. Run the command "npm install <module> --save" for each module you want to use.

Running the install command will actually add a line in your package.json file. Also, the first time you run it in a directory, a folder named node_modules is created. If you look in that folder, you will see all of the module code. When you run Node with code that requires one of these modules, Node will be able to find them.

You can try the steps by creating the package.json file with the following lines in it:

```
{
  "name": "test",
  "dependencies": {
  }
}
```

Now run the following command:

```
npm install async --save
```

After running this, a node_modules folder is created, with the async module files installed. Your package.json file will look as follows:

```
{
  "name": "test",
  "dependencies": {
    "async": "^1.5.2"
  }
}
```

There is a convention in the package.json file for listing modules. The name is listed followed by the version you desire. You can specify an exact version or you can specify just the major number and have NPM get the latest minor version. The '^' character for our usage of the async module means that if any time you refresh your usage with an "npm install" command you would bring in any minor or patch updates. For our example above, that will not bring in a 2.x.x version, as that would be a major version update.

If you add dependency modules in your package.json file by hand and specified the exact version, you can then run the "npm install" command. NPM will install the latest version for what version numbering you specify. The following command is what you run if you edit package.json first and then want to install all the modules you specified:

```
npm install
```

It is typical for developers to stick with a known good working version for each of their modules and not update to any new major versions even when they become available. When you feel you need some new capability or security patch of a newer version, then do an update and do extensive testing of everything again. Major versions have new functionality.

Here is another code sample you can run. It makes use of the async module that was installed. Put this code in your server.js file:

```
var async = require('async');
var fs = require('fs');

async.eachSeries(['package.json','server.js'], function(file, callback)
{
  console.log('Reading file ' + file);
  fs.readFile(file, 'utf8', function read(err, data) {
    console.log(data);
    callback();
  });
}, function(err){
    if( err ) {
      console.log('A file failed to load');
    } else {
      console.log('All files have been successfully read');
    }
});
```

You will learn more about the async module later. This is basically using the async module capability to sequence through an array of values and sequentially do what processing you want on each entry.

Note that you are using the fs module. You don't need to run an install or even list the fs module in the package.json file. This is because this module is part of Node.js, but you still need the require statement. If you now run "node server.js", you will see the contents of your files printed out.

*Note: Deployment of your Node.js application is easy. You can just copy everything, including the node_modules directory, to a machine. When you use a PaaS environment install, such as with AWS Elastic Beanstalk, you don't need to copy the node_modules directory. AWS will run the npm install for you. If you want version numbers locked, set the specific versions or use what is called a shrinkwrap file or use NPM version 5 or later.*

# Callbacks and concurrent processing

To start with, you need to understand that there is just one main thread of execution in your program. This is the thread that starts up your application and begins execution of your JavaScript code. From there, Node.js sends all your code to the V8 JavaScript VM engine, and OS abstracted calls to begin its execution.

Everything at the highest level of your JavaScript places processing time on a single thread. Only one thing can happen there at a time, so write your code so it is not compute intensive.

***Note:*** *VM stands for Virtual Machine and is a concept that V8 uses to isolate JavaScript execution. Don't confuse this definition of a VM with a VM that you find hosted in a cloud data center.*

You have already seen Node code can use the callback style of coding. This style is prevalent, as well as the async/await and Promise style usage. The Node.js library provides for these non-blocking asynchronous callbacks. Your code never blocks, but returns immediately and then at some later time, the callback function is executed. This gives you the concurrent execution capability that Node is famous for.

***Note:*** *There are other mechanisms for doing asynchronous code such as using Promises and Async/Await. These are also popular and are basically a syntax preference. Async/Await will not be covered in this book.*

# Code execution flow

The code execution path is interesting to trace through. I will now explain a bit of how this all works. Look at the following code that will be used to help understand the execution flow for your JavaScript code:

```
var x = 7;
var s = "Hi there";
var fs = require('fs');

function blah(n, s) {
   if (n == 0) { return "No"; }
   return s.toUpperCase() + " " + Math.pow(n,3);
}

console.log(blah(x, s));
fs.readFile('package.json', 'utf8', function(err, data) {
   console.log(data);
});
```

If this code is in your server.js file and you execute it on the command line as node server.js, your execution looks as shown in the upcoming figure. You can see the code boundary crossings. You can also see that almost everything is non-blocking. At a lower level, there is a thread in Node that does end up being blocked, but this thread does not affect you at all. Your code still has the callback that is asynchronously called, so you are not blocked by it. In the illustration, the execution time moves left to right. I have illustrated the boundary between your main JavaScript thread and the Node.js framework with Libuv, a multi-platform library with a focus on asynchronous I/O (more information on Libuv in a later chapter). The upper part is your code being executed in the V8 VM.

# PART II: The Service Layer (Node.js)

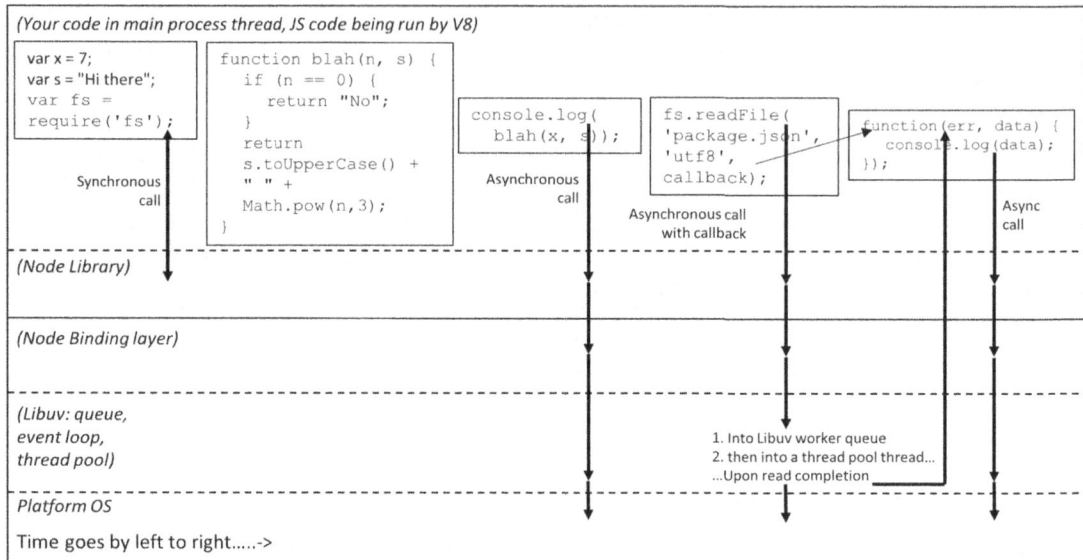

*(Your code in main process thread, JS code being run by V8)*

```
var x = 7;
var s = "Hi there";
var fs =
require('fs');
```
Synchronous call

```
function blah(n, s) {
   if (n == 0) {
      return "No";
   }
   return
   s.toUpperCase() +
   " " +
   Math.pow(n,3);
}
```

```
console.log(
   blah(x, s));
```
Asynchronous call

```
fs.readFile(
   'package.json',
   'utf8',
   callback);
```
Asynchronous call with callback

```
function(err, data) {
   console.log(data);
});
```
Async call

*(Node Library)*

*(Node Binding layer)*

*(Libuv: queue, event loop, thread pool)*

1. Into Libuv worker queue
2. then into a thread pool thread...
   ...Upon read completion

Platform OS

Time goes by left to right.....->

*Figure 35 - Code execution flow*

Follow left to right in the figure above and you can see how each bit of code is run. To start with, the line of code that does the require() will block code execution until it completes. The console.log() call does not block. The lower layer will asynchronously print out the value. The function blah() is called and in that, the JavaScript Math.pow() object function executes synchronously in V8 and not in the Libuv layer.

The fs.readFile() call uses an asynchronous callback to keep your upper layer code non-blocking. You can see the third parameter to the readFile() function is a function callback. The Node.js framework starts executing readFile() for the first bit of code, but the Node.js code is just calling into Libuv to hand off the request to be executed on the Libuv thread pool. This then immediately returns, and your execution continues in the code.

The Libuv execution thread for the file I/O eventually returns and then the callback function gets called and runs on the main thread. You cannot get access to the thread pool processing directly from your JavaScript code.

Filesystem calls go to the thread pool of Libuv. Network calls are processed differently than filesystem calls as will be explained later. In either case, Libuv does all the work for you to make things work across platforms and in a non-blocking way.

# Continuous processing with Node

Another concept to study, is how Node can be running in a continuous processing loop. As shown, the previous code sample runs to completion and then the Node process exits. This is because Node knows if there is any more work to execute and, if there isn't any, it just exits.

You can understand that there are certain modules you can use that will basically keep the Node process running forever. If this is the case, you can stop the Node process as you would normally do on your operating system.

You earlier saw code that had a while loop that never had any way to exit. That example would be a little odd, since it runs all the time, and completely blocks the single processing thread. A more reasonable piece of code would be something that uses an interval timer to do some periodic computation. Here is some simple code that keeps Node running:

```
setInterval(function () {
  console.log('Hello again!');
}, 5000)

console.log('Hello World!');
```

This is actually something similar to what the NewsWatcher code does to periodically look for news stories. NewsWatcher will need to run some processing on a periodic basis.

Another thing that will keep your process running forever would be the use of modules such as HTTP, Net, or Express. When you set up the code to listen for TCP connections and listen on a socket, you set up code that will run in the lower level of Libuv. Take the following example that was used before:

```
var http = require('http');

var server = http.createServer(function (request, response) {
  response.writeHead(200, {"Content-Type": "text/plain"});
  response.end("Hello World\n");
});
server.listen(3000);
```

What happens here is that Libuv is set up to use the low-level OS socket capabilities to listen and respond to incoming connections and requests. Libuv then has a loop to respond to any of these events. When they happen, your callback code can run, such as the one above. The main Node.js process loop actually checks with Libuv to see if it needs to be running because of work it has in its queue or is listening for. If so, then the process is kept alive.

*Note: The Libuv thread pool is not involved in socket listening. This is because the low-level OS capabilities handle the async non-blocking processing and generates the notification events upon completion.*

# 8.4 Node.js Module Design

Node itself is composed of various modules that run as part of the core service. Modules are what enable all the functionality in Node besides what is provided with the JavaScript language. The simple "Hello World" example demonstrated this. That code made use of the **console** module. Other modules, such as the **express** module, are third-party modules you install to bring in additional functionality. You use the NPM to get all your external modules installed on your machine and then use a `require()` statement to use them in code.

You will write many of your own project code as modules for your own consumption. If you are ambitious, you may want to write a module and make it available as a download from the NPM repository for others to benefit from.

I will now show you how a module is constructed, and you will see how Node exposes the functionality of a module. You don't necessarily need to know how this works internally to Node.js, but I include this information for those who are curious.

## A module can return a JavaScript object

Modules contain JavaScript code that is typically set up to return an object. The object is then exposed in a special way, so that a client can make use of it through the `require()` function shown in a previous example. Node.js does the work to take your module code and surface it through its internal exposure for other code to call. Node keeps track of all the loaded modules and manages loading, configuring, running, and caching of the modules. All require statements throughout your code will return the same object (e.g., HTTP) as it is cached once and used for every require (singleton pattern).

All you really need to know is that module code is exposed through a special Node object named `exports`. The `exports` object is created for you by Node in each and every module file. Then, when code calls the `require()` function, Node returns the `exports` object with whatever functionality was placed in it, such as a function. The following example shows a simple module you can write that exposes a single function:

```
// mymodule.js that holds your module code
module.exports.welcome = function(name) {
   console.log("Hi " + name);
}
```

As I mentioned, you provide functions and properties in a module file. These are then exposed outside of that file. You can add properties to the exports object such as the `welcome()` function in the previous example. What Node did for you in the above code was to create the exports object when it ingested your file.

Node has code for the require() function that sets everything up to be exposed. Node takes your code from your file that it opens and parses, and does something similar to the following:

```
function require(file) {
  module.exports = {};

    ...The file parameter is used, and that file is opened
    ...and code is parsed and taken and placed below.

  // Your extracted code
  module.exports.welcome = function(name) {
    console.log("Hi" + name);
  }
  // End of your extracted code

  return module.exports;
}
```

An empty object is created in the first line of the function. That object then has properties added to it, such as the function you see. Finally, the object is returned so other code can call this function off the object.

For the code to use the function in the module, it needs to use the `require()` function. The `require()` function takes the name of the file that has your module code in it. You reference mymodule.js as follows, and call the function you have exposed.

```
// file server.js that uses the sample module
var w = require("./mymodule.js");
w.welcome("Bob");
```

Node actually has an internal `module` object that it creates for each module exposed. There is a lot more that is going on behind the scenes than that, but you don't need to know the details. You can learn more of the internals if you wish, by reading through the actual source code, since it is an open-source project.

# A more complicated module

You can hang multiple properties on the exports object, such as objects, strings, numbers, arrays, etc. In one of the previous examples, I showed you the use of the HTTP module. It has a `createServer()` function attached to it. This is a design pattern known as the factory design pattern. You don't use the function directly, as was done with the `welcome()` function in the previous example, but call it to get an object that you can then use.

You can also expose a class through a constructor function that clients take and construct themselves or you can go further and provide a function that does the creation for them like the factory pattern I just mentioned. If you have multiple classes to expose, then you would want to use the factory pattern. If you only have one class, then you can expose that with a constructor function. If you want to provide a constructor function in a module, it should look as follows:

```
// mymodule.js
var a = require('http');

function Welcome(nameIn) {
  this.name = nameIn;
}

Welcome.prototype = {
  this.name = null,
  showName: function () {
    console.log("Hi " + name);
  },
  updateName: function (nameIn) {
    this.name = nameIn;
  }
};

module.exports = Welcome;
```

The following example shows how this module can be used:

```
var Welcome = require('./mymodule.js');
var w = new Welcome("John");
w.updateName("blah");
w.showName();
```

The `welcome()` function acts as the constructor in this example. The `prototype` keyword in JavaScript allows you to set properties that exist for all instances and are thus not created again for every instance. Modules are single instance cached anyway, so that might not be important. Node keeps references to each that are used and gives the same instance back every time it is required.

Note how I included the usage of the HTTP module to be used by the sample module above. This shows how modules can require other modules for their own functionality and do so with the standard `require()` function. These included modules are not visible outside of the internal code. The code using the module can't actually get access to the HTTP module, unless it also has a `require('http')` statement as well.

## Import instead of require with Node version 10

The JavaScript language itself has releases and is an implementation based on a standard called ECMAScript. Node has an implementation of JavaScript that is on a path to adopt more ECMAScript syntax as that evolves. In the ES6 release of ECMAScript, a new module system was introduced that uses the 'import' keyword. Version 10 of Node takes a step towards implementing this, so you would no longer need to use 'require' statements if you prefer to use the 'import' keyword instead. People writing HTML browser applications are already familiar with this new standard. Here is what the old way looks like compared with the newer way:

```
// Older CommonJS syntax that existed in Node
const sm = require('./somemodule')

// ESM syntax for getting the default export
import sm from './somemodule'
```

# 8.5 Useful Node Modules

If you go to https://nodejs.org/api/, you will find the official documentation on the core Node.js modules. Glance through what is there so you can keep it in mind if you need to reference it in the future.

Besides the modules that come with Node, there are plenty of other installable packages from NPM. You might want to Google for articles that lists the most popular ones.

Keep in mind that many of the NPM downloads are for code modules you use inside an application and others are tools that you download and run. Some code modules are also only for use as express middleware. Some are only for other frameworks such as React or Angular.

Here is a small list of some core npm modules you may find useful in Node.js code that you write. A few come with Node itself and the rest you can download from NPM.

## PART II: The Service Layer (Node.js)

| Module: | Purpose: |
|---|---|
| async | Used to force your code to run in a workflow. Instead of doing things like nesting callbacks, you can use async to set up sequential calls. Also used to run parallel functions with the ability to know when all have finished. |
| child_process | For child process spawning and management. |
| cluster | For setting up a cluster of Node.js processes to distribute the load across. |
| events | A primitive module that many others are built on, to emit or listen to events. |
| express | Web server functionality for configuring routes, serving up static files and providing template data binding functionality. |
| fs | Standard OS filesystem functionality. |
| helmet | For mitigating different types of HTTP security vulnerabilities. |
| http | This module serves a dual purpose. You can use it to set up a listening service for incoming HTTP requests. You can also use it to make outgoing HTTP requests. |
| joi | For performing validation on HTTP request JSON body properties. |
| lodash | A collection of useful helper functions. (see https://lodash.com/docs/) |
| mongodb | Used to interact with a MongoDB database. |
| net | Standard low-level networking functionality for servers and clients. |
| os | Basic utility functions for accessing OS information. |
| process | The standard type of functionality for working with processes on an operating system. |
| request | Simplifies making HTTP/S client calls. |
| response-time | Displays the response time for HTTP requests. |
| socket.io | A higher level way to build bi-directional communications between clients and servers. |
| stream | A primitive module that many others are built on to provide readable and writable streams on top of data. |
| url | Utility functions for working with URLs. |
| util | Internal utility functions that Node itself takes advantage of that are exposed for you to use. |
| zlib | For compression and decompression. |

114

# Chapter 9: Express

The E in the MERN acronym stands for Express. Express is a module commonly used in a Node.js application. It is not a part of the Node.js installation but can be downloaded from NPM and installed separately and integrated into your application. Express is popular and consistently one of the most downloaded Node.js modules on NPM. The Express module allows you to implement the functionality of a web server. It provides a way to specify the route handling for incoming requests and simplifies response generation.

One of the things Express does for your project is to remove the need for the HTTP module that comes with Node. The HTTP module that is built into Node.js requires you to write a lot of code to set up the routing and responses. Express makes all that a lot easier.

The following are the key capabilities of Express:
- Specifies the route handling of incoming HTTP requests.
- Has mechanism to inject middleware into a request to modify it as it gets passed along.
- Easily integrates third-party middleware to provide extended capabilities for request processing.
- Provides a `request` object with properties and methods to look at everything connected with the incoming request.
- Provides a `response` object to use to set everything up for a response.
- Configuration for JSON payload serving.
- Configuration to serve up static files.
- Pairs up with server-side template engines to set context and return HTML with data binding.

# 9.1 The Express Basics

As with any external Node module, you must do an NPN install and require Express in your code. You can do this at the top of a file such as your server.js file. You can then make the Express listener active. Here is some sample code to set up and use Express:

```
var express = require('express');
var app = express();
app.listen(3000);
```

I will later walk you through the construction of the NewsWatcher sample application and show you how to set everything up for that. Before that happens, I will cover the basics of how Express is used.

# Express configuration settings

As part of using the Express module, you will need to configure some settings in your code to determine how it works, and these are settings you need at startup. You use the Express object `set()` function to do this. The `app.set(name, value)` syntax is used for setting a value for predefined values that Express uses as configuration. The `set()` method will configure values such as what port to listen on. Here is an example:

```
var express = require('express');
var app = express();
app.set('port', 3000);
```

To disable a setting, you can call `app.set(name, false)`. You use the `app.get(name)` method to retrieve any set value. There are `enable` and `disable` functions, but they are the same thing as calling set with a true or false. The following are some of the settings that can be used for configuring the mode of operation of Express:

**Settings with Value options:**

| | |
|---|---|
| case sensitive routing | A Boolean to determine route interpretation as to case sensitivity. For example, if set to false, then "/News" and "/news" are treated the same. |
| env | A string value that sets the environment mode such as "development." This is purely for your use to set and read. For example, you would have logic to determine which URL endpoints, database connections, etc., depending on if you are running the code to try out new development code or if the code is running in production. |
| etag | Used to set the ETag response header for all responses. Set it to `strong`, `weak`, or `false` if you want to disable it. You can also pass in a custom function. The default value is `weak`. |
| jsonp callback name | A string to specify the default JSONP callback name such as `?callback=`, which is the default. JSONP to bypass the cross-domain policies in web browsers. (Not necessarily needed.) |
| json replacer | A string to specify the JSON replacer callback. This is a function you define to decide on a property-by-property basis if they are returned on a JSON route response. |
| json spaces | Specifies the number of spaces to use for JSON indenting for readability. |
| port | The port to listen on. |
| query parser | You can set this to `simple` or `extended`. `simple` is based on the query parser from Node and `extended` is implemented by the qs module and is the default. |
| strict routing | A Boolean to set to `true` if you want routes like "/News" and "/News/" treated as different paths. The default setting is `false`. |

| subdomain offset | A number that defaults to 2 for how many dot-separated parts to remove to access the subdomain. |
|---|---|
| trust proxy | To be used if you have a front-facing proxy. You will need to set this up to be trusted. This is `false` by default. |
| views | The directory(s) to use for template lookups. |
| view cache | A Boolean that enables template caching. If the `env` setting is set to `production`, then this value defaults to `true`; otherwise, it defaults to `false`. |
| view engine | A string to specify the template engine to use with your provided templates. |
| x-powered-by | A Boolean (defaults is true) to enable the "X-Powered-By: Express" HTTP header to be returned. You set it to `false` so as not to return it to prevent any hackers from knowing too much about your implementation. |

The Express object has a property named `locals` on which you can place your own custom properties that you may want to associate with your Express application. You use the `app.locals` object as shown in the following example:

```
app.locals.emailForOps = 'Help@myapp.com';
```

# Listening

With code in place to manage your settings, all you need to do next is make a function call to allow the Node.js runtime to run its platform-specific code and set up a socket connection for the given port on the server machine to listen on. Any incoming connections to the IP address of the machine at that specified port number will be bound through to your code to respond. Here is the code to do that:

```
var server = app.listen(3000, function() {
  debug('Express server listening on port ' +
server.address().port);
});
```

With some understanding of the initialization and settings, you can look at implementing a fully functional web service with HTTP request route handling.

# 9.2 Express Request Routing

When an HTTP request comes into Node, the request will make its way to the Express code you have written to service it. Your Node.js instance will be running on a machine that is hosted and exposed on the internet. Node will be executing in a process on that server, and through the Express code it will be listening on a port socket for incoming connections.

HTTP requests can come to your server to render a browser page or fulfill REST web service API requests that deal with JSON payloads. A typical URL request is as follows:

```
http://newswatcherscale-env.us-west-2.elasticbeanstalk.com/news?region=USA
```

If you use query strings as shown above, you can get those values in the Express request handlers. The following diagram illustrates Express routing:

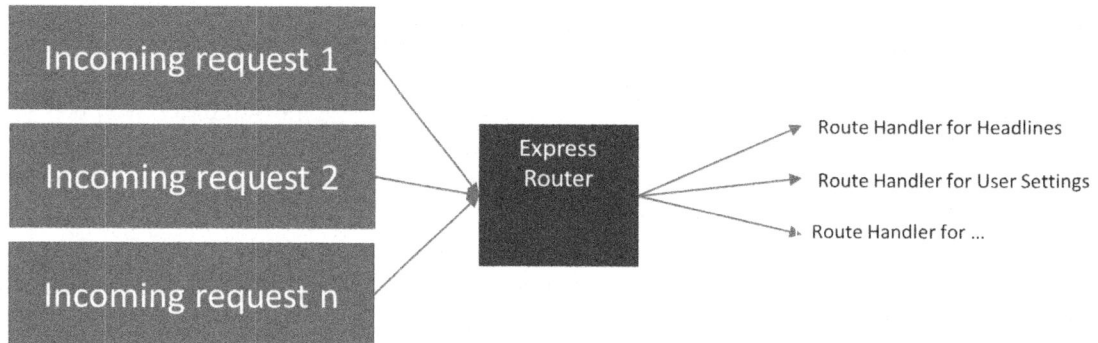

*Figure 36 - Express routing*

You can set up the code to service lower-level TCP or UDP types of connections if you decide it is needed. You can research the Node Modules Net and UDP. This chapter will only be concerned with servicing HTTP requests with Express.

# Routes

Once you have the Express JavaScript object through the require statement, you can use methods that set up the servicing of HTTP requests. Express will then listen for connections on the specified port and it can understand the various HTTP verbs and break down what is being passed in as part of the URL path, query string, and the HTTP headers and JSON body.

The Express object gives you functions to use for handling the various HTTP verbs (get, put etc.). Here are the standard verbs that are used for CRUD type operations to expose an API in your service layer:

```
app.get(path, MyCallback);   // Read item(s)
app.post(...);               // Create a new item
app.put(...);                // Replace an item
app.patch(...);              // Update an item
app.delete(...);             // Delete an item

function MyCallback(req, res) {
  res.send("Send something back");
};
```

Express supports the following HTTP methods:
- `checkout`
- `copy`
- `delete`
- `get`
- `head`
- `lock`
- `merge`
- `mkactivity`
- `mkcol`
- `move`
- `m-search`
- `notify`
- `options`
- `patch`
- `post`
- `purge`
- `put`
- `report`
- `search`
- `subscribe`
- `trace`
- `unlock`
- `unsubscribe`

See the Express documentation for the complete list of supported methods. The general signature looks as follows:

```
app.<METHOD>(path, callback [, callback ...])
```

`<METHOD>` is replaced by one of the methods such as `get`. The first parameter is the URL path. This is not the complete URL, but the portion of it after the domain name.

The second parameter is the callback function that gets called for that request. You can provide multiple callbacks and they will each get called sequentially. I will cover more on this later.

Route paths can be strings, string patterns, or regular expressions. They can also be an array that combines any of the mentioned formats. As a string, a route path can be things such as "/books". Be aware that the query string portion is not considered part of the path.

You can use "/" to specify that all paths should be picked up. Alternatively, all paths will be picked up if you omit the path parameter completely.

The callback function you provide has at least two parameters in its standard form. The first parameter is the request object, which contains information about the incoming request. The second parameter is the response object and is used to send back a response, such as serving up an HTML file or sending back a JSON payload.

The following example shows the use of a pure REST style URL. You just need to provide the routing path that occurs after the domain portion of the URL. Here is the URL and the way you will specify the path.

```
// http://mysite.com/news/categories/sports
app.get('/news/categories/sports', callback);
```

For each verb, such as get, you may have different routes for different resources that are being retrieved. Here is some code that sets up multiple path routes for the get HTTP verb, each returning something unique:

```
app.get('/about', function(req, res) {
  res.send("About page");
});

app.get('/news', function(req, res) {
  res.send('News page');
});
```

Be aware that the ordering of your route handling code is extremely important. Any incoming request is basically consumed by the first path that is found to handle it.

# A single path for multiple verbs

If you find you have several verbs that all respond to the same path, you can specify them together by using the route() method. This may help you alleviate typing in the path multiple times. An example of this is:

```
app.route('/customer')
  .get(function(req, res) {res.send('Get a customer');})
  .post(function(req, res) {res.send('Add a customer');})
  .put(function(req, res) {res.send('Update a customer');});
```

# All verbs at once

Besides standard verbs, there are also the methods all or use to respond to all verbs of the incoming requests. The first route handler below is for all verbs that are for the path `/test`. Then the second is set up for everything else and returns a 404 - Not Found code. The `use` function here is not using a path, so it is for all verbs and all paths not serviced yet.

```
app.all('/test', function(req, res) {
   resp.type('text/plain');
   resp.send('This is a test.');
});

app.use(function(req, res) {
   res.type('text/plain');
   res.status(404);
   res.send('404 - Not found');
});
```

# Advanced path specification

You have seen the simplest of cases with route handling paths up to now. There is a technique that can give you more advanced parsing capability for a URL path. Below is one of the URLs from a previous example:

```
// http://mysite.com/news/categories/sports
app.get('/news/categories/sports', callback);
```

The problem with this example is that you may need to service 20 different categories of news stories. For example, what about the paths `/news/categories/science` or `/news/categories/politics`? It can get very monotonous to set up routes for each and every one. To make this easier, Express allows you to set up placeholder parameters in the path that you can get at later in the callback code and then have the code handle it.

To be able to retrieve parts of a path, you use a special syntax in the `path` parameter by placing a colon character in the string. This then sets up a JavaScript property you can later access in the handling function as part of the request object. The code below sets the path up with a colon. A property will then be available on the request object as shown.

```
// http://mysite.com/news/categories/sports
app.get('/news/categories/:category', function(req, res) {
  console.log('Your category was ' + req.params.category');
});
```

With the above code, you create one route that can service all requests for news stories and can feed that category into the backend retrieval mechanism.

The `req.params` property can be used as an array, in case there are multiple in a row. The following example shows this, for when you have more than one of these parameters set up to use:

121

```
app.get('/products/:category/:id', function(req, res) {
  console.log(req.params[0] + req.params[1]);
});
```

You can also construct your URLs so they contain query strings. For a query string that you want to process such as /news?category=sports, you don't need any special syntax in the path parameter. Specify the path up to the start of the query string and stop there. The name-value pairs of the query string will automatically be parsed out and made available to you as part of the query property of the request object. For this example, there would be a property named category, with a value of sports:

```
// http://mysite.com/news?category=sports
app.get('/news', function(req, res) {
  console.log(req.query.category);
});
```

# 9.3 Express Middleware

At this point, you know how to set up callback functions that get executed for a given incoming request. Now I'll introduce the concept of middleware as a means of inserting route processing code that will happen before your ending route handler runs. Some middleware code will be provided by Express, other middleware can come from modules downloaded from NPM, and other middleware can originate from what you write.

The concept of middleware is that you can have code run as an inserted step that is placed before your actual route callback gets run. Middleware code gets chained together in a series of calls that you specify. Express Middleware can then act upon and modify the request object that is being passed along the way. You can do this to reuse code across multiple routes.

Some inserted middleware terminates the request, as it takes care of everything, and the requests never even go to any of your route handling. An example of this is the Express static file serving middleware. You also may have some middleware that intercepts calls to verify a user's authorization and does not continue if the request is determined to be invalid.

Additional examples of middleware may be something that caches content for you or is a function that logs all operations.

Many available middleware modules are simple to add, yet powerful in what they provide. I'll show you several of them in this chapter. Conceptually, you can take the Express routing diagram previously shown and modify it as follows to show middleware being injected that passes functionality down the chain:

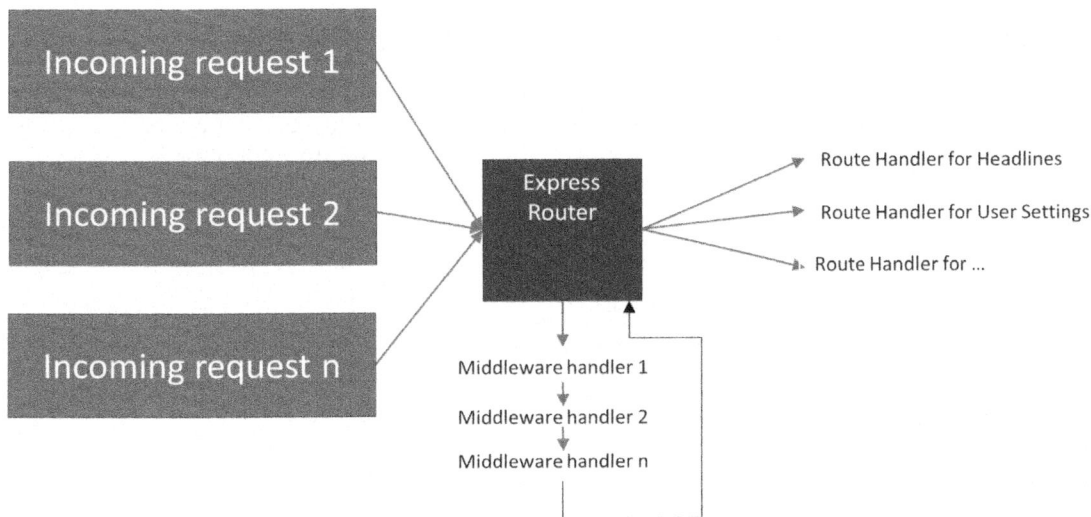

*Figure 37 - Express routing with middleware*

Middleware functions look almost identical to what you have seen already as callbacks. They are just callback functions with the same signature you have seen and thus have access to the request and response objects. This means that the request object can be modified before being passed along. For example, the request body in the response object can be modified to have additional data added to it before it gets to its final destination-handling callback.

Middleware extends the capabilities of Node.js beyond its core functionality. For example, let's say you are writing a web server that will serve up static files, such as image files. Node.js allows you to do that — if you write the code to do so. Instead, there is a module that acts as middleware in Express that makes it incredibly easy to implement. The module that does this is the `static` module and it comes as part of the Express install. It can be hooked up to serve up static files with very little code.

Read the Express documentation to learn more about available modules you can download from the NPM repository. Middleware can accomplish things like authentication, caching, logging, session state, cookies, etc. These act as shared pieces of code that you can use across all or just certain routes.

**Note:** *Don't confuse the concept of a middleware module with the general concept of a module in Node.js. They still use a* `requires` *function, but are used differently than regular Node modules.*

# Hooking up your own custom middleware

Normally, route processing stops at the first match that is found for a URL path and then a callback is run. However, if you make one minor modification to your code, you can string together multiple callbacks for the same route. Notice the one modification made below:

```
// First handler for the route
app.get('/test', function(req, res, next) {
  ...
  console.log("Got here first");
  next();
});

// Second handler for the same route
app.get('/test', function(req, res) {
  ...
  console.log("Got here second");
});
```

The difference is that the first route callback function has a third parameter named `next`. This parameter is a function, and its usage tells Express that you want the callback to act as middleware code to be injected before the actual end route is called. The order is important as stated before.

`next()` is a function that you call when your middleware code is done with all processing. You must call `next()` or the request will be abandoned and not make it to your end handler. In the example, the first handler runs and then, because of the `next()` function call, execution continues to the second handler.

The previous example code also can be structured so that the callbacks are not separated out. You then just list out callbacks one after another. You would need to have the code in the `cb1()` function that calls `next()` or `cb2()` will not be run.

```
app.get('/test', cb1, cb2);

function cb1(req, res, next) {
  console.log("Got here first");
  next();
});

function cb2(req, res) {
  console.log("Got here second")
});
```

124

# Universal middleware

You can hook up middleware that will get inserted into every single route and for every single verb. To do this, you simply use `app.use()`. This sets up Express to use this function across all incoming requests. You can leave off the path in this case, as "/" is the default path if you don't provide one. Yet you may want to provide a path so that the middleware only gets run for a certain path.

```
app.use('/', function(req, res next) {
  ...
  next();
});
```

You can insert as much middleware as you need for your routes. The order in which you list them in your code will be the order in which they are sequenced through. Be aware that certain third-party middleware from NPM are required to be placed before others. Refer to the middleware's documentation for more information.

I will now show you a practical example of some custom middleware you may want to implement. Let's say that you have a special path that you only want administrators to have access to. You can create a function that does validation before allowing the request to proceed for further processing. Here is how you do that:

```
// Middleware injection
app.all('/admin/*', doAuthentication);
app.get('/admin/stats', returnStats);
app.get('/admin/approval', approval);
```

With the `app.all()`, the authentication happens for all verbs and acts as middleware. Inside the `doAuthentication()` will be code to determine the authenticity of the request. If it is detected to be invalid, then you will not call next() and the other two route handlers will never be called. You will see something similar being done in the NewsWatcher application.

If you call `next()` and pass an error object parameter, then that route terminates from normal execution and error handling middleware then gets invoked. This type of error handling will soon be explained.

# Parameter middleware

Express allows you to set up a middleware callback function for a given parameter property that you defined in other route handlers. You do this with the `app.param()` function. This callback is called before any route handler. Here is an example:

```
// Using the global param handler
app.param('id', function(req, res, next, value) {
  console.log("someone queried id " + value);
  if (value != 99)
    next();
});

app.get('/products/:category/:id', function(req, res) {
  console.log(req.params[0] + req.params[1]);
});
```

If you don't have any routes with "id" in them, then the `param()` callback will never get called. The `param` callback will be called before any route handler in which the parameter occurs. You still need to call `next()` to continue the processing.

# Router object

If you have a lot of routes and want to subdivide them for better organization, you can use the `router` object and keep each in their own modules. In that way, you can isolate your logic and not have it affect the other routes you have set up.

You set the `router` objects up independently. Until you activate them with `app.use()`, they will not be functional. The following is an example of setting up some middleware and some routes for two different routers and then activating them in the app:

```
var routerA = express.Router();
var routerB = express.Router();

// Set up routes that will end up with news
routerA.get('/weather', function(req, res) {...
});
routerA.get('/sports', function(req, res) {...
});

// Set up routes that will end up for blog
routerB.get('/tech', function(req, res) {...
});
routerB.get('/art', function(req, res) {...
});

// Sets up /news/*
app.use('/news', routerA);

// Sets up /blog/*
app.use('/blog', routerB);
```

# Middleware error handling

Any middleware function can return an error. It does so by calling next(err), where err is an Error object. Execution of the middleware and any subsequent routing is ended and processing of the error takes place. To process the error, Express has a default function that it calls that writes the error back to the client. You have the option of providing your own function or chain of functions for handling middleware errors. If you provide a function or chain of functions, then the default one will not be called. In the following example, note how there are four parameters on the error handling middleware function:

```
app.get('/test/:id', function(req, res, next) {
  if (req.params.id == 0)
    next(new Error('Not Found'));

  next();
};

// A middleware error handling function. err is an error object.
app.use(function(err, req, res, next) {
  console.error(err);
  res.status(500).send('Something bad happened');
});
```

Once execution has shifted to the error handling middleware, you can return back to regular route processing if you do a next('route') call. Doing that will jump you to whatever is the next defined route handler. The diagram of Express routing can be further expanded to add in middleware error handling.

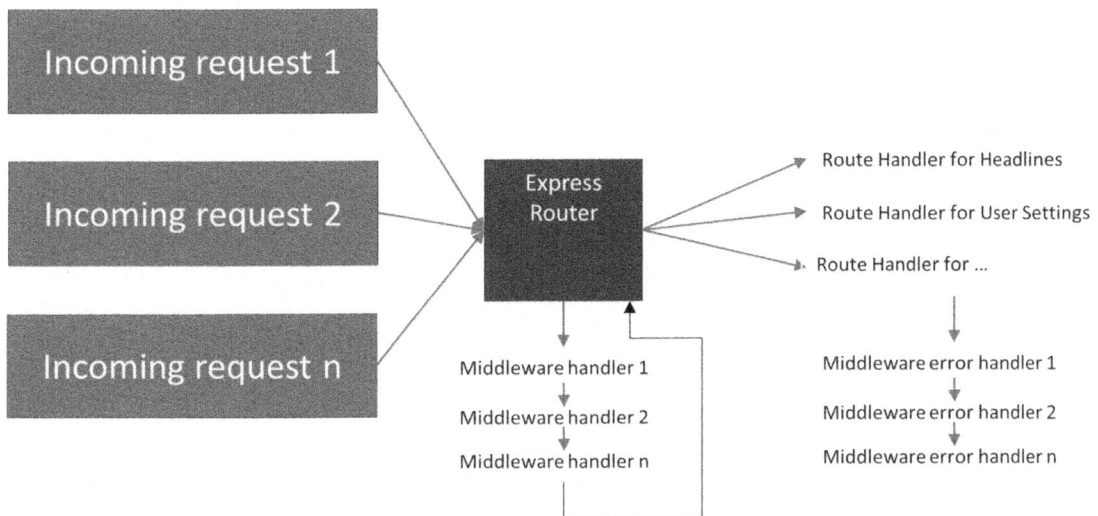

*Figure 38 - Express routing with route handing and middleware error handling*

# Using next() even if your handler is not middleware

To be clear, just because you add a `next` parameter on a route handler does not mean you are implementing some middleware. For example, you will most likely need `next()` as a parameter on all your end route handlers in order to do central error processing.

As an example, the following code is an end route being handled by the Express router object. As it is written, you assume nothing will go wrong and it will just carry out some operation:

```
var router = express.Router();

router.delete('/:id', function (req, res) {
  res.status(200).json({ msg: 'Logged out' });
});
```

What if you want to detect an error and use middleware error handling as explained in the previous section? This is where you will need to add the `next` parameter to be called if there is an error. If there is no error, the route will complete by calling the `res.status()` function to send back a response. Nothing else gets in the way, and if running normally, you don't call `next()` because there is nothing else to chain.

In order to use the error handling middleware, an end route itself must be able to pass control to the error handling chain. To do that, the code needs to be modified to add in the `next()` function to be called. As mentioned, the ending route handler would call `next()` with an error parameter if something other than a successful return happens. The following example shows the additional error handling. The error handler then can send a response with an error code and message.

```
var router = express.Router();

router.delete('/:id', authHelper.checkAuth, function (req, res,
next) {
  if (req.params.id != '77')
    return next(new Error('Invalid request'));

  res.status(200).json({ msg: 'Logged out' });
});
```

# Static file serving middleware

One middleware module that comes with Express is the `static` middleware that allows interception of requests for files and returns them. This alleviates the need for you to provide an end route handler of your own. For example, if you want to serve up jpg image files, you can use the following code:

```
var express = require('express');
var app = express;

// Middleware injection
app.use('/images', express.static('images'));
app.listen(3000);
```

`app.use` is setting up the middleware route. Instead of the callback function being provided by you, you insert the call for using `express.static()`. This takes care of everything for you to send a response back. In your Node.js project code, you will need to provide the folder of images. An HTML page could access an image as follows:

```
<img src="http://yoursite.com/images/someimage.jpg"/>
```

As a second parameter to `express.static()`, you can pass in an options object that can have the following properties on it: dotfiles, etag, extensions, fallthrough, immutable, index, lastModified, maxAge, redirect, setHeaders.

For example, `setHeaders` is a function you use to set headers to send with the files. Another example will be setting the `lastModified` property to true and the `Last-Modified` header value is set to the date of the file being sent. For more information, refer to Express documentation.

# Third-party middleware

A growing number of third-party modules provide Express middleware for your applications. Go to the Express site (http://expressjs.com/resources/middleware.html) to find a list of modules you can download.

To use the middleware, you typically call `app.use(<middleware>)`. This means that all paths and verbs will flow through it. Most third-party middleware that intercepts routes will call the `next()` function so that processing will eventually reach your code, if you have that need.

Here are a few useful Express middleware components for your reference:

| | |
|---|---|
| passport | Used to authenticate requests. You can set this up to log a person in using OAuth (e.g., through Facebook), or federated login using OpenID. There are more than 300 strategies available through the passport module. |
| body-parser | This middleware intercepts any HTTP Post verb requests that have body data, such as from a form submit. The middleware code runs and |

| | |
|---|---|
| | then by the time your handler code runs, the response object has a body property with sub-properties off it for each of the body values.<br><br>```<br>var bodyParser = require('body-parser')<br>app.use(bodyParser());<br>```<br><br>With later versions of express you do not need this and can replace it as follows:<br><br>```<br>app.use(express.json())<br>```<br><br>```<br>// In your handler, you can look at the values<br>app.post('/', function(req, res) {<br>  console.log(req.body);<br>}<br>```<br><br>Query string values can also be placed into the body object for you. |
| compression | This will compress requests that pass through the middleware. You will place this statement before any other middleware or routes, unless you only want certain routes to be compressed. If you look at the headers of a returned response, you will see the headers have an entry for Content-Encoding.<br><br>```<br>var compression = require('compression')<br>var express = require('express')<br>var app = express()<br>app.use(compression())<br>``` |
| cookie-parser | This middleware does all the work to make available a cookie property on your request object.<br><br>```<br>var cookieParser = require('cookie-parser')<br>app.use(cookieParser());<br>```<br><br>```<br>// look at all of the properties on body<br>app.post('/', function(req, res) {<br>  console.log(req.cookies);<br>}<br>``` |
| errorhandler | This accomplishes the sending back of stack traces to the client when an error occurs. Only use this when running in a development environment.<br><br>```<br>var errorhandler = require('errorhandler')<br>app.use(errorhandler());<br>``` |
| express-session | Server-side session data storage. |
| express-simple-cdn | Usage of CDN for static asset serving with multiple host support. |
| helmet | Helpful for mitigating several HTTP security vulnerabilities. |

130

| response-time | Response time tracking to add the X-Response-Time header. The value inserted is in milliseconds. You can use this to track your SLA over time and be alerted as to what needs further investigation for performance optimization. |
|---|---|
| morgan | This is for request logging. This frees you up from writing any of your own `console.log` statements. For example, incoming HTTP requests go through this middleware and it logs those requests to the console window. You can also specify the format of the logging and direct the output to a file.<br><br>```\nvar fs = require('fs')\nvar morgan = require('morgan')\n\n// create a write stream (in append mode)\nvar accessLogStream = fs.createWriteStream(__dirname +\n'/access.log', {flags: 'a'})\n\napp.use(morgan('combined', {stream: accessLogStream}))\n``` |
| multer | Multi-part form data or uploading files in chunks. |
| serve-favicon | For customizing the icon in the browser. |
| timeout | For setting a timeout period for HTTP requests. |
| express-validator | For validation of incoming data. |
| connect-redis | Session store using Redis cache. |
| connect-timeout | For routes that may run into some backend processing issues and need to be limited in the amount of time they take, you can use this to cut them off and return an error. You still need to determine what the right approach is for termination and resubmission of requests.<br><br>```\nvar cto = require('connect-timeout')\n\napp.get('/some_questionable_route', cto('5s'),\n  function(req, res, next) {\n    ...some possibly long running code...\n    ...check req.timeout to see if it is ever true and\n    ...then return false\n    return next();  // finished processing in time, go\nto next function\n  },\n  function(req, res, next) {\n    res.send('ok');\n  }\n);\n``` |
| vhost | For routing by hostname different sub-domains (e.g., www.mysite.com versus api.mysite.com). |
| express-stormpath | User storage, authentication, authorization, SSO, and data security. Will work with the Okta API. |

# 9.4 Express Request Object

Let's look more in-depth at the usage of the request object in Express route function handlers. The request object contains all the information you need for digesting the incoming request. For example, you have seen how the properties `req.params` and `req.query` are used. You have also seen some third-party middleware that adds more properties to the request object. Here is a reference to some of the properties that are available. You can refer to the Express documentation to find the complete list.

The request object is a parameter of your express route callback function. You can name it anything you like, though "req" is a good name for it. The following example shows how to get the complete URL that this request came in from.

```
app.get('/', function(req, res) {
  console.log(req.url);
});
```

Here is a reference to the some of the properties that are available on the `request` object. Refer to the Express documentation for the complete list.

**Request object properties:**

| | |
|---|---|
| `accepts(types)` | For content negotiation on a return.<br>```if (req.accepts('text/html') == 'text/html')<br>  res.send('<p>Hello</p>');<br>} else if (req.accepts('application/json') ==<br>'application/json')<br>  res.send({ message: 'Hello' });<br>} else {<br>  res.status(406).send('Not Acceptable');<br>}``` |
| `app` | A reference to the instance of the express application object. |
| `params` | This is used to access route parameters. You need to first set the route specification string and then you can use the `params` property of the request object.<br><br>```app.get('/user/:id', function(req, res) {<br>  res.send('user' + req.params.id);<br>});``` |
| `query` | Used to get the URL querystring. A property will exist on the query object for each.<br><br>```// For URL "/users/search?q=Smith",<br>app.get('/users/search', function(req, res) {<br>    console.log(req.query.q);});``` |

| body | Contains properties of key-value pairs of data submitted in the request body. You need to add the body-parser middleware for it to work. `var bodyParser = require('body-parser'); app.use(bodyParser.json()); app.post('/', function (req, res) { console.log(req.body);})` |
|---|---|
| route | This is an object that has properties of the current route, such as `path`, `keys`, `regexp`, and `params`. |
| cookies | When using the cookie-parser middleware, this property is an object that contains cookies sent by the request. Each cookie that is attached is a property on the cookies object. `req.cookies.someName` |
| signedCookies | Exists if the cookies have been signed and protected from tampering. `req.signedCookies.someName` |
| ip | The remote IP address of the incoming request. |
| protocol | Such as `http`, `https`, or `trusted` if setup with a trusted proxy. |
| secure | This has a value of `true` if SSL is in effect. |
| headers | The HTTP headers you can access. `req.headers['x-auth']` |
| url | The URL of the request. |
| path | The path part of the request without the query string. |
| route | An object that contains the matched route and lots of other properties, such as the method and function that handled it. |
| hostname | The host from the HTTP header, e.g., example.com |
| subdomains | The subdomain part that is in front of the hostname, e.g., ["blah", stuff"] if from stuff.blah.example.com. |
| xhr | This is set to `true` if the request came from a client call such as from XMLHttpRequest, which had set the X-Requested-With field. |

Here is a reference to some of the methods that are available on the `request` object. Refer to Express documentation for the complete list.

**Request object methods:**

| get(field) | To get the request header fields. `req.get('content-type'); // e.g. "text/plain"` |
|---|---|

| `accepts(types)` | To check if a certain type is available, based on the Accept header field. If what you send in as a parameter does not match one of the values, then you will receive an undefined return. `req.accepts('html');` |
|---|---|
| `is(type)` | To find out what type the incoming request is. `req.is('text/html');   // returns true` |
| `acceptsLanguages(lang [, ...])` | Based on the Accept-Language field of the header, it returns the first language on a match, or false if none are accepted. Similar calls are `acceptsCharsets` and `acceptsEncodings`. ```var lang = req.acceptsLanguages('fr', 'es', 'en'); if (lang) {     console.log('The first accepted is: ' + lang); } else {     console.log('None accepted'); }``` |

# 9.5 Express Response Object

Requests are routed to your callback because of a routing path you have set up. You will eventually return a response back to the requester. The Response object is what you use to do that with. You should at least send back an HTTP status code. You also may return some HTML, text, or (better yet), a JSON payload.

Methods on the response object are combined and have a cumulative effect on the returned response. The example code below sets the status of 200 OK for a successful HTTP request:

```
var express = require('express');
var app = express();

app.get('/', function(req, res) {
  res.status(200);
  res.set({'Content-Type': 'text/html'});
  res.send('<html><body>Some body text</body></html>');
});

app.get('/test_json', function(req, res) {
  res.status(200);
  res.set('json spaces', 4);
  res.json({name:'me', age: 37});
});

app.listen(3000);
```

In the first route, the Content-Type is set as "text/html" and then the send() method is used to finish the returned response with some returned HTML. The second route returns some JSON.

The response object has many useful properties and methods. Each usage of the response object is used inside a function callback. You can name it anything you like, though "res" is a good name for it.

Here is a reference to the properties and methods that are available on the response object. Refer to the Express documentation for the complete list.

**Response object properties:**

| App | A reference to the Express application. |
|---|---|
| headersSent | A Boolean value that is true if HTTP headers have been sent. |
| locals | Local variables scoped to the request, might be identical to app.locals. A template can use these for its data binding. |

**Response object methods:**

| status(code) | Used to set the status of a return in the case of an error. `res.status(404);` |
|---|---|
| set(field [, value]) | For setting the fields of the response header. `res.set({'contentType': 'text/plain', 'ETag': '123'});` |
| get(field) | Retrieves what the setting is for a header field. `res.get('contentType');` |
| redirect([code,] URL) | The path to redirect to instead of the one it came in at. You can provide an optional status code. A 302 "Found" is the default value of the code. You can also redirect relative to the current URL of the service. `res.redirect('http://example.com');` |
| cookie(name, value, [options]) | Sets a cookie. The name is the identifier of the cookie. The value parameter can be a string or object converted to JSON. The options parameter can set up things such as domain, expires, httpOnly, maxAge, path, secure, and signed. `res.cookie('rememberthis', '1', { maxAge: 900000, secure: true });` |
| json([body]) | Send a JSON body back. |

| | |
|---|---|
| | `res.json({ msg: 'Hello' })` |
| `jsonp([body])` | Sending JSON with JSONP support.<br><br>`res.jsonp({ msg: 'Hello' })` |
| `attachment([path to file])` | You can set an attachment to be returned. If you pass a parameter, it is expected to be a file. The `Content-Disposition` and `Content-Type` are set for you.<br><br>`res.attachment('path/to/logo.png');`<br>`// Content-Disposition: attachment;`<br>`filename="logo.png"`<br>`// Content-Type: image/png` |
| `sendFile(path [, options] [, fn])` | Transfers a file.<br><br>`res.sendFile('me.png', {maxAge:1,`<br>`root:'/views/'}, function(err){});` |
| `end([data][,encoding]` | Use this to end the response without any data being returned.<br><br>`res.status(404);`<br>`res.end();` |
| `format(object)` | Use this if you are going to receive requests for content of more than one type. You can line up multiple pieces of code for each Accept HTTP header type.<br><br>`res.format({`<br>`  'text/plain': function(){`<br>`    res.send('Hi');`<br>`  },`<br>`  'text/html': function(){`<br>`    res.send('<p>Hi</p>');`<br>`  },`<br>`  'application/json': function(){`<br>`    res.send({ message: 'Hi' });`<br>`  },`<br>`  'default': function() {`<br>`    res.status(406).send('Not Acceptable');`<br>`  }`<br>`});` |
| `append(field [, value])` | Adds the specified text and value to the header.<br><br>`res.append('Warning', '199 Miscellaneous warning');` |
| `send([body])` | Sending of an HTTP response.<br><br>`res.send({ message: 'Hello' });` |

| sendStatus(code) | Sets the response code for the return.<br><br>`res.sendStatus(200);` |
|---|---|
| render(name, [,<br>data][, callback]) | Template response sending. See the next section of this book for more information.<br><br>`res.render('user', { name: 'Tobi' },`<br>`function(err, html) {`<br>`    ...`<br>`});` |

# 9.6 Template Response Sending

One of the things that Express enables is sending server-side HTML, which has been generated from templates that have data bound to them. There are several popular template languages similar to HTML markup that are supported through Express. I will highlight just one, but you can investigate others.

Using the Express response object, you can formulate a response to send back with a function named `res.render()`. This function takes a file that contains the template as one parameter, referred to as the `view`. As a second parameter, you can provide the data object that binds to the template. The template that you have loaded through Express binds the data and produces the resulting HTML as the output to pass back on the request.

You can pass a third parameter as a callback function to get the rendered string and process any errors that may have occurred. Here is what the code looks like that utilizes a template to send back as a response to a request:

```
app.set('views', path.join(__dirname, 'views'));
app.set('view engine', 'jade');

app.get('/test', function(req, res) {
  res.render('test.jade', {
    title: 'My News Stories ',
    stories: items
  });
});
```

First, you need to have incorporated the NPM jade module into your project. Then you need to make calls to tell Express what directory the template files are in and what template engine you are using. Express will then internally use the Jade module you have included in your project.

# PART II: The Service Layer (Node.js)

Notice the two `app.set()` calls used to configure the use of Jade. The `app.get()` sets up the request route with a handler. It is in that handler function where you have the call to render the template with the given data to bind to it.

Refer to Jade documentation to learn about the template syntax. It uses a curly brace syntax to bind to properties. Here is a Jade template that can take the passed in data context and bind those values to elements:

```
// test.jade file content for the template view
doctype html
html
  head
    title my jade template
  body
    h1 Hello #{title}
    div.newstbl
      each story in stories
        div.storyrow
          a(href=story.link)
            img.story-img(src=story.imgUrl)
            h6 #{story.title}
```

Be aware if you use template rendering that you are relying on server-side rendering or SSR of your HTML. If you prefer to utilize a SPA architecture for a client-side browser application, you will not want to do this. In Part Three, I will describe how to return HTML that uses React component Document Object Model (DOM) rendering for the client-side SPA application. You can also do server-side rendering or SSR with React, either to return all your rendered pages, or just a few of them.

# Chapter 10: The MongoDB Module

This chapter is one of the most important in Part Two of this book. One of the main purposes of a service layer is to provide access to the data layer. To do that, you will be utilizing a Node.js module that has been created to interact with MongoDB on the backend. You will be learning how to use the "mongodb" module from the NPM repository.

You need to first include this module in your package.json file so it is available in your project. The usual `require()` statement is then used to make functionality available in your code. I will cover that again when you construct the NewsWatcher sample application.

*Note: The mongodb NPM module API is quite massive and it would take a huge book to document it all, so this small book cannot make you an expert in all its usage. For example, I have chosen to perform the functions to create and delete collections and to perform other administrative duties through the Atlas management portal. You should make a quick pass through the API to see what other capabilities it has that you may want to take advantage of. Look for the mongodb module on the NPM site and find a link to the documentation from there.*

## The MongoClient object

Your JavaScript code needs to establish a connection to a MongoDB server. To do this, you use the `connect()` function of the mongodb `MongoClient` object. After establishing a connection, you will find a lot of useful functions for interacting with a MongoDB collection. The function signature for `connect()` looks as follows:

```
connect(urlConnectionString, [options], [callback])
```

The first parameter of the function is the URL of the service endpoint for your MongoDB instance. The second parameter contains options that can be used for settings on the server, such as for a replica set, etc. The last parameter is your callback function, where you will receive the client object that then allows you to be connected to a database and a collection.

If you have an incorrect URL, you will get an error returned. You can find the actual URL connection string to use by opening the Atlas management portal. Click the **CONNECT** button for the cluster. Click **Connect Your Application**. You will see the connection string listed. You can click the **COPY** button to capture it. URL-encode your username and password if they contain any characters that are not readily used in a URL without converting them, such as '#' or '@'.

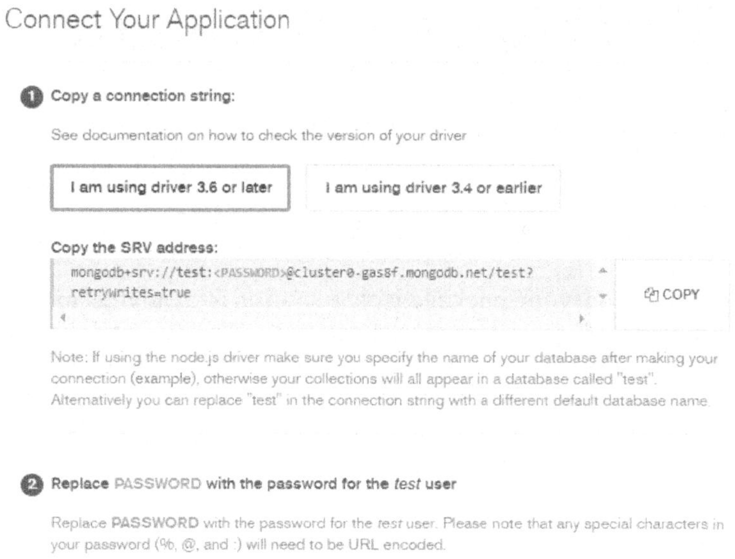

Connect Your Application

**1** Copy a connection string:

See documentation on how to check the version of your driver

| I am using driver 3.6 or later | I am using driver 3.4 or earlier |

Copy the SRV address:

```
mongodb+srv://test:<PASSWORD>@cluster0-gas8f.mongodb.net/test?
retryWrites=true                                                    ⧉ COPY
```

Note: If using the node.js driver make sure you specify the name of your database after making your connection (example), otherwise your collections will all appear in a database called "test". Alternatively you can replace "test" in the connection string with a different default database name.

**2** Replace PASSWORD with the password for the *test* user

Replace **PASSWORD** with the password for the *test* user. Please note that any special characters in your password (%, @, and :) will need to be URL encoded.

*Figure 39 - Atlas database page with connection URL*

In more readable text, this is something like:

```
mongodb+srv://test:pwd@cluster0-
gas8f.mongodb.net/test?retryWrites=true
```

There are placeholders for a password in the string. You should also go to the **Security** tab above the cluster and create a user login that you will use in your code. This is different from the account login you use with the administrative Atlas portal. You need to choose a setting to give the user account read and write access when you create the user.

Here is some example code calling connect() to establish a connection using the URL from the above example.

```
var db = {};
var MongoClient = require('mongodb').MongoClient;
MongoClient.connect("<Connection string>", { useNewUrlParser: true,
useUnifiedTopology: true, minPoolSize: 10, maxPoolSize: 100 }, function
(err, client) {
    db.collection = client.db('newswatcherdb').collection('newswatcher');
}
```

The callback code uses the database connection and calls the collection() function. This gets the "NewsWatcher" collection object which can then be used to perform CRUD operations.

You are now ready to learn about a few of the methods exposed with the mongodb module. The focus for you is on learning the functions necessary to perform the CRUD operations.
140

# 10.1 Basic CRUD Operations

With the collection object now obtained, you are ready to learn about the CRUD operations. In the terminology of MongoDB, for single document interactions, CRUD translates to the following functions:

- `insertOne()`
- `findOne()`
- `findOneAndUpdate()`
- `findOneAndDelete()`

I'll now walk you through each of the four functions and show you how to use them. Later, you will put together the NewsWatcher sample application and use them all again, plus a few others. At that time, you will make your code more robust and include error handling.

*Note: There are variations of the CRUD functions listed above. For example, to find multiple documents, it is necessary to use the `find()` function. When reading any of the documentation, pay attention to any text stating that it is deprecated. Deprecated items may work in a current version but may not function in future updates.*

*Note: The driver supports callbacks, Promises, and async/await. We will use callbacks.*

## Create

The following is the function signature used for creating a document in a collection:

```
insertOne(doc, [options], [callback] ) -> {Promise}
```

The first parameter is the JavaScript object that you want to have inserted. This then gets created as a BSON document. If you do not provide the _id property in your JavaScript object, MongoDB will generate one for you when it stores the document.

Your callback function will have, as the first parameter, an error object that you can check to see if something went wrong on creation. If you don't provide a callback function, a Promise is returned for you to use. The following code passes an object and uses the callback function.

```
// This example does not use the options object parameter
db.collection.insertOne({property1: 'Hi', property2: 77}, function
(err, result) {
  if (err) console.LogError('Create error happened');
  else console.log(result.insertedId);
});
```

The options parameter of the `insertOne()` function is an object that has a set of properties on it. This same object is used for many of the calls in the API. I will describe it here for your reference as you will see this used again. All the properties of this object are optional. Here is a table that describes some of the properties of the `options` object:

**Request Options Object Properties:**

| Property: | Purpose: |
|---|---|
| `writeConcern` | The write concern (how the acknowledgment works). |
| `maxTimeMS` | The timeout you want for the return. |
| `bypassDocumentValidation` | To bypass schema validation. |
| `serializeFunctions` | Bool to serialize functions on any object. |
| `forceServerObjectId` | Bool for server assignment of _id values, not driver. |
| `retryWrites` | To retry failed write. |
| `readPreference` | What machine to read from – primary, secondary etc. |

The `result` parameter in the callback function has very few properties on it. The driver has evolved and done away with the previous ability to get back the stored document in the result. In theory this makes sense, since you were the one putting the document in and only might want to know the id of the created document. The `result` properties are as follows:

| Property: | Purpose: |
|---|---|
| `acknowledged` | If the write was accepted. |
| `insertedId` | The generated `ObjectId`. |

# Read

To retrieve a single document from a collection, use the `findOne()` function. The following is the function signature for retrieving a document:

```
findOne(query, [options], [callback]) -> {Promise}
```

The first parameter is the query criteria. You can go back to Chapter Three to review what that looks like. What this function does is to simply return the first document that matches the query criteria and no more. The callback has an error object followed by the document returned. Here is an example (not using the options parameter) of `findOne()`:

```
db.collection.findOne({ email: "nb@abc.com"}, function (err, doc) {
  if (err) console.LogError(err);
  else console.log(doc);
});
```

142

`Options` is an optional object with 20 optional properties you can use. Refer to the mongodb module's documentation for a complete list of properties. The following is a description of just a few of them:

| Name | Type | Description |
|------|------|-------------|
| `fieldsAsRaw` | object | The projection criteria to specify the fields to return as unserialized raw buffer. |
| `hint` | object | To tell the query what indexes to use. |
| `explain` | Boolean | Return an object with query analysis and not the result. |
| `raw` | Boolean | Return the BSON. |
| `readPreference` | ReadPreference \| string | Which machine in replica set to read from, such as the primary or secondary. |
| `maxTimeMS` | Number | How long to wait in milliseconds before aborting the query. |
| `session` | ClientSession | Optional session to use for the operation. |

`findOne()` returns a JavaScript promise if no callback was provided.

You can use the `explain` option to look at how well your indexes are working. You will see some JSON returned that gives useful data about the query. Make sure to remove this option afterward as it prevents your actual result from being returned.

# Update

The following is the function signature for updating documents in MongoDB:

```
findOneAndUpdate(filter, update, [options], [callback]) -> {Promise}
```

The first parameter is the filter parameter, which is the query criteria needed to identify the document. The second parameter is for the update operators to use. Your callback function would have as the first parameter an error object that you could check to see if something went wrong on the update. Here is an example of usage that uses an option to have the updated document returned in the callback:

```
db.collection.findOneAndUpdate(
  {email: "nb@abc.com"},
  { $set: { name: "Charles" }},
  { returnDocument: "before" },
  function (err, result) {
    if (err) console.log(err);
    else if (result.ok != 1) console.log(result);
    else console.log(result.value);
});
```

The following is a description of a few of the `options` object properties:

| Name | Type | Description |
|---|---|---|
| projection | object | The projection criteria to specify the fields to include or exclude in the return. |
| sort | object | Specifies a sorting order for multiple documents that are matched. |
| maxTimeMS | Number | How long to wait in milliseconds before aborting the query. |
| upsert | Boolean | Create the document if it did not exist. |
| returnDocument | String | Set this to "after" if you want the updated document returned. |
| session | ClientSession | Optional session to use for the operation. |

`findOneAndUpdate()` returns a promise if no callback was provided.

# Delete

The following is the function signature for deleting a document in MongoDB:

```
findOneAndDelete(filter, [options], [callback]) -> {Promise}
```

The first parameter is the query criteria needed to identify the document. Your callback function will have as the first parameter an error object that you can check to see if something went wrong on deletion. Here is an example of usage:

```
db.collection.findOneAndDelete({email: "nb@abc.com"}, function(err,
result) {
  if (err) console.log(err);
  else if (result.ok != 1) console.log(result);
  else console.log("User Deleted");
});
```

Here is a description of a few of the options object properties:

| Name | Type | Description |
|---|---|---|
| projection | object | The projection criteria to specify the fields to include or exclude in the return. |
| sort | object | Specifies a sorting order for multiple documents matched. |
| maxTimeMS | Number | How long to wait (milliseconds) before aborting the query. |
| Session | ClientSession | Optional session to use for the operation. |

`findOneAndDelete()` returns a promise if no callback was provided.

144

# 10.2 Aggregation Functionality

In the data layer chapters where the MongoDB capabilities for querying were covered, I intentionally omitted one specialized type of query - the capability of MongoDB to perform aggregation over the data. Aggregation gives you the ability to report on summarizations of data such as grouping or finding the sum, min, or max across a value.

Because this gets rather involved, I held off introducing it until now. You need to focus some serious study on this topic to master all that is possible with the aggregation capability.

Here is a simple example to give you a feel for how it works. Imagine that for the bookstore used in prior examples, you want to find out the number of customers living in each state. You will use the `aggregate()` function for this. The following example is the signature of the `aggregate()` function:

```
aggregate(pipeline, [options], callback) -> {null|AggregationCursor}
```

The `pipeline` parameter is an array of MongoDB supported aggregate commands. Think of these as stages that data is being piped through from one to the next. You can string quite a few aggregate operators together.

Here is an example that gives the count of people by state. Notice how we chain the function to go to the `toArray()` function.

```
db.collection.aggregate(
  [
    { $group: { "_id": "$address.state", "count": { $sum: 1 } } }
  ]).toArray(function(err, result) {
    console.log(result);
});
```

The result may be as follows:

```
[{ _id: 'UT', count: 54 },
 { _id: 'KS', count: 988 },
 { _id: 'FL', count: 1259 }]
```

The callback has an error object that you can check.

You can insert operators like `$match` and `$project` that help narrow down the documents and what properties are passed through the pipeline.

# 10.3 What About an ODM/ORM?

You have learned how to directly connect to MongoDB with a node module created specifically for that purpose. Yet there is another NPM module you can use to approach interfacing with MongoDB in a completely different way. This is the "mongoose" module.

Those of you from a relational database background will understand there are such things as Object Relational Mappings (ORMs) for connecting to a SQL Server. Perhaps you are familiar with Entity Framework for .Net, or Hibernate for Java? The equivalent in a document-based database is called an Object Data Mapping (ODM).

Look up "mongoose" on NPM or GitHub and you will find an ODM that sits on top of MongoDB. This module can be used instead of the mongodb one that we previously covered in this book. The mongoose module adds an additional abstraction. With this, you get features such as schematization of the data and validation.

Here is what code looks like that uses the mongoose module to save and query a document. I will use the sample of the customer document from the online bookstore example, but I will cut back on the number of properties for a shorter example. Here is what some code may look like that uses mongoose:

```
var mongoose = require('mongoose');
mongoose.connect('<The usual connection URL>');

var Customer = mongoose.model('Customer', { name: String, age:
Number, email: String });

var c = new Customer({name:'Aaron', age:32, email:'ab@blah.com'});

c.save(function(err) {
  if (err) console.log(err);
});

mongoose.model('Customer').find(function(err, customers) {
  console.log(customers);
});
```

There are ways to accomplish each of the ODM capabilities on your own. For example, to add in a simple module that helps you do server-side input validation, you can use "express-validator" or "joi", but why go to all the work if an ODM already exists? An ODM module may be the way to go to give you more robustness with your application.

# 10.4 Concurrency Problems

Once your application starts to have multiple concurrent users, you may run into issues updating documents in MongoDB. Let's take an example where you have a single document that contains a list of high scores for an online game and you want to keep the top five scores across all players. As users finish a game, a call is made to submit the score they achieved and insert it into the list of top scores. High scores are best, so if a new score is posted that should be added because it is higher than other entries, the lowest score is dropped off the list.

If multiple players all submit their scores at the same time, you can understand that some contention may arise with updates to this single document. Here is a diagram of how this might look:

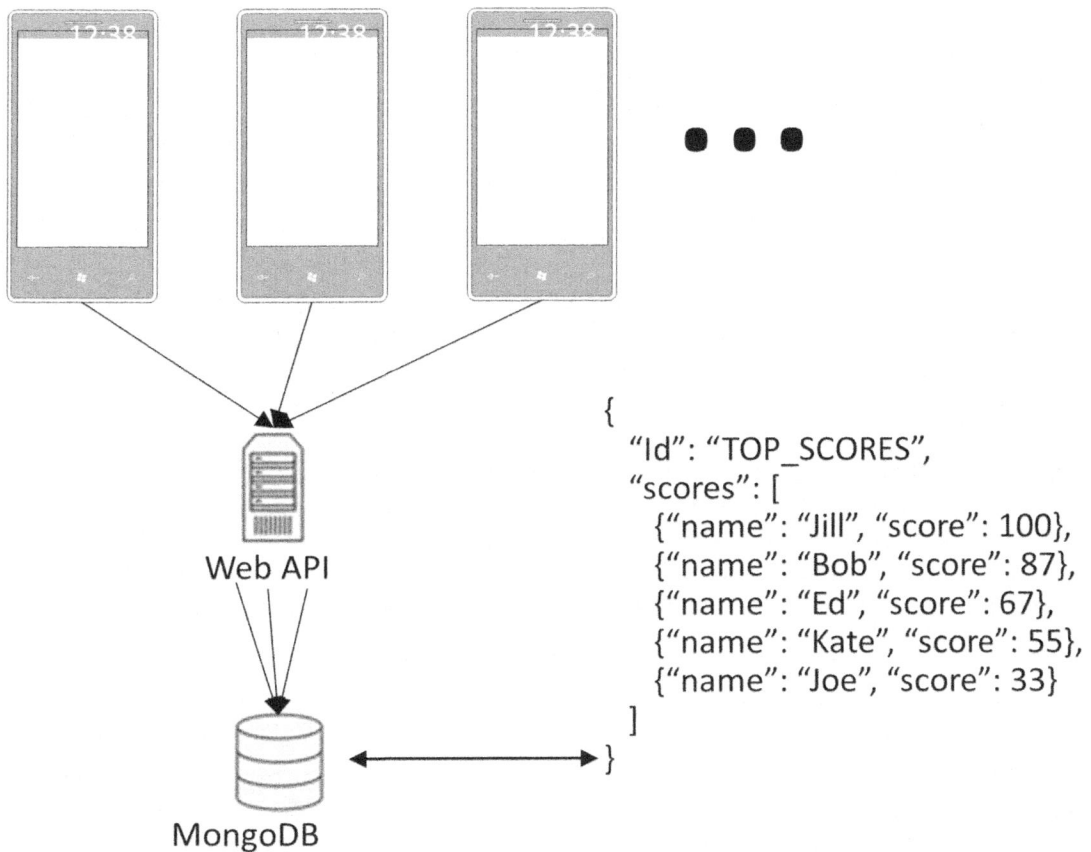

```
{
  "Id": "TOP_SCORES",
  "scores": [
    {"name": "Jill", "score": 100},
    {"name": "Bob", "score": 87},
    {"name": "Ed", "score": 67},
    {"name": "Kate", "score": 55},
    {"name": "Joe", "score": 33}
  ]
}
```

*Figure 40 - Top score contention illustration*

# PART II: The Service Layer (Node.js)

All these users are causing score insertion code to run concurrently. Each request gets sent to a Node.js Rest API endpoint route to allow multiple simultaneous calls to read and update the single document in the MongoDB DBMS.

What happens if two scores are submitted for processing at the exact same time? Each request would first read the document and then insert a score if it is higher than the lowest number in the list. The document then is sent back for replacement in the collection. Here is a sequence diagram that illustrates the problem with time flowing top to bottom:

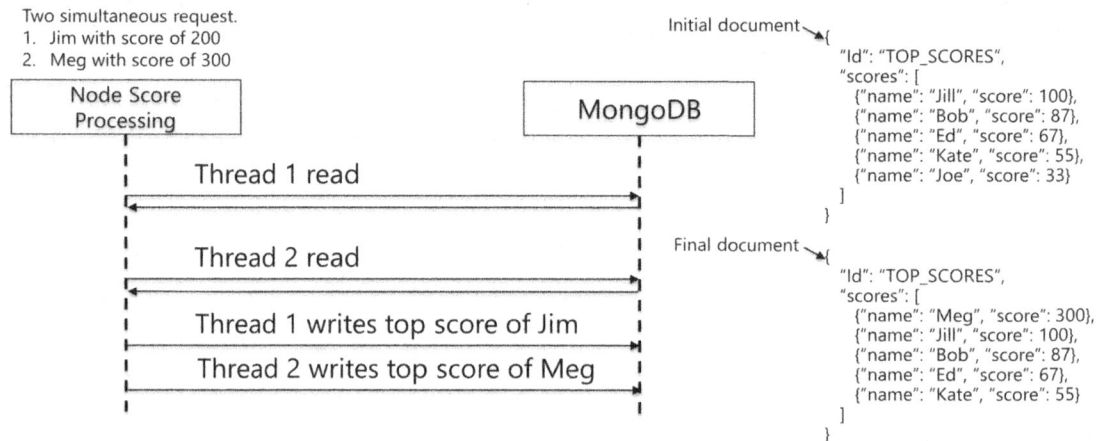

*Figure 41 - Simultaneous update*

If you look carefully, you can see that the final document is not what it should be. You want to see both Meg and Jim inserted at the top and then Kate and Joe dropped off from the bottom of the array. Instead, Meg is at the top and only Joe drops off. Jim is really going to be disappointed when he checks the high scores and does not see his name listed.

Neither of the updates knew the other one was happening and so the last write "wins." This is a classic problem and not something unique to MongoDB. There have indeed been many solutions invented over the years to solve the problem of simultaneous submits. For example, some database technologies offer locking. The problem with this is that locking can be tricky and requires you to implement detection of stale locks. MongoDB does not support API level locking. It does, however, do some locking on its own internally for certain calls you make.

The solution that MongoDB and its underlying storage engine (WiredTiger) implement for you is termed optimistic concurrency. If you use the findOneAndUpdate() function, this will happen at the document level. However, both processing calls use a read followed by an update function, simultaneous reads will still get the same data and one write will prevail. So the second write will overwrite the first.

148

The MongoDB driver offers `findOneAndUpdate()`. If two calls come in, the first one pauses the second one from doing anything until the first call has finished on that document. The second call will not even do its work until the first one has completed its write of the data. The second call then keeps retrying for several times until it is successful.

Let's go back to the initial sequence diagram and draw this out one more time with retries put in:

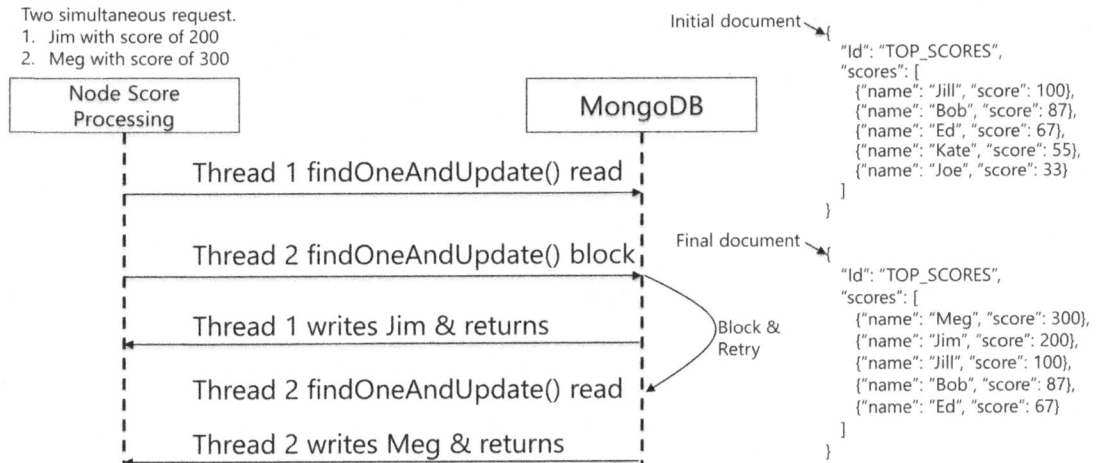

*Figure 42 - Simultaneous writes with optimistic concurrency*

Optimistic concurrency retry logic can be inefficient if your database is always being locked and retries are constantly happening. Imagine if four users consistently update their scores at the same time every second, over and over. If four attempts are made at the same time, only one can succeed and then the other three must try again. The second round has three attempts, and one works, and so forth.

The point is that you might be wasting MongoDB compute time. This may cause your MongoDB service to perform poorly. You can always look at your design and see if you need to have one single document that is always being contended for. You may run into cases where you simply have a "hot" document with lots of simultaneous updates all the time as illustrated with the game scores.

Imagine multiple independent Node.js processes that are all hitting a single backend MongoDB database. This means you will have to come up with a way to coordinate across independent processes.

If you want to implement the ultimate architecture to handle massive scaling and avoid conflicts you get with optimistic concurrency, you need to create a single queue that is outside of all your node processes.

# PART II: The Service Layer (Node.js)

Here's how this works. You place your update operation requests in an external queue, such as an AWS SQS resource for global access across all node processes. You then create a single unique node process that watches the AWS queue and processes the requests. Or even better, use an AWS Lambda function. The following example shows this solution. The dots next to the Scaled Web API represents multiple machines with load balancing across node processes:

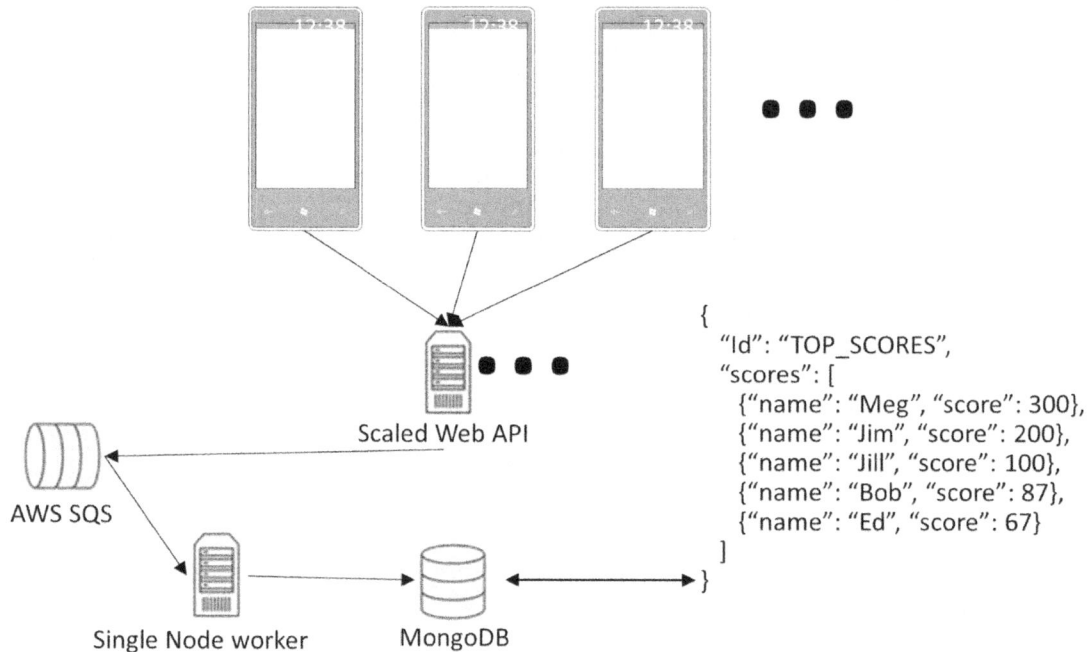

*Figure 43 - Queue serialization solution*

This also is the safest solution if you need to ensure consistency and repeatability. There is one main drawback with this solution — you need to deal with the issue of returning to the initial caller if the operation is successful. This might not be important, but it could be.

You can decide to take the "fire-and-forget" approach and always return success. Perhaps, losing a top score for some reason can be considered acceptable if later the worker role fails to insert a score for some reason.

Alternatively, what if you have a bank transaction that requires a notification one way or the other back to the user? You need to implement some way to do an asynchronous return. There are available notification mechanisms that would solve this. For example, the request would be queued, and an id returned that the UI could use to poll with to get the completion status. You could also use WebSockets, gRPC or HTTP2 capabilities to push back to the browser the result.

150

# Chapter 11: Advanced Node Concepts

Some particularly difficult issues to be aware of require careful coordination across the Node ecosystem. Certain intricacies must be handled properly. Done incorrectly, and it can be disastrous. Done correctly, everything will hum along. This chapter will cover a few of the subtleties you might need to address.

# 11.1 How to Schedule Code to Run

A timer function can be used to schedule code to run at a later time. This can be done as a one-time request, or it can be set up to happen on a recurring interval. You can also investigate using the cron NPM module to schedule the running of periodic Node code. The following code schedules a function to run in five seconds and shows how to pass in a parameter:

```
setTimeout(myFcn, 5000, "five");

function myFcn(param1) {
   console.log("Hi %s", param1);
}
```

The callback is not going to happen exactly at 5000 milliseconds, but Node will do its best to fit it in when it is time, and the callback is then given to V8 to run.

***Note:*** *It is best to split out backend batch or timer based processing code to scale better as wells as. You would use AWS EventBridge to fire a Lambda or run something in ECS/EKS. This way you do not over burden your main node.js server transaction processing backend.*

You can use the function `setInterval()` to schedule a recurring function. If you want to cancel the recurring timer, you can cancel it at any time with `clearInterval()`. The following example shows how this would look:

```
var id = setInterval(myFcn, 5000, "five");
...
clearInterval(id);
```

Be aware that Node will run your function over and over, even if the previous call has been blocked and has not yet completed. If you want a callback executed only if the previous one has completed, you can write your own implementation of `setInterval()` to prevent that behavior.

Another function named `setImmediate()` is available that does not have a timer go off, but places the callback to be executed after I/O before `setTimeout()` and `setInterval()` events are processed.

If you are interested in scheduling code to run at an even higher priority, you can use the process object `nextTick()` function. This call will place your callback to be executed above all other processing in the event queue, even before I/O callbacks are executed. Be careful to not use this unless you are using it responsibly. If you do this too often, at some point you will completely cut out all I/O processing.

# 11.2 How to Be RESTful

It is up to you to design the HTTP API endpoint that your Node application will process. You can create a RESTful service and even support OData or GraphQL. You are also free to include support for query strings if that is your preference. If you go the RESTful route, here are a few things to keep in mind.

The first thing is that RESTful web services are intended to be stateless. If you are scaling your server-side Node process and using the cluster module or have scaling through Elastic Beanstalk, then you most likely want stateless servers. Any of your horizontally scaled servers can process any incoming client call. If you want state information kept around, you can store state information with the client or have it cached for server-side retrieval in something like a Redis cache.

Make sure to design your REST API URLs so they make sense. There are standard things to consider such as using nouns and not verbs in your paths. You may run into some dilemmas, but there is probably an answer for your challenge in a forum you can research.

When you come out with a new version of your REST API, you need to make that clear and perhaps support multiple versions for a while. You can strive to keep your API as backward compatible as possible. You can insert a version number into your path to move clients from one version to another. For example, if you had "/api/users" as a path, you could have clients start using "/api/v2/users" for new functionality. You can always tack a query string on the end to specify the versioning such as "?Version=2015-12-22". An HTTP header setting is also possible such as "x-version: 2015-12-22".

# 11.3 How to Secure Access

You should conduct a review of all your data connection points and scrutinize all data that is being transferred and stored. Make sure to take appropriate precautions with sensitive data for which you are required to safeguard for your customers. Information such as a home address can be used to identify a person and should never be leaked. Financial and medical records many times have laws and regulations concerning their storage and transmission. You are not just trying to ensure your business interests are safe, you are responsible to safeguard your customers from harm.

One vital detail you need to work out is how users will identify themselves and be allowed to access your Web API from a client application. The other security concern you need to solve is how to prevent any eavesdropping or man-in-the-middle type of hacks as information flows back and forth from the client to the Web API service. This section will explore these and other related topics and present a solution for each.

# Access token

Once a person is identified by logging in with their password, the Web Service needs to recognize and allow access to data associated with the account. One possible means of user interaction authorization is to have a customer sign in and then use a session cookie with every request coming in.

Another similar mechanism is to generate an access token and have that passed in with every client request. There is a standard way to do this with something called a JSON Web Token (JWT). A JWT is something that can be generated on the server-side in response to a client login request when presented with a username and password. The JWT is basically an encoded set of information about the user that is signed to make sure data is not tampered with. Here is the sequence diagram for how a JWT is created and passed from layer to layer:

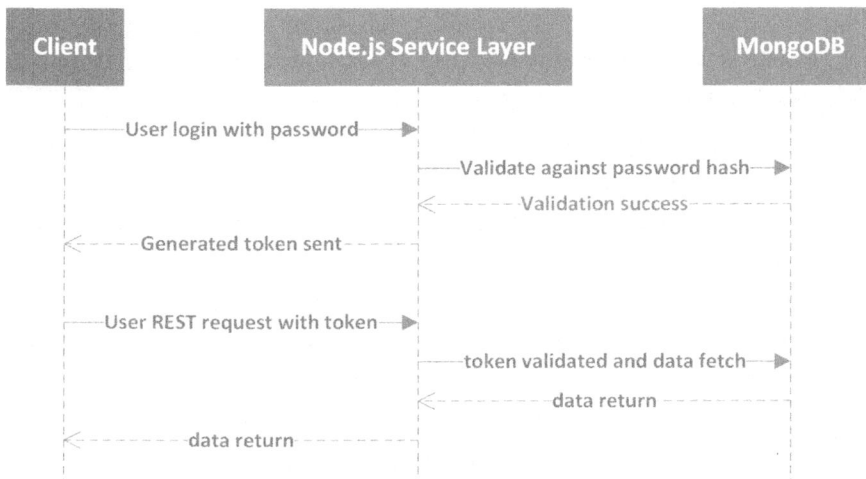

*Figure 44 - JWT token passing*

You can embed whatever you want into the JWT. Put in information that will help you in your processing back and forth. Add in the IP address as well as the HTTP user-agent header value for additional security. This can then be validated on later calls to make sure the same person is using the token. You can also set an expiration time on the token so that a person is required to log in every so often.

To offload user authentication, you can use the Oath 2 standard. This allows you to have a person redirected to some other web presence that the person already trusts and log in there first. Users may prefer this, as they only need to manage one single sign-on credential.

With the NPM passport module, you can implement Oath 2 to delegate the user authentication to an externally trusted site such as Google or Facebook. A token of authenticity is passed back that has information about who the user is.

*Note: The JWT should always be transferred using HTTPS because it easily can be decrypted. Make sure it doesn't contain anything that compromises your security. The token is in the header such as "x-auth: <token>" or "Authorization: Bearer <token>" if you want to mimic standards such as Open ID Connect (OIDC).*

# All data traffic should be encrypted

Once you have sufficient momentum on your code, another priority is to implement certificate-based authentication and encryption using the HTTPS standard. This will enable your site to be viewed as legitimate and ensure that data is transferred using the encrypted TLS/SSL protocol.

If you search on the internet, you will find code samples that show you how to configure Express to require HTTPS. If your Node.js service is to be hosted on a machine directly exposed on the internet, this is what you need to do:

```
const https = require('https');
const fs = require('fs');

const options = {
   key: fs.readFileSync('keys/agent-key.pem'),
   cert: fs.readFileSync('keys/agent-cert.pem')
};

https.createServer(options, (req, res) => {
   res.writeHead(200);
   res.end('hello world\n');
}).listen(8000);
```

If you choose to use a PaaS solution, the above code is not necessary. This is because your Node.js service is hidden behind the server that acts as the reverse proxy and load balancer.

154

However, this means that you need to configure HTTPS for the AWS Elastic Beanstalk service external-facing load balancer. You will see how this is set up when the sample application is put together.

# 11.4 How to Mitigate Attacks

When you expose a service on the internet, it will be vulnerable to attacks of all kinds. Some attacks may be intentionally malicious and others just annoying. All threats should be taken seriously. At a minimum, attacks can disrupt your service, which is unacceptable. Beyond that, cyber-attacks can steal sensitive information and do damage to your customers and to your own reputation.

Obviously, a simple native mobile phone game of tic-tac-toe would not have as large an attack surface as a three-tier e-commerce or financial application. The more infrastructure and code that you have, the larger your attack surface will be. Hackers will look for the vulnerability that is easiest to exploit. Your security is only as good as your weakest point of attack.

To get an accurate look at all possible avenues of attack, you need to draw out a data flow diagram that shows all the processes, interactions, data stores, and data flows. Each process represents those that you own or ones that you rely on. Some of those processes can be classified as completely external and possibly out of your hands, but they should still be on the diagram.

For each of the elements in the diagram, you will want to do some analysis to determine what threats can exist. For example, if a SQL database is in your diagram, you will determine what it is storing, how data gets in and out, and what configuration and administration are occurring. You may discover that your SQL database is vulnerable to a SQL injection attack.

With your analysis completed, you will mitigate each of the vulnerabilities. In some cases, you simply may need to write a few lines of code; in other cases, you may need to re-architect parts of your system. To fully address this topic, you will want to buy a book specifically dedicated to this topic. I will now present a few security concerns associated with Node.js and Web API interactions.

## Never trust ANY input!

One basic strategy is to never trust any input. Always do what you can to validate all data before using it. Look into using the node modules "joi" or "validator." Using a data layer ORM/ODM can give you these data validation capabilities, but you still need to sanitize your data from script injections. Here is some validation using the joi module:

```
var schema = joi.object({
    displayName: joi.string().alphanum().min(3).max(50).required(),
    email: joi.string().email().min(7).max(50).required(),
    password: joi.string().regex(/^[a-zA-Z0-9]{3,30}$/)
});
schema.validateAsync(req.body).then(value => {
    //...
}).catch(error => {
    return next(err);
});
```

You can see that only alphanumeric characters are allowed for the user display name. The joi module is also making sure no extra properties exist on the body. These types of restrictions are extremely important, so don't underestimate their usefulness.

Another issue you'll face is data transmission size. What if someone starts sending large JSON packages or ones that contain extra objects or properties in them? You can set up your body-parser middleware to turn down requests that are too large and thus may be suspicious. The default size is 100kb so you can make that smaller just to be safe. Here is how you set that up:

```
app.use(bodyParser.json({ limit: '10kb' }));
```

Let's now look at some of the types of attacks that can occur on your exposed Web API. None of these are specific to Node.js. and exist because of the fundamental way that browsers and HTTP work.

# DOS/DDOS attack

The denial-of-service (DoS) or distributed denial-of-service (DDoS) attack is where traffic is thrown at your web app to try to bring it down. It might be possible to overwhelm it so that others are prevented from using it. If successful, the attack denies service to the legitimate people who are intended to use it. In some cases, an attack may lead to incorrect behavior of your application to exploit it for other gains.

The distributed version of this attack means that the attacker is employing multiple distributed machines at once. The term bot is commonly used, which means that these machines are set up to run scripts or programs to carry out the attack and to constantly enlist other machines to participate. It acts like a virus and replicates.

A DoS attack is purely malicious and rarely happens by accident. You need to be prepared for it and mitigate this risk regardless of intention. You can be sure that big e-commerce sites like Walmart.com, Amazon, and eBay often see these kinds of attacks and take them seriously. An attack like this might even cause your scaling infrastructure to unnecessarily kick in and start costing you more money in cloud operating costs.

156

At this point, I must repeat – *never* do anything compute intensive on the main Node.js thread. If you have a lot of requests coming in that trigger some intensive synchronous code, your process basically will be unable to respond. This makes it easier for a real DoS attack to occur if you allow your main thread to get overloaded. Let's now look at ways to mitigate a DoS attack.

When you initially created your Elastic Beanstalk Node.js application environment, it set up Nginx to act as the reverse proxy and load balancer. Nginx can be configured to limit the rate per IP address as well as limiting connection count per IP address.

Another approach you can take using AWS is to set up an AWS API Gateway in front of your service layer. That gives you a lot of what you need for defense, such as throttling per connection to head off a DoS attack. You also get authorization, reporting, and API consumption of your web API contract in a central shared way.

You also can do some type of IP blocking on your own, such as tracking the access time per IP and then limiting each IP to once-per-second access. Check out the NPM module express-rate-limit. You can use a cloud-mitigation provider that applies its expertise to track patterns of attack and identify DDoS attacks and disable them.

# XSS – Cross-Site Scripting

A Cross-Site Scripting (XSS) hack is where a foreign script gets injected and run as part of your web-rendered site. A likely vulnerability is where you accept input from the user and then later re-display that input back to them and others who view the site. The browser does not stop JavaScript that was maliciously inserted as it cannot tell the difference. Your application allowed the user to enter something in the first place. Normal users will not be typing in malicious scripts to be run — it is hackers who love to do this for fun and profit.

Let's take an example of something that can affect the NewsWatcher application. In that application, people can comment on a shared news story. This is an occasion where input from the user is accepted and later displayed back to them. Let's say that a malicious user enters the following text as a comment on a news story instead of some nice comment:

```
<script>alert("Hi");</script><img src="smiley.gif">
```

Now all other users looking at the comment will be affected. In the UI code, you may have some HTML that displays all of the comments and end up looking like this:

```
<ul><li>
    <p>'<script>alert("Hi");</script><img src="smiley.gif">'</p>
</li></ul>
```

Everyone will now see an alert box and a cute little smiley face staring back at them. You never intended for this to happen, but you did nothing to stop it. The savvy hacker can even hack the client-side JavaScript code and mess with the JSON before it gets sent back to the server. Web APIs simply can never trust the data that is sent to them.

Imagine if the hacker references some script across the internet that will really wreak havoc. If the hacker understands what your API is on the backend, the hacker can run any command virtually as a logged-in person and really do some damage. This means the hacker will have hijacked the user session.

This goes back to the simple statement that you should never trust user input. To mitigate this, you can take action to validate the input as you collect it. For example, in the NewsWatcher code, you can validate things like the username and don't allow anything but alphanumeric characters. Your sanitization can also scan all characters and change a character like '<' into "&lt;".

Fortunately for the NewsWatcher application we are using React to render the UI. React does the work to disable any scripts from running. React simply displays the actual text without letting the browser interpret it. In the example above, you would actually see the literal string "<script>alert("Hi");</script><img src="smiley.gif">" in the comment list and all would be good.

# CSRF – Cross–Site Request Forgery

A Cross-Site Request Forgery (CSRF) hack is where a request is made to a site you are currently logged into with your browser. You already are authenticated and have an authorization cookie stored by the browser. The attacker tricks you into viewing a page he or she has set up, and as that page is loading in the browser, it runs a script that sends a request to the site you are already logged onto.

As an example, let's say you are logged on to your banking website. While still logged on, you open an email that is from a malicious attacker that says: "Click here and win a million dollars!" When you click on the link, the destination URL directs you to the hacker's malicious site that then loads a page that runs a script that sends requests to your banking site and transfers money to them and changes your password at the same time. Since you are already logged on, the browser happily sends the authentication cookie along with the request and you are hacked. The banking site has no idea that this request was not valid.

Our NewsWatcher sample application does not use cookies, so it is not vulnerable to this attack. The JWT token sending is under the control of your client code and is not automatically sent by the browser like a cookie is.

If you do end up implementing some design that is vulnerable to a CSRF hack, you can use the "csrf" NPM module to implement a mitigation. This will create a secret token that only your site knows that is only sent from your pages.

# The NPM Helmet module

I have discussed each of the security concerns and discussed mitigations. In this section, you will look at the NPM Helmet module that would give you the ability to further mitigate possible attacks.

The Helmet module tweaks your HTTP headers to set things up to utilize certain best practices for security risk mitigations. The Helmet module works as Express middleware by injecting itself into the request-response chain. It does not do anything that you can't do by hand, though I recommend it because you would write a lot more lines of code to accomplish what the Helmet module does with a single line of code.

I will show you some code that will be the starting point when using the Helmet module. This code will set up things such as enforcing HTTPS and mitigating clickjack attacks, certain XSS mitigations, and attacks based on MIME-type overriding attacks.

You can take the defaults and further specify any deviations from there that you prefer. Refer to the documentation for any of the specific HTTP headers you want to individually control on your own. The following code shows you how easy it is to use the Helmet module:

```
var express = require('express');
var helmet = require('helmet');

var app = express();
app.use(helmet()); // Take the defaults to start with
```

The one usage you do need to control on your own is the setting of a Content Security Policy (CSP). This basically lets the browser be aware of where resources legitimately can come from. This then prevents resources that are unknown to you from being injected. The basic usage of the Helmet module with CSP settings added becomes the following code:

```
var express = require('express');
var helmet = require('helmet');

var app = express();
app.use(
  helmet({
    contentSecurityPolicy: {
      useDefaults: true,
      directives: {
        "default-src": ["'self'"],
```

```
        "script-src": ["'self'", "'unsafe-inline'", 'ajax.googleapis.com',
'maxcdn.bootstrapcdn.com'],
        "style-src": ["'self'", "'unsafe-inline'",
'maxcdn.bootstrapcdn.com'],
        "font-src": ["'self'", 'maxcdn.bootstrapcdn.com'],
        "img-src": ["'self'",  'https://static01.nyt.com/', 'data:']
      },
    },
    crossOriginEmbedderPolicy: false
  })
);
```

If you want to have violation notifications sent back to your service, you can use a setting and then handle that Express route in your code. Refer to the documentation for some sample code for that. The Helmet module cannot be your only mitigation for security threats. You need a thorough analysis of your data flow diagram to come up with every threat and start building your security plan.

## The Node Security Project initiative

An ongoing initiative to audit the code of NPM modules and keep a database of known vulnerabilities is by no means comprehensive, but it is good that this initiative is underway. Starting with NPM version 6, any install you do will automatically get verified to tell you if there are any known vulnerabilities with the NPM modules you are using. You can also run a command to have an audit run:

```
npm audit
```

Make sure to execute it in the folder where your package.json file is. The npm snyk cli is also a very valuable tool to use to identify out of date components and vulnerabilities.

# 11.5 Understanding Node Internals

You now know enough to use Node without knowing more about its internal workings. You are thus free to skip this section if you wish. You may, however, want to come back to this section later as it may help clear up some of your advanced questions that may arise.

*Note: I wrote this section from what I learned by reading through the actual source code for Node.js found on GitHub. Go to https://github.com/nodejs/node if you are interested in the internal workings of Node.js as I was. Libuv code also is included in the Node.js source code. You can find out more about its API if you go to http://libuv.org/.*

# The layers of Node.js

You can see from the following block diagram that a main Node.js codebase exists with dependencies below that. The top two layers of code are what make up the Node.js framework. The bottom two layers are dependencies that Node.js relies on.

The very top layer is the JavaScript library that you will be using directly from your code. Any time your code does something that is outside of the standard JavaScript calls, you will end up using something that is found in this layer. For example, all of the following code is made possible by the library layer:

```
var http = require('http');

var server = http.createServer(function (request, response) {
  response.writeHead(200, {"Content-Type": "text/plain"});
  response.end("Hello World\n");
});

server.listen(3000);
```

This top layer is what is documented on the Node.js official site https://nodejs.org/en/docs/ as its API and is the only layer you are required to know anything about to get started using Node. All core Node modules are exposed through this library layer.

Here are the four layers that comprise the operation of Node as a framework runtime. Keep in mind that this diagram may change as Node.js updates over time.

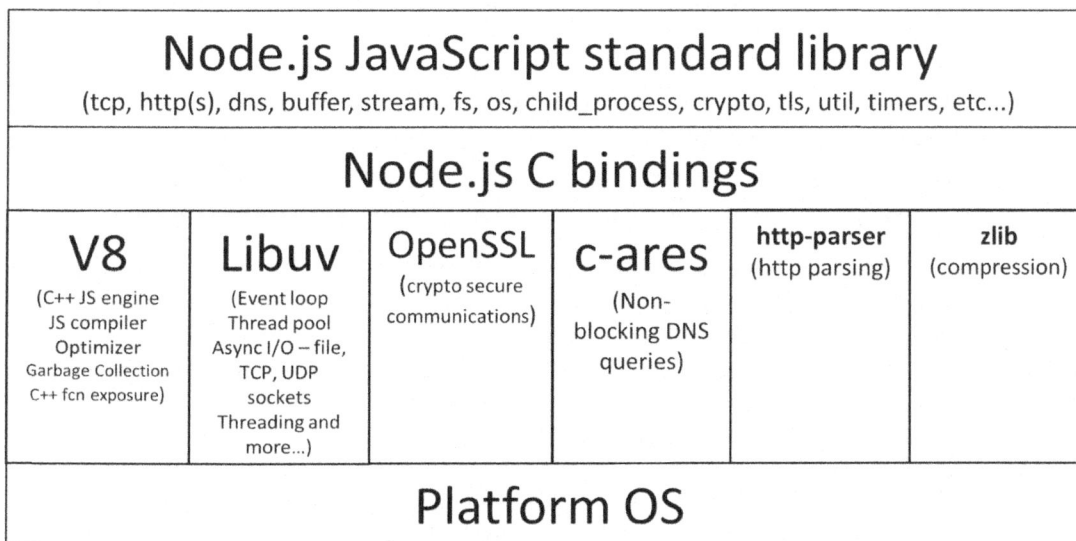

| Node.js JavaScript standard library | | | | | |
|---|---|---|---|---|---|
| (tcp, http(s), dns, buffer, stream, fs, os, child_process, crypto, tls, util, timers, etc...) | | | | | |
| Node.js C bindings | | | | | |
| V8 (C++ JS engine JS compiler Optimizer Garbage Collection C++ fcn exposure) | Libuv (Event loop Thread pool Async I/O – file, TCP, UDP sockets Threading and more...) | OpenSSL (crypto secure communications) | c-ares (Non-blocking DNS queries) | http-parser (http parsing) | zlib (compression) |
| Platform OS | | | | | |

*Figure 45 - Node.js platform*

# PART II: The Service Layer (Node.js)

While it would be great if the OS itself could understand JavaScript, neither MacOS, or Windows has JavaScript libraries exposed as they do for C/C++ usage. Thus, you need some translation from your code, to code that the operating system can understand. Therefore the bindings layer is needed.

The Node.js C bindings layer is made up of C++ code. This is what takes your code that is in JavaScript and allows it to call down to code libraries like Libuv that are written in C.

Node uses V8 as a dependency and does so for two main purposes. V8 makes it possible for your JavaScript code to call through to C++ code running in the Node.js process. Take the following example of some Node.js JavaScript application code you may have. This code displays the size in bytes of your package.json file:

```
var fs = require('fs');

fs.stat("package.json", function(error, stats) {
  console.log(stats.size);
});
```

The way it works is that Node has exposed the fs module in the library layer that is being used here. Node uses some capabilities of V8 to actually take the fs.stat() call and have that make a call to a C++ function in the bindings layer called Stat(). This C++ function in the binding layer then makes a call to Libuv, which in turn calls OS appropriate low-level code. Eventually, the call makes its way to a Unix flavor OS library call of stat() or, on a Windows system, the call made is NtQueryInformationFile() that is exposed in ntdll.dll. The callback then makes its way back to your JavaScript where the asynchronous operation completes.

V8 acts as a Virtual Machine in the sense that it can isolate and execute some code and be independently hosted many times on one machine. It provides all of the aspects necessary for a language runtime. You can read an introduction to V8 JavaScript engine by going to https://developers.google.com/v8/intro#about-v8. Here is some text from the Google site:

*"V8 is Google's open source, high-performance JavaScript engine. It is written in C++ and is used in Google Chrome, Google's open source browser...V8 compiles and executes JavaScript source code, handles memory allocation for objects, and garbage collects objects it no longer needs...V8 does, however, provide all the data types, operators, objects and functions specified in the ECMA standard. V8 enables any C++ application to expose its own objects and functions to JavaScript code."*

Besides Libuv and V8, there are a few other dependencies that Node uses for operations as noted in the diagram. To actually look at what the dependencies are of Node.js, you can go to the GitHub project and look in the "deps" folder. Node.js is very portable and its dependencies have been ported to many platforms.

162

You now have seen how all the layers fit together. You see that JavaScript code can be executed and that function calls can make their way through Node modules. Some of those modules invoke code through the binding layer in C++ down into Libuv.

The next thing to understand is how the processing loop comes into play. I held off explaining this thus far, but it is now important to understand in context. Understanding the processing loop helps you to be aware of where the processing takes place and how vital it is for you to keep your async callback code as performant as possible, to keep the main thread free.

# Operation of Node

Do not believe everything you may read about Node.js. Some people are under the misconception that Node.js only has a single thread and can only do one thing at a time. This is not exactly as some represent it. I will teach you exactly how Node.js executes and dispel any misconceptions.

Now that you have seen the layers of code that your application sits on top of, you are ready to see how these layers operate to orchestrate the execution flow of a Node application. The main operation of Node is illustrated in the following diagram. Not every dependent component has been included in the figure, just the main ones concerning processing flow:

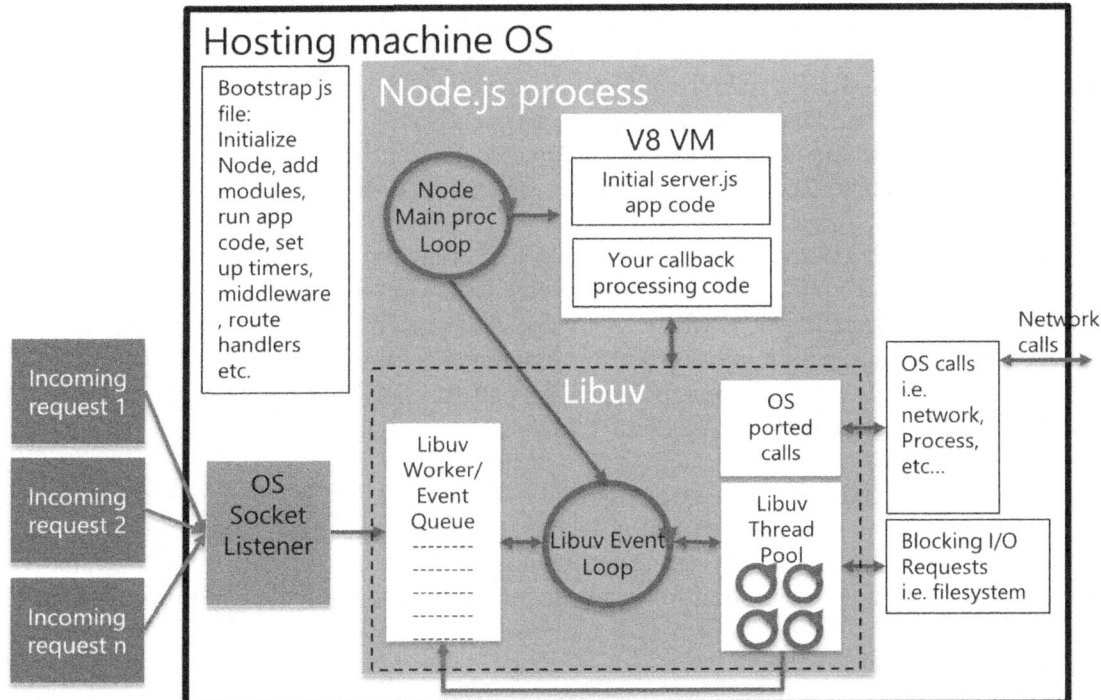

*Figure 46 - Node.js conceptual architecture*

163

# PART II: The Service Layer (Node.js)

It is true that Node.js has a single-threaded architecture for its main JavaScript V8 processing. What is mostly executing on that thread are your JavaScript functions that Node executes in response to events in the form of asynchronous callbacks. This is what gives you concurrent processing in your app. The reality is that *more* than one operation is executing in the guts of Node and on the operating system in the background, but only one return can be processed in your code at a time.

Internally, Node can use multiple threads to shuttle work off to. Node.js utilizes a callback code pattern that frees up the developer from having to manage the threading and polling. For example, you make your call for retrieving data from a backend database and, as part of that call, you provide a callback function. Your code immediately returns and then Node.js takes care of knowing when the data is returned and schedules your callback to run later. Your callback is thus running asynchronously.

Node.js does not have any sleep, mutex lock, or any similar functions that you may see in other runtimes. On the lower layers of Node, such as in Libuv, the threads of Node can be executing in parallel, such as taking advantage of running on a multi-core machine. The OS will be able to take the executing threads of Node and spread those out across the cores that are available. Thus, things like file system operations and network requests do occur in parallel, even though your processing of the results can only occur serially.

Your JavaScript callback code is executed asynchronously, meaning you make the call, and it is not blocking and code execution can continue elsewhere, and you don't know when the return will be processed. It is concurrent because your JavaScript code can call many different asynchronous APIs and have them all in action or queued without you worrying about them. It is parallel because the OS will take the many processing threads or low-level calls that Node makes and spread them out across the available CPU cores.

If you go back and consider the code sample that used the fs module, you saw that it eventually made it all the way down to the lowest level OS `stat()` function. That call is obviously a blocking call! This is where Libuv does the work for you to queue up that request to take it off the main V8 thread and return immediately. Libuv will then run the `stat()` call on a thread in its own thread pool. When the call finishes, Libuv knows the callback to call and sends it back up to be executed in V8.

The thread pool threads of Libuv are being used over and over. While there are four of them by default, you can increase the number of threads if you find a need for it. The NewsWatcher sample does not use the filesystem directly, so it does not alter that number. In Newswatcher, the Net module ends up being used for database interactions and that does not use the Libuv thread pool.

# Bootstrapping of Node

If you look back at the previous figure, you can see the upper left portion has what is termed the bootstrapping of Node. This is something that Node sequences through to get up and running. Here is the general sequence and explanation of the diagram:

1. The Node process is run from a command line and has a `main()` function that is called as its process entry point in its C++ code.
   a. The main function creates the runtime environment and then loads it.
   b. An object called "process" is created that has properties and functions on it. This object is very important and is used throughout the code.
   c. A JavaScript file is now run in the V8 VM that bootstraps the whole process.
      i. Some internal Node modules are parsed and made available.
      ii. The file (e.g., server.js) that node was started with as an argument is read and run in V8.
2. Your server.js file runs and at this point can use the full capabilities of Node to do things like use JavaScript, set up timers, set up HTTP listeners, make HTTP requests, set up middleware, etc.
3. After your server.js code has finished being processed, the Node process enters its perpetual processing loop. At this point, the process keeps running as previously stated as long as there is work to process.
   a. Libuv processes work on its queue with its process loop and threads.
   b. Callbacks make it back to the main thread to be run by the V8 VM. V8 will queue calls that come in and process them.

For those of you who are still skeptical about my stating that there are two processing loops, here is the proof. Here is the V8 code from node.cc that the `main()` function eventually enters and keeps executing. This is really what can be called the main processing loop of Node:

```
bool more;
do {
  v8::platform::PumpMessageLoop(default_platform, isolate);

  more = uv_run(env->event_loop(), UV_RUN_ONCE);
  if (more == false) {
    v8::platform::PumpMessageLoop(default_platform, isolate);
    EmitBeforeExit(env);

    // Emit `beforeExit` if the loop became alive either after emitting
    // event, or after running some callbacks.
    more = uv_loop_alive(env->event_loop());

    if (uv_run(env->event_loop(), UV_RUN_NOWAIT) != 0)
      more = true;
  }
} while (more == true);
```

The Libuv processing code is run while it has work to process, and is called and is under the control of the main loop of the node process. The main Node process loop calls Libuv to let it run with the `UV_RUN_NOWAIT` flag. The main loop thus is continually calling to have the Libuv loop run over and over. Here is the Unix ported version of the Libuv processing loop:

```c
int uv_run(uv_loop_t* loop, uv_run_mode mode) {
  int timeout;
  int r;
  int ran_pending;

  r = uv__loop_alive(loop);
  if (!r)
    uv__update_time(loop);

  while (r != 0 && loop->stop_flag == 0) {
    uv__update_time(loop);
    uv__run_timers(loop);
    ran_pending = uv__run_pending(loop);
    uv__run_idle(loop);
    uv__run_prepare(loop);

    timeout = 0;
    if ((mode == UV_RUN_ONCE && !ran_pending) || mode == UV_RUN_DEFAULT)
      timeout = uv_backend_timeout(loop);

    uv__io_poll(loop, timeout);
    uv__run_check(loop);
    uv__run_closing_handles(loop);

    if (mode == UV_RUN_ONCE) {
      uv__update_time(loop);
      uv__run_timers(loop);
    }

    r = uv__loop_alive(loop);
    if (mode == UV_RUN_ONCE || mode == UV_RUN_NOWAIT)
      break;
  }

  /* The if statement lets gcc compile it to a conditional store. Avoids
   * dirtying a cache line.
   */
  if (loop->stop_flag != 0)
    loop->stop_flag = 0;

  return r;
}
```

You can see that the while loop has code that causes a break statement to happen in the case of Node calling it.

Callback functions in your JavaScript usage of modules such as with `fs.reafFile()` are kept by Node in a structure to keep track of them. When the low-level Libuv call returns, the callback happens on the C++ main thread of Node.js. If you look at the diagram again, you can see there is a two-way arrow from Libuv to the V8 VM.

The call coming down from JavaScript through the C++ bindings uses the V8 `FunctionTemplate` class to accomplish this. The callback going up from Libuv makes it back to the C++ bindings layer code and uses the V8 `Function::Call()` function to have the V8 VM execute the actual JavaScript callback that you provided.

This callback is executed in the V8 VM execution mechanism. Node simply hands this off to V8 for the guts of the JavaScript execution.

# The effect of compute intensive code

The main event thread should only be used to do fast, less intense processing for inbound or outbound results. Take the following example that shows what could be termed as *unacceptable*:

```
var http = require('http');
var bcrypt = require('bcryptjs');
var count = 0;

var server = http.createServer(function (request, response) {
  for (i = 0; i <= 10; i++) {
    bcrypt.hashSync("hjkl5678jhg", 10);
  }

  response.writeHead(200, { "Content-Type": "text/plain" });
  response.end("Hello World\n" + count++);
});

server.listen(3000);
```

Every web request that comes in will cause this compute-intensive code to run. This would basically devour all the compute time as requests keep coming in. Your event loop would not have any time to process incoming requests.

The point here is that you should never do CPU intensive calculations in your main Node thread callback code. If you do, it will prevent the entire system from working correctly and efficiently.

You can solve this problem in a simple way with Node, which has built-in capabilities to send expensive processing to separate forked processes that do not affect your main Node.js process. Processing can be forked to child processes or even to a separate executable running on your computer that you have written in another language. You can also queue up work in a central queue and have Node workers pull from that that are running separately.

You can also create an add-on in C++ for Node.js. You may consider some remote service calls if you have capabilities accessible that way. You can do many creative things to solve this problem. Another possibility is you can use AWS Lambda to accomplish the offloading of processing or otherwise compute intensive code.

In the NewsWatcher sample application, we will be forking off code that does the fetching of the global list of news stories. I only do it this way to simplify the provisioning and deployment for now. The code that is running in the forked process could be taken out and run using AWS Lambda.

# 11.6 How to Scale Node

Can the Node.js process ever get overloaded? Let's say 1,000 HTTP Get requests come in to your Node.js server all at once. This, in turn, may require 1,000 files being read from disk. The answer is that Node is set up to be able to manage a sizable workload, so it may do just fine. You will need to do some profiling for your workload scenarios.

With few exceptions, Node.js makes your requests in a way that is non-blocking. For file system interactions, the requests are shuttled off to threads in the Libuv thread pool to run in parallel. Each of those do their work and return to the event loop for callback execution scheduling. As thread pool threads get freed up, they are given more work.

Eventually, all the 1,000 file read requests will complete. This happens in the least amount of time possible and with the least number of resources. All of this can be handled quite efficiently by one single node process.

You can visualize this queue handoff as shown in the following diagram. Note that the event loop of the Node process simply sequences through everything in its queue. You can configure through code to have more than four threads if you like, and you can experiment to see what performance improvements you can make.

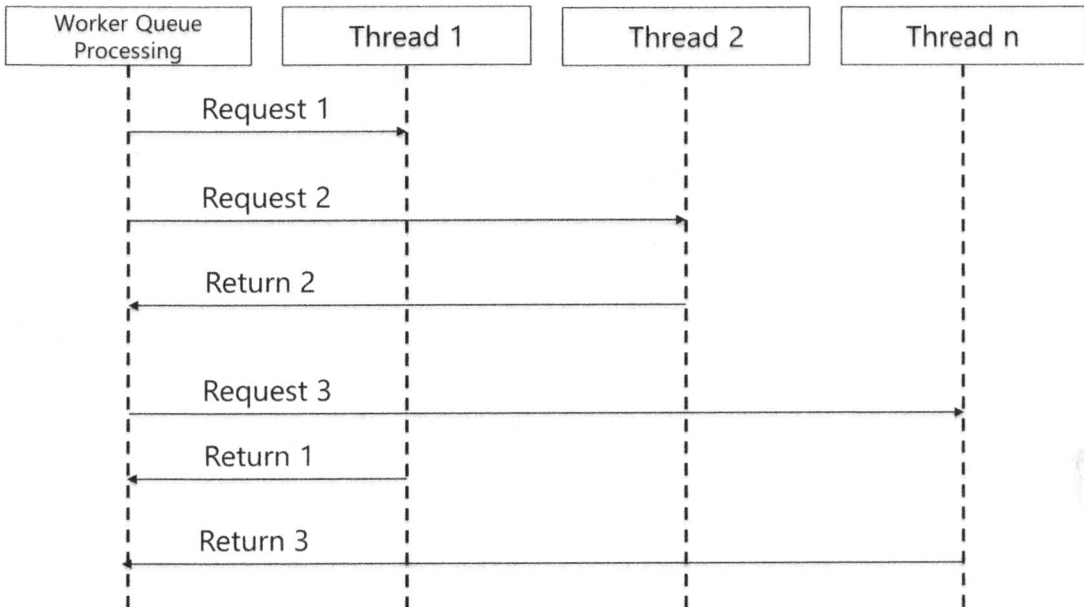

*Figure 47 - Thread pool handoff for file system requests*

Results are always interleaved and processed one by one on your single main thread. As always, the performance is affected by how your JavaScript code processes the results. That is where the bottlenecks almost always happen.

Remember that network I/O is handled differently than file system calls and does not use the Libuv thread pool queue as was discussed for file system usage. Any HTTP calls in your Node code are handled by low-level OS mechanisms that can scale.

Be aware that the NewsWatcher sample application will connect to a MongoDB database in the data layer. This module is written to use low-level TCP calls to the MongoDB service. The module you go through to make those calls uses the Net module of Node which will make platform network calls to the MongoDB infrastructure. The MongoDB module will pool database connections. This means you have a certain number of connections to use for all the calls in your Node.js code, and calls are queued and rotated through as connections are freed up from previous calls. You can configure the number of connections in the pool.

There is obviously a limit to how many operations you can achieve per second before performance starts to degrade. Perhaps you have network I/O requests going off to some slow-returning MongoDB call. The requests could grow out of control if they are not handled fast enough, and the DB connection pool requests start to queue up. Of course, you also need to provide for scalability in your data layer or that will become the bottleneck.

---

*Note: At some point, Node must rely on the OS and hardware (disks and network cards). Remember, that you have other system contentions to be aware of such as disk controller contentions from other processes running on a given machine.*

# Multiple Node processes per machine

Eventually, you can move to other architectural variations to handle your load. To scale to a greater capacity, you can start up multiple Node.js processes on the same machine and distribute the load across those. If you have lots of cores on a machine, you can make use of all of them with a Node process for each core.

To do this scaling of Node across the cores of a single machine, you can use the cluster module of Node.js itself to do the load balancing. There is also a process manager named PM2 that you can download from NPM to do the load balancing. PM2 has other capabilities such as monitoring, restarting of Node processes, and running rolling deployments.

If you deploy your Node.js application through an AWS service, such as Elastic Beanstalk, you can utilize the power of PaaS to scale everything for you by scaling up the number of cores on an EC2 VM (known as vertical scaling) or by scaling up the number of EC2 instances (known as horizontal scaling). Eventually, you will not be able to handle all requests and processing on a single machine. This is when you must employ horizontal scaling.

# SOA Architecture

SOA principles can be understood by contrasting a three-tiered service-oriented architecture with what can be considered a monolithic architecture. Perhaps you have seen Java-based or ASP.Net web servers, with code that renders the HTML on the server and sends it back to the client. It can have a lot of intertwined code to spit out this HTML.

Monolithic applications have many shortcomings. For example, if a change is made to one single part of the code, the entire application needs to be deployed again. Monolithic code is typically tightly coupled and very fragile. The monolith ends up being hard to test, enhance, and maintain.

Moving to a SOA means providing a layer that the UI can access to get at functionality as a set of services. This UI either can be server-side or from a client-side SPA. A great practice evolved to create these services as HTTP/REST web services that use JSON as the data transfer format.

SOA web services are small units of code that do one thing and are testable. A Node.js application can use Express and have a nice RESTful approach through its exposed paths in the URL that goes to the route handlers. This ends up being a SOA with all these benefits. You get the benefits of splitting out the capabilities into separate code bases to be individually

170

developed, tested, and maintained. Multiple people can independently work on each individual service as each route handler is independent. Each service has its own logic and can go to its own database if necessary. More than one individual service can actually go to the same database but will have separate document types within that shared database.

The code patterns for this type of SOA are what is implemented in the NewsWatcher sample application and can achieve superior quality and scaling. You can deploy the same Node.js application across a horizontally scalable cluster of machines. You even can argue that you still have at least one of the characteristics of a monolith since a single codebase with multiple services still needs a complete deployment for each change in an individual service. The nice thing that makes this SOA different is that each service is independent and not affected by a change made in another service, so it also is unaffected by a new deployment of the collection of routes.

# Scaling horizontally

Let's now discuss horizontal scaling. Let's say you have four cores per machine and have your main Node process on one core. You can count on the OS making use of all cores for you for operations such as file system operations that use the Libuv thread pool.

You can employ the technique of forking processes from the main Node process to offload heavier computations and to do periodic batch processing. If you are doing this, however, then you don't have the spare cores to use through cluster or PM2 usage. The better approach is to horizontally spread out across machines. You must do this anyway for high scale needs.

Scaling across machines is easy to configure in AWS. If you use Elastic Beanstalk, it sets up a load balancer for your Node.js application. Elastic Beanstalk lets you configure how many machines to load balance between. You can also set up Elastic Beanstalk to use an auto-scaling group instead of using individual EC2 instances.

Each EC2 machine also will be in a different availability zone to give you redundant failover as needed. This is in addition to the load balancing benefit that gives you more capacity. One advantage with AWS Elastic Beanstalk is that it works with one machine at a time when machine OS upgrades or patches need to happen. Machines are taken out of rotation, updated, and then put back into rotation. When you need to update your Node.js application, your deployment also happens in this exact same way with a "rolling update." This way you achieve what is called a "zero downtime" deployment.

*Note: Be aware that with any of these scaling mechanisms, you must ensure that you are running in a completely stateless way and avoid any affinity settings to truly be able to distribute calls across all Node processes. To manage a connection requiring state, you can insert the use of something like Redis to fetch state if needed or keep state only on the client or in a token passed back and forth.*

# PART II: The Service Layer (Node.js)

The following topology diagram shows you exactly what we are going to deploy to in AWS to get the security, scaling, and high availability we want to achieve. Added security is achieved by creating a private subnet for the Elastic Beanstalk EC2 machines. That way they are not exposed to the external internet. They can be reached only through the secured ALB and in addition, they can call out to the internet in code by way of placing a NAT gateway in the public subnet that they can route through.

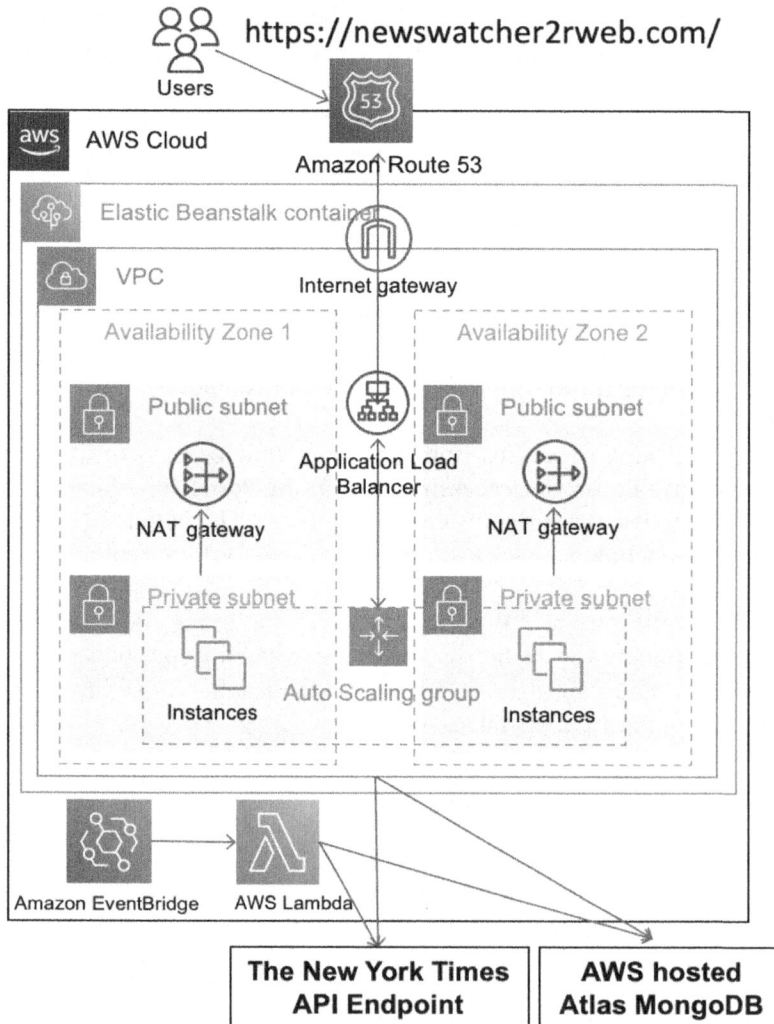

*Figure 48 - AWS hosting topology*

# Static resource serving

In the NewsWatcher sample application, the Node.js application acts to serve up the HTTP requests for the API routes. These are the RESTful web services that the React UI consumes. Node.js serves up the React web app. These are static resources because they don't change. Thus, all files required for the React application — such as images, css, html, js, etc., — are contained in a directory that Node knows about and serves up. For example, in the next figure, here is the logging that happens when the React application is requested. You can see the JavaScript bundled file for React that is being sent. A 304 code means that a resource was not sent since the version requested is still the latest, and the browser has it up to date.

```
$ node server.js
Express server listening on port 3000
{ msg: 'FORK_RUNNING' }
Fork is connected to MongoDB server
Connected to MongoDB server
GET / 304 17.208 ms - -
GET /static/js/main.91212db2.js 304 1.824 ms - -
GET /static/css/main.86667ca2.css 304 9.239 ms - -
GET /api/homenews 200 1410.983 ms - 18504
GET /poweredby_nytimes_30b.png 304 1.758 ms - -
```

*Figure 49 - Static resource returns*

There are more efficient ways to serve up the React application and free up the Node.js application from that work. For example, a Nginx reverse proxy load balancer can know about those requests and serve them up. Another way is to use something like Amazon S3 storage and set up AWS Route 53 to go through AWS CloudFront that pulls the React application from the S3 storage. This is what is referred to as a Content Delivery Network (CDN).

In the case of Nginx or CloudFront, they both support caching. Since the React application resources are not updated often, requests are fetched from the cache and are extremely fast. Express does not do any caching, but you can use client-side caching with `ETag` or `Max-Age`. By default, the static middleware has `ETags` enabled.

# A Docker Container for your Node.js process

It is possible to use Docker containers as a means of deploying and running your Node.js application. Elastic Beanstalk can use a self-contained Docker container. You get the same provisioning, load balancing, and scaling you get by just deploying a straight Node.js application. You could also deploy to AWS ECS/EKC with a docker container.

You may want to take this approach if you find that you have other things to install alongside the Node.js application in the Docker container. If all you have is your Node.js code to deploy using Elastic Beanstalk, you may not gain any benefit in using Docker. If you have a Docker container all running and tested on your development machine, it is more likely to function in production or at least not be missing anything that needs to be installed alongside it.

# Microservices architecture

We discussed that implementing SOA was a great idea and that we achieved that with our RESTful approach. Also as mentioned, the NewsWatcher sample application keeps all the RESTful routes in the same application. Now think about what will happen if each individual HTTP/REST route (e.g., auth, user, etc.) is split into its own Node.js application and separately deployed. This gets us into the topic of realizing an architecture called a Microservices architecture. The following figure shows the before and after visualization of how our NewsWatcher application can be made into a Microservices architecture.

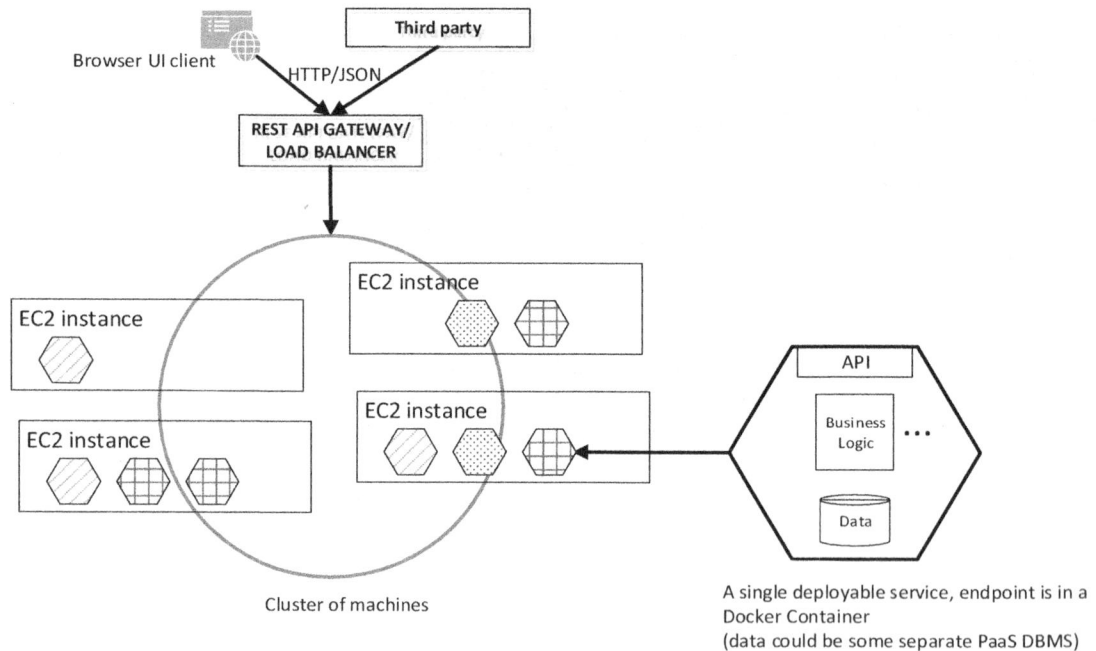

*Figure 50 - Microservices cluster deployment hosting*

Each different background fill pattern in the diagram represents a different hosted service endpoint, such as, Billing, User Profile, Inventory, etc. Each endpoint can be independently scaled, and the load is balanced across available machines in the cluster. Amazon has ECS for this and EKS and each of those can use Fargate for serverless operation.

***Note:*** *You can keep your Node.js application Docker images in AWS ECR (EC2 Container Registry).*

Microservices are independently deployed and run, and each has their own API, middle business logic, and backend data layer as necessary. The data can be documents in a shared MongoDB collection. Microservices can be thought of as a specific way to implement SOA.

# Chapter 11: Advanced Node Concepts

The term SOA for service-oriented architecture was a concept that surfaced many years ago. SOA implementations, at that time, were quite formal and burdened by specifications (such as SOAP WS-*, XML/WSDL), which were sometimes implemented on top of an Enterprise Service Bus (ESB). That has evolved over time, yet the original concepts are still valid in today's modern HTTP REST and JSON.

Before embracing a Microservices architecture, you should be sure you really need one in the first place. Implementing a full build and deploy system for microservices is complicated. Always keep things as simple as possible for your needs. Let's look at a few reasons people give for why Microservices architecture is so great to understand under what circumstances you may want to attempt this architecture.

One benefit of Microservices architecture is that services are split out into discreet, small pieces. The mindset is that smaller pieces are easier to build and test. You may have heard of companies that manage thousands of microservices at once. Small microservices provide all the backend servicing for any number of consumers on the front end. Amazon is one such company that utilizes Microservices architecture. Chances are, at this point, that you are a much smaller company than Amazon and your applications are much simpler.

Remember to do what is best for your needs. Build your backend services at a granularity that makes development and management easiest for your operation. It is unfortunate that the term Microservices starts with "micro", as it makes one think that the services within the architecture must be small. Indeed, some people advocate microservices that are 100 lines or less. If that is a fit for you, then I would advise you go to a serverless architecture and not do a full Node.js application, and to look at things such as AWS Lambda Functions.

If you are writing a backend billing service, you can imagine that it could be a large amount of code. You can initially try to split it out into smaller consumable units so each can be used independently. If the units cannot be used independently, then you should develop one cohesive service. A statement was made in a presentation by Amazon on its Microservices architecture that the main amazon.com page makes 100 to 150 backend calls to microservices at initial load time. These are very small consumable services.

Another benefit is that you can scale better with microservices. Not all services will be deployed across all machines in the same amounts. Services such as billing may not get called as often and can be deployed on a few machines, while a product lookup service may be called a lot more often and need to be deployed onto lots of machines.

The following issues would cause you to split out and run separate services:
1. You have completely autonomous teams (geographically separated) that don't know about the other services or teams. Each is creating and deploying independently.
2. Each service gets deployed on the number of machines required to handle the load it needs to support.

The point of a sophisticated PaaS solution for a Microservices architecture is that it will handle this distribution and shuffling of many services to balance them across machines.

# Microservices frameworks

If you split up your Node.js application into several smaller applications, each needing its own "npm start" command to run, then you will want to consider using Docker for each of these discreet Node.js applications. To do that, you will want to use something like Amazon ECS, EKS, or Fargate. There is also OpenShift Online, Kubernetes, Docker Swarm, Cloud Foundry, or NGINX Unit (with Application Server and Service Mesh). These frameworks manage the deployment of your Docker containers to a cluster of machines.

Sophisticated Microservices architectures may even employ some type of service discovery registry. Frameworks such as Scenica are available to do this. You will need to justify that you need this complexity and overhead.

You can download a Node.js package named Hydra from NPM that may be of benefit to you if you are creating a more complex implementation. Hydra provides things like service discovery, distributed messaging, message load balancing, logging, presence, and health monitoring.

AWS Lambda functions are actually a great way to manage backend processing. This can be everything form the actual transaction backend of Node or spot processing that you need to run from time to time like a batch ETL process. Just be aware that Lambda functions have their place. They can scale to many thousands of concurrent executions and can run for up to 15 minutes each.

# Queues and worker processes

A common approach in backend services is to take requests that require longer running processing and immediately accept that request and return success, but just queue it up. Imagine if you have a photo storage service that has an HTTP Post endpoint API that accepts images. You may want to create thumbnail images and do some image processing to do facial recognition and add meta data to identify people.

To keep your service scalable, you can have your web service accept the image, copy it to S3, and then place an entry into MongoDB or AWS SQS to be looked at later. Then you can have separate EC2 machines or AWS Lambda Functions that scale, and independently read the work requests, and do the processing. This is how to manage the work and keep the web service highly responsive. You can set up how many concurrent Lambda functions are allowed to run and SQS then does the throttling of feeding jobs as allowed.

# Chapter 12: NewsWatcher App Development

It is now time to begin constructing a Node.js service layer that integrates with the MongoDB data layer from Part One of this book. This chapter takes the concepts you have already learned regarding Node and applies them in a real project. What you will be creating is a RESTful web API that your presentation layer will be able to integrate with. You will learn how to implement everything needed for a fully functional cloud-hosted web service. You will utilize best practices for testing and DevOps in the chapters that follow.

*Note: Don't forget that you can access all the code for the NewsWatcher sample project at https://github.com/eljamaki01/NewsWatcher2RWeb.*

# 12.1 Install the Necessary Tools

One of the amazing things about Node is how simple and quick it is to set up a server. It can be hosted in AWS as a PaaS offering using Elastic Beanstalk or other hosting infrastructure, or even be manually deployed and managed. Node.js runs on many different operating systems.

To get started, install the following:
- ✓ A code editor such as Visual Studio Code, Sublime text, Vim, etc.
- ✓ Node.js from https://nodejs.org/. Get a stable version. This installs the node executable for you and also installs the NPM executable.

*Note: Visual Studio Code is not the same as Visual Studio. VS Code is a completely new tool that offers a rich editing environment as well as integrated features for source code control and debugging. VS Code has the capability to launch tasks through the means of tools like Gulp with no need to jump out to a command line. These tools can be used to automate, build, and test steps that you need to run frequently. With Visual Studio Code, you will be able to create a project and run it locally on your machine and have access to IntelliSense, debugging, and web app publishing through Git/GitHub.*

# 12.2 Create an Express Application

Start by creating a folder for your application and naming it. I named mine "NewsWatcher2RWeb" and will use that name herein to refer to the application folder. You can now create the minimum amount of code required for a Node.js application. To help maintain your sanity, you should start with the smallest amount of code possible and push it all the way to deployment. This will eliminate unneeded investigation time into issues that are unrelated to being able to just get the basic service up and working.

Launch Visual Studio Code and click **File->Open Folder**, then select the folder you just created and named ("NewsWatcher2RWeb"). With VS Code open, you see the **EXPLORE** view open, and in there you will find two subfolders. You can also use a command prompt and type **"code ."**. to open up VS Code in the project directory.

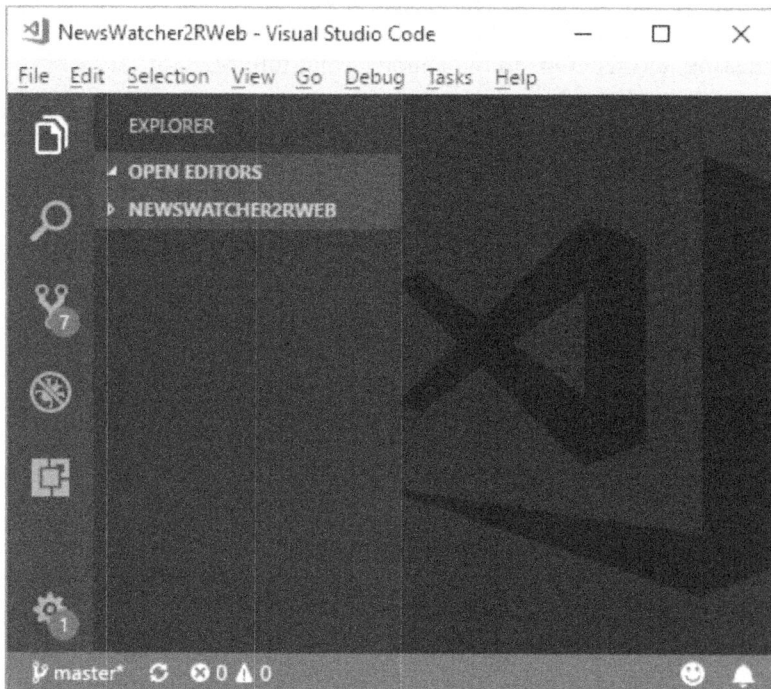

*Figure 51 - VS Code UI*

The **NewsWatcher2RWeb** folder shows you all files and subfolders.

You can now get started and create a simple Node.js application, and then deploy it to an AWS Elastic Beanstalk EC2 instance. Once that is verified and running, you can add more code to fill out the full functionality of the web service.

On the NEWSWATCHER subfolder, click on the icon to create new files. Start by creating these three files:
- .gitignore
- package.json
- server.js

The .gitignore file will not be used right now, but will come into use when you use Git and GitHub. You will list files and directories you want to exclude from Git source code control.

At this point, there is no node_modules folder yet. That folder will automatically be created when you install the node modules with the "npm install" command.

Add the following code to the server.js file:

```
var express = require('express');
var app = express();

app.get('/', function (req, res) {
  console.log('Send message on get request');
  res.send('Hello full-stack development!');
});

app.set('port', process.env.PORT || 3000);

var server = app.listen(app.get('port'), function () {
  console.log('Express server listening on port:' +
              server.address().port);
});
```

Add the following lines to the **package.json** file:

```
{
  "name": "NewsWatcher",
  "version": "0.0.0",
  "description": "NewsWatcher",
  "main": "server.js",
  "author": {
    "name": "yourname",
    "email": ""
  },
  "scripts": {
    "start": "node server.js"
  },
  "dependencies": {
    "express": "^4.17.3"
  }
}
```

Save all the files, then open a command prompt window and navigate to your project's directory. At the command prompt, type npm install. This will look at your package.json file and install the standard Node modules along with the Express module that is listed as a dependency. A new directory will be added with the name node_modules.

You can now try running your Node project locally. Once it is proven to function, you will work on getting it deployed to AWS. At the command prompt, type "npm start". You can also type "node server.js" to run it. If the project runs successfully, you will see the following console output:

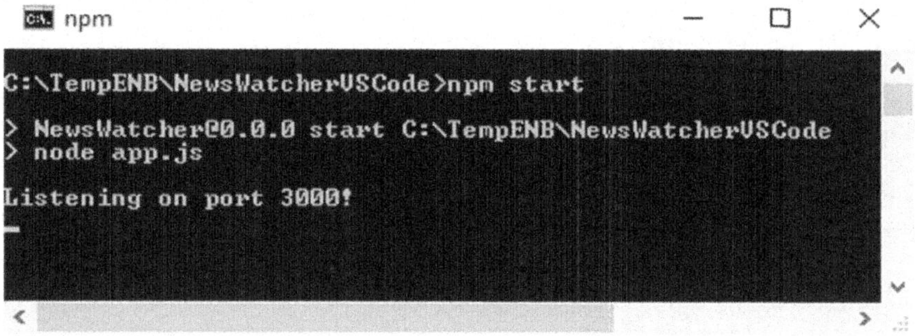

*Figure 52 - Local console output*

Open a web browser and navigate to http://localhost:3000/. In the browser window, you will see your message:

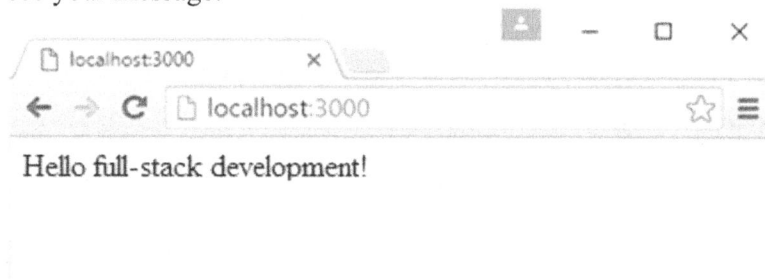

Hello full-stack development!

*Figure 53 - Project message in a browser window*

# 12.3 Deploying to AWS Elastic Beanstalk

It is now time to create your Elastic Beanstalk app through the AWS Management Console. To do so, you must already have an AWS account. *Be aware that you may incur some costs at this point if you are not running with a free account with AWS!*

To create the app:
1. Open a web browser and navigate to https://console.aws.amazon.com/console/.
2. In the upper right corner of the web page, for **Region**, select a region such as **US East (N. Virginia)** as the region you want your services to be running in.

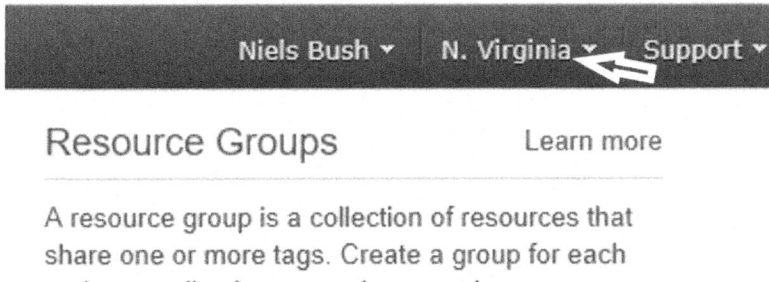

*Figure 54 - Region selection*

3. From the selection of services, click **Elastic Beanstalk**:

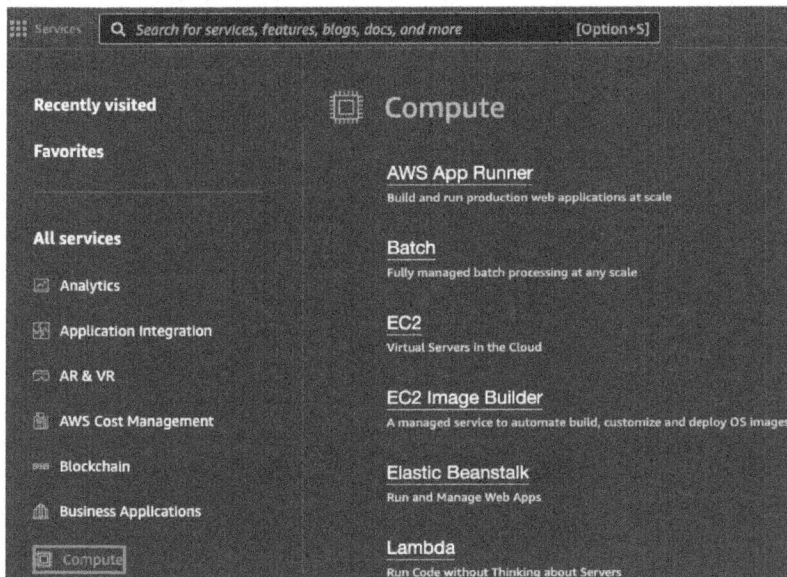

*Figure 55 - Select Elastic Beanstalk*

4. Click **Create New Application**. Enter a name and description. I gave it the name "newswatcher".
5. Click to create a new environment by clicking **Create one now**. Select a **Web server environment**. I chose to run with t4g.micro instance sized machines.
6. Select **Node.js** as the platform and click **Create environment**.

AWS changes the UI from time to time, so you might have to consult the AWS documentation on the exact steps to follow. Once the site is ready, the Elastic Beanstalk portal looks as follows:

## PART II: The Service Layer (Node.js)

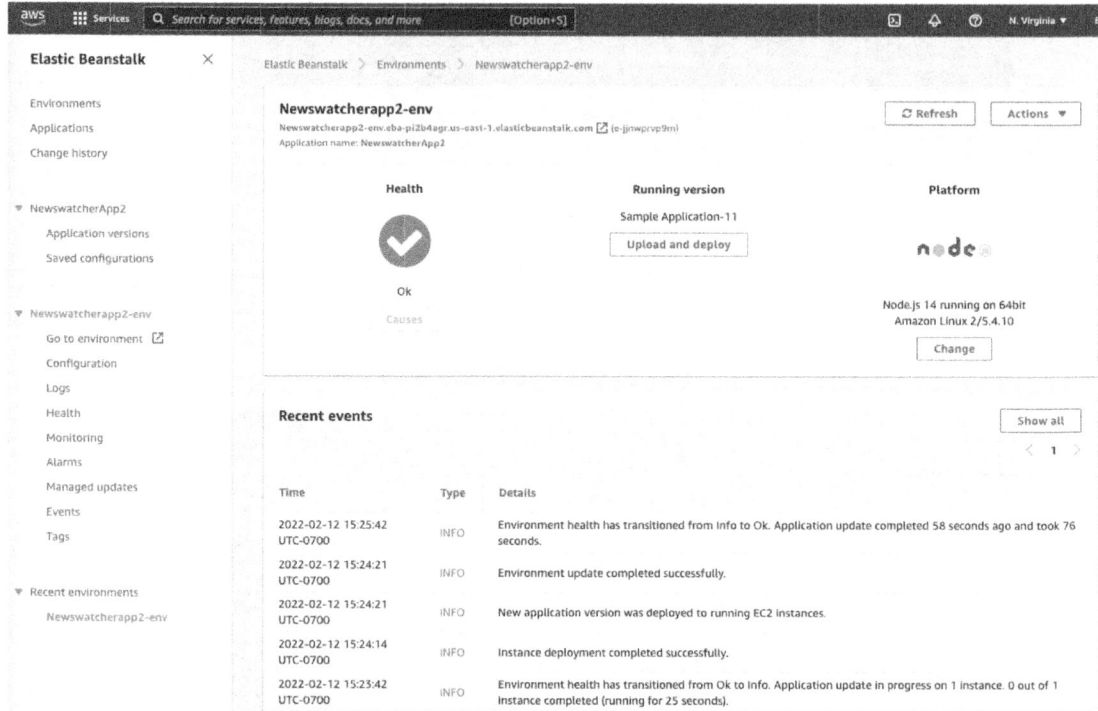

Figure 56 - Elastic Beanstalk portal

Click the **URL:** link at the top of the page to see your site working.

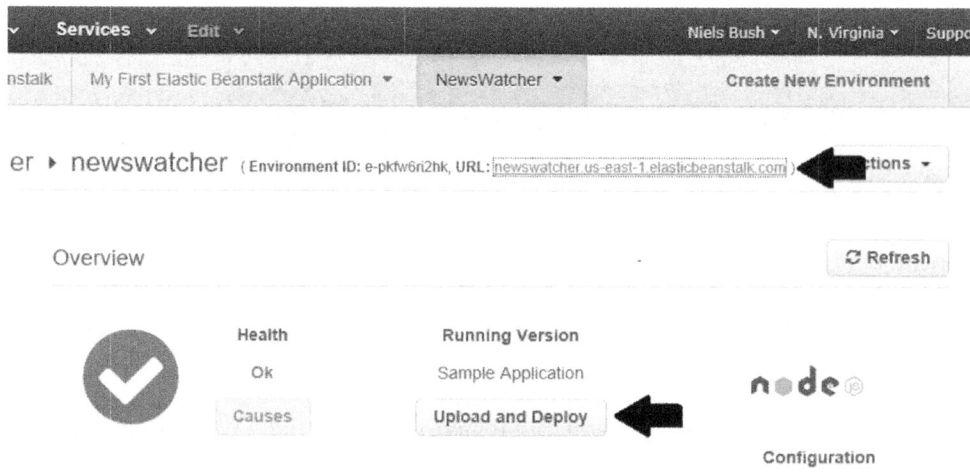

Figure 57 - Launch the site

Now you can deploy your own simple Express application. On a Microsoft Windows machine, this is as follows:

1. Using Windows File Explorer, navigate to your project folder.
2. Select the package.json and server.js files together, then right-click and select **Send to -> Compressed (zipped) folder**. Give the zip file a name and save it.
3. In the Elastic Beanstalk dashboard for the NewsWatcher application, click **Upload and Deploy** button and select your zip file.
4. Wait for the confirmation that the deployment is ready and click the URL again.

Your Node.js application is now working for everyone to see.

You do not need to zip and send the node_modules folder. The deployment to Elastic Beanstalk will run "npm install" for you on the EC2 instances. One thing you should do is to set an environment variable through the Elastic Beanstalk management console so that the node install becomes "npm install –production". Set the environment variable in the **Configuration -> Software Configuration -> Environment Properties**. as:

```
Property name:NPM_CONFIG_PRODUCTION   Property value:true
```

This will make the install go much faster as it will not deploy any npm modules that are needed in test or development environments.

*Figure 58 - Upload and deploy the zip file*

You just performed a manual deployment. While a few manual deployments might be tolerable, eventually you will want full continuous-integration scripts that run tests and deployments for you.

# 12.4 Basic Project Structure

You can now add the rest of the code for the NewsWatcher application. You can start by adding in the rest of the Node.js dependencies that you will need. Edit your package.json file to be as follows, then save it. It should look like the following. Go to the GitHub project for the complete file.

```json
{
  "name": "newswatcher",
  "version": "0.3.0",
  "main": "server.js",
  "scripts": {
    "start": "node server.js",
    ...
  },
  "dependencies": {
    ...
    "async": "^3.2.3",
    "async": "^3.2.3",
    "bcryptjs": "^2.4.3",
    "bootstrap": "^5.1.3",
    "dotenv": "^16.0.0",
    "express": "^4.17.3",
    "express-rate-limit": "^6.3.0",
    "helmet": "^5.0.2",
    "ignore-styles": "^5.0.1",
    "joi": "^17.6.0",
    "jwt-simple": "^0.5.6",
    "mongodb": "^4.5.0",
    "morgan": "^1.10.0",
    "react": "^18.0.0",
    "react-bootstrap": "^2.2.3",
    "react-dom": "^18.0.0",
    "react-redux": "^7.2.8",
    "react-router-dom": "^6.3.0",
    "react-scripts": "5.0.1",
    "redux": "^4.1.2",
    "response-time": "^2.3.2",
    "web-vitals": "^2.1.4"
  }, ...
}
```

Now open a command prompt window and type npm install at the command prompt. You will add code to the rest of the files later in this chapter. You can, of course, go to the GitHub project and get all the code for the NewsWatcher sample application. You are now ready to put your service layer web application RESTful API together.

# 12.5 Where It All Starts (server.js)

The starting point of your Node application can be in a file named whatever you wish; I prefer server.js or index.js. This file instructs Node what to do to be initialized and what to subsequently execute, once up and running. This file does things like establish a MongoDB connection, set up error handling, and establish the Express listener for the HTTP request handling.

At the top of the server.js file is where you will place the `require` statements that load the needed modules. If you recall, module references inside a file are internal to that file. However, subsequent require statements in other files reference what was brought into memory from the same require that might have happened elsewhere.

Some of the modules you will specify as being required are used as middleware for Express. This means you set them up with a `require` statement and then don't use them directly. They act to intercept calls through an `app.use()` setting. If needed, you can refer to the chapter on middleware to review this concept. Here is the first section of code in the server.js file.

```
var express = require('express'); // Route handlers and template usage
var path = require('path'); // Populating the path property of the request
var logger = require('morgan'); // HTTP request logging
var responseTime = require('response-time'); // Performance logging
var helmet = require('helmet'); // Security measures
var rateLimit = require('express-rate-limit'); // IP based rate limiter
var compression = require('compression'); // Traffic compression
```

The first require statement provides the object that will be needed for leveraging Express. The path module is basically used to provide a helper object that will be used to manipulate strings for specifying the file paths in your project. The rest of the require statements are for setting up modules that act as middleware.

These next lines to add will set up environment settings for the code that can have things such as passwords and other secrets and configurations that are needed. When run locally, this is read out of a file named ".env". When run in production, the file is not used, but the values are set in the AWS Elastic Beanstalk (and Lambda) environment settings and the code sees those instead.

```
if (process.env.NODE_ENV !== 'production') { // reading in of secrets
  require('dotenv').config();
}
```

These next lines pull in code from modules that provide the route handlers for incoming HTTPS Post, Put, Get and Delete calls:

```
var users = require('./routes/users');
var session = require('./routes/session');
var sharedNews = require('./routes/sharedNews');
var homeNews = require('./routes/homeNews');
```

You now make the Express application object available and create a setting on it that will be needed for when it is run in AWS. Since the app is behind a load balancer with Elastic Beanstalk, you don't want the load balancer IP address sent in the header requests but want the IP address of the actual machine that it is acting on behalf of. This is what the trust proxy setting does. Compression is easy to set up with this one single line which will accomplish compression of all HTTPS traffic.

```
var app = express();
app.enable('trust proxy');
app.use(compression());
```

## Middleware
Next, you can see how the Express middleware is hooked up for some of the modules you are incorporating. Here are those lines:

```
const limiter = rateLimit({ // DDoS attack protection
  windowMs: 1 * 60 * 1000, // 1 minute
  max: 2000, // limit each IP address per window
  delayMs: 0, // disable delaying - full speed until max limit is reached
  message: { message: 'You have exceeded the request limit!' },
  standardHeaders: false, // Disable info in the `RateLimit-*` headers
  legacyHeaders: false // Disable the `X-RateLimit-*` headers
})
app.use(limiter);

app.use(
  helmet({
    contentSecurityPolicy: {
      useDefaults: true,
      directives: {
        "default-src": ["'self'"],
        "script-src": ["'self'", "'unsafe-inline'", 'ajax.googleapis.com',
'maxcdn.bootstrapcdn.com'],
        "style-src": ["'self'", "'unsafe-inline'",
'maxcdn.bootstrapcdn.com'],
        "font-src": ["'self'", 'maxcdn.bootstrapcdn.com'],
        "img-src": ["'self'", 'https://static01.nyt.com/', 'data:']
      },
    },
    crossOriginEmbedderPolicy: false
  }));
```

186

```
// Adds an X-Response-Time header to responses to measure response times
app.use(responseTime());

// logs all HTTP requests. The "dev" option gives it a specific styling
app.use(logger('dev'));

// Parses a JSON body request payload and provides the 'body' property
app.use(express.json())

// serving of the React app and other static content like images
app.get('/', function (req, res) {
  res.sendFile(path.join(__dirname, 'build', 'index.html'));
});
app.use(express.static(path.join(__dirname, 'build')));
```

This code takes the modules brought in through the require() statements and inserts them as middleware by calling app.use().

The first piece of middleware is what gives protection against DoS attacks. You can look up the module in GitHub to see how it works as middleware.

The next piece of middleware is Helmet. I covered its use earlier. Helmet is a security mitigation module that tweaks the HTTP headers.

There are five other uses of middleware that are documented in the code to tell you what they do. Each is useful and you will benefit from them all.

You can see the use of the path module to provide functionality to manipulate path strings with the join function. The __dirname variable is provided by Node so that you can use it to get the name of the directory that the currently executing code resides in. It will return the directory of the server.js file. It will be the local path if you are running it locally or whatever it is on the AWS production machine if it is running in the deployed environment.

The coding of the React Web SPA React is covered in the third part of this book. You can see the code here that serves up this static content that is using the Express static module. More explanation will be given later.

### The MongoDB Data Layer Connection
Most of the code in the service layer deals with interactions with the backend data storage layer. You initialize your MongoDB connection by utilizing the mongodb module that is an NPM download. You use the connect() function and then set up the usage of the "newswatcher" collection. The connect() function takes the MongoDB connection URL.

# PART II: The Service Layer (Node.js)

```
var db = {};
var MongoClient = require('mongodb').MongoClient;

//Use connect method to connect to the Server
MongoClient.connect(process.env.MONGODB_CONNECT_URL, { useNewUrlParser:
true, useUnifiedTopology: true, minPoolSize: 10, maxPoolSize: 100 },
function (err, client) {
  if (err == undefined || err == null) {
    db.client = client;
    db.collection = client.db('newswatcherdb').collection('newswatcher');
    console.log("Connected to MongoDB server");
  } else {
    console.log("Failed to connected to MongoDB server");
    console.log(err);
    process.exit(0);
  }
});
```

We save the connection as a property on an object named **db**. This db object will be used later with your Express routes through middleware injection. Watch for that code coming up soon. As noted earlier, then first parameter to connect() is the URL to connect to that you get in the Atlas portal. The second parameter is the configuration object. In here is where we set up the pool size, meaning that multiple requests can come in at the same time from multiple users of NewsWatcher and it each can make their own connection through the pooling of those connections by the lower level MongoClient module. Here is where we are telling that code to reserve a minimum of 10 connections and maximum of 100. Connections are reused after each request is made, so in essence if 1000 requests came in at the same time 100 could start to process and the others would queue up and be services in order until all completed.

*Note: The call to connect() is an asynchronous call. The code following this gets run before the call returns and thus before the* db *object properties are set. This is not the best, but works as we don't ever reference it till much later in the process execution.*

The last thing to note about the above code is that you have a configuration file (.env) for storing settings that you want to have in a central place. Some of the values in that file are ones you want to keep secret. **Don't** post that file for anyone to see. As mentioned previously, you keep needed configuration values there, but these values are also set as environment name/value pairs in an Elastic Beanstalk environment. The .env file of mine is not something that is available in the GitHub project. You will need to create your own at the same directory level as the server.js file. I give you the content later, with necessary redaction of anything I need to keep secret.

## Sharing Objects

The db variable are needed in your routing code. The db object database connection will be used for all CRUD operations, so you need to make that available. You expose these variables through middleware injection. This means that Express allows us to inject objects into the

188

request processing chain by adding them as properties on the request object. You will see how this is done.

We place a middleware function right at the top of the Express chain which every request will have to pass through first. In there, you attach new properties on the request object that is being passed along. This then is a global call that happens and everything in the further Express calling chain can use it.

The third property of the callback is `next`. This is a function you will need to call. As required with Express, you call `next()` to move the execution along to the rest of the processing chain for the request. Remember that the `use` function applies across all requests, so that a `get`, `put`, or any other request is routed through here first, as we have places it before the other `app.X()` calls. By doing this, in any future express handler function, you have access to the `req` parameter and as part of that you have the `db` object attached, so that you can use the database connection as follows - `req.db.collection.insertOne(...)`. This is how we interact with MongoDB. You will see plenty of examples of this coming up. Here is the code to intercept all incoming routes and add the database object reference.

```
app.use(function (req, res, next) {
  req.db = db;
  next();
});
```

**Express Route Handlers**
We are almost done with the main server code that sets everything up. The following code defines the routes for the HTTP/Rest API. This sets up what the endpoint paths will be, and in other code we provide servicing of the Post, Put, Get and Delete verbs each will support. Inside each of the supporting modules, you will find the HTTP/Rest verbs in functions that handle each. These are modules you write that use the Express Router object, as explained earlier. The individual modules have a require statement to pull them in.

```
// Rest API routes
app.use('/api/users', users);
app.use('/api/sessions', session);
app.use('/api/sharednews', sharedNews);
app.use('/api/homenews', homeNews);
```

There is also an error handling route that handles invalid URLs that come in. This returns a 404 code to signal that the resource was not found. Basically, if none of the routes match, then this one will. For example, if a request comes in for /api/blah, it will go here. This code activates an express error handler because it calls `next(err)`.

```
// catch everything else and forward to error handler as a 404 to return
app.use(function (req, res, next) {
  var err = new Error('Not Found');
  err.status = 404;
  next(err);
});
```

Here are the error handling routes for when you have an error returned in the code. You get here when `next(err)` is called in any routing code. There is also a handler that only kicks in when running in your development environment. You want to do this so that you can add in what the stack trace is. The second handler is the one that kicks in within the production environment.

```
// development error handler that will add in a stack trace
if (app.get('env') === 'development') {
  app.use(function (err, req, res, next) {
    if (err.status)
      res.status(err.status).json({ message: err.toString(), error: err });
    else
      res.status(500).json({ message: err.toString(), error: err });
    console.log(err);
  });
}

// production error handler with no stacktraces exposed to users
app.use(function (err, req, res, next) {
  console.log(err);
  if (err.status)
    res.status(err.status).json({ message: err.toString(), error: {} });
  else
    res.status(500).json({ message: err.toString(), error: {} });
});
```

The final lines in server.js contain the necessary code that tells Express to be listening for HTTP requests. In production, it picks up the port necessary to run in that hosted environment. The last line export the server for our testing framework to use.

```
app.set('port', process.env.PORT || 3000);

var server = app.listen(app.get('port'), function () {
  console.log('Express server listening on port ' + server.address().port);
});

server.db = db;
console.log(`Worker ${process.pid} started`);
module.exports = server;
```

# 12.6 News Stories Document

You need to create a MongoDB document to store the contents of the shared global news stories. I covered this previously in Part One of this book, so you may have already created this document. If not, you can create it now using the Atlas portal. This functions as the holder for the master list of current news stories and top stories found on the home page. This is then used by all users to do matching with their news filters. This way, each user does not need to fetch all the news individually. This document will have a distinctive value set for the _id property so you know what it is for. Be careful — do not accidentally delete it at any point. This document looks as follows:

```
{
  "_id": "MASTER_STORIES_DO_NOT_DELETE",
  "newsStories": [],
  "homeNewsStories": []
}
```

# 12.7 Central Configuration (.env)

To run your Node application locally, you will need certain environment variables set that your code can reference. These are values such as those to establish the connection to the MongoDB database. The value for the MongoDB connection URL can be found in the Atlas portal as explained.

Be aware that I cannot divulge the actual contents of my .env file, as you would then have access to services I need to protect. Replace values as needed in your own .env file. Create this file at the same level as your server.js file. Do not check it into GitHub! The .gitignore file is in place to prevent this, as you can find that file listed as one to ignore.

```
MONGODB_CONNECT_URL=mongodb+srv://<account>:<password>@cluster0.jdl4x.mongo
db.net/<database>?retryWrites=true&w=majority
JWT_SECRET=<yoursecretkey>
NEWYORKTIMES_API_KEY=<yoursecretkey>
GLOBAL_STORIES_ID=MASTER_STORIES_DO_NOT_DELETE
MAX_SHARED_STORIES=30
MAX_COMMENTS=30
MAX_FILTERS=5
MAX_FILTER_STORIES=15
USE_CACHE=TRUE
AWS_ACCESS_KEY_ID=<key>
AWS_SECRET_ACCESS_KEY=<key>
USE_LOCAL_LAMBDA=FALSE
USE_SSR=FALSE
```

# 12.8 HTTP/REST Web Service API

It is now time to fill in the REST Web Service API. The API will accept and pass back JSON payloads in response to HTTP requests. Eventually, you will create the SPA web page that calls the web service being designed here.

If you think about the REST API that you want to expose, it becomes clear that you need to create all the CRUD operations for each resource that is necessary. Your resources are sessions, users, sharednews, and homenews.

Here is a table that lists everything the REST API supports. The `id` parameter is the identifier of individual resources for a user being accessed. The `sid` parameter is the identifier for stories.

| Verb and Path | Result |
|---|---|
| POST<br>`/api/sessions` | Create a login session token. |
| DELETE<br>`/api/sessions/:id` | Delete a login session token. |
| POST<br>`/api/users` | Create a user with the passed in JSON of the HTTP body. |
| DELETE<br>`/api/users/:id` | Delete a single specified user. |
| GET<br>`/api/users/:id` | Return the JSON of a single specified user. |
| PUT<br>`/api/users/:id` | Replace a user with the passed-in JSON of the HTTP body. |
| POST<br>`/api/users/:id/savedstories` | Save a story for user, content of which is in the JSON body. |
| DELETE<br>`/api/users/:id/savedstories/:sid` | Delete a story that the user had previously saved. |
| POST<br>`/api/sharednews` | Share a news story as contained in the JSON body. |
| GET<br>`/api/sharednews` | Get all the shared news stories. |
| DELETE<br>`/api/sharednews/:sid` | Delete a news story that had been shared. |
| POST<br>`/api/sharednews/:sid/comments` | Add a comment to a specified shared news story. |
| GET<br>`/api/homenews` | Get all the homepage news stories. |

You may have noticed that some verbs are missing that you might have expected to find. For example, you will not see a GET /api/users to get the list of all users. You don't really want other people to see everyone that is a user of the NewsWatcher application, so don't offer that. You certainly could decide to offer it, but you would then want to place a middleware restriction on it that only allows logged in administrators to have access to it.

An example of a restricted API route that you have is DELETE /api/users/:id. A user can only delete themselves, so you restrict that to just the logged-in user for deleting their own account and not an account of anyone else. You can write some code to allow admins to be able to delete anyone.

Remember that the token is useful for restricting access. It is up to you to define what roles and access you will need and then enforce them. In this case, each call only works for a single account to access that user's own data for whatever is authorized. Perhaps administrators who log in can be identified and allowed access to everything.

# Visualizing the code

As you may recall, the lines below are found in the server.js file and are used to set up your Express route handling. The first two Express application calls are used for sending the files back that the client browser application will need. The last four Express application calls are for route handling of everything listed in the REST API resource table. Here are the relevant lines from the server.js file again:

```
// Main file to serve up that is built by React build process
app.get('/', function (req, res) {
  res.sendFile(path.join(__dirname, 'build', 'index.html'));
});

// Serving up of static content such as HTML for the React SPA, images,
// CSS files, and JavaScript files
app.use(express.static(path.join(__dirname, 'build')));

// Rest API routes
app.use('/api/users', users);
app.use('/api/sessions', session);
app.use('/api/sharednews', sharedNews);
app.use('/api/homenews', homeNews);
```

The following is a pictorial representation of how the routing code all fits together. This does not contain all the files and details, but it gives you an idea of the routes that are being serviced. I have included here the UI rendering React code, though that has not been covered yet. You can see that I divide up the code by the architectural layer it resides in.

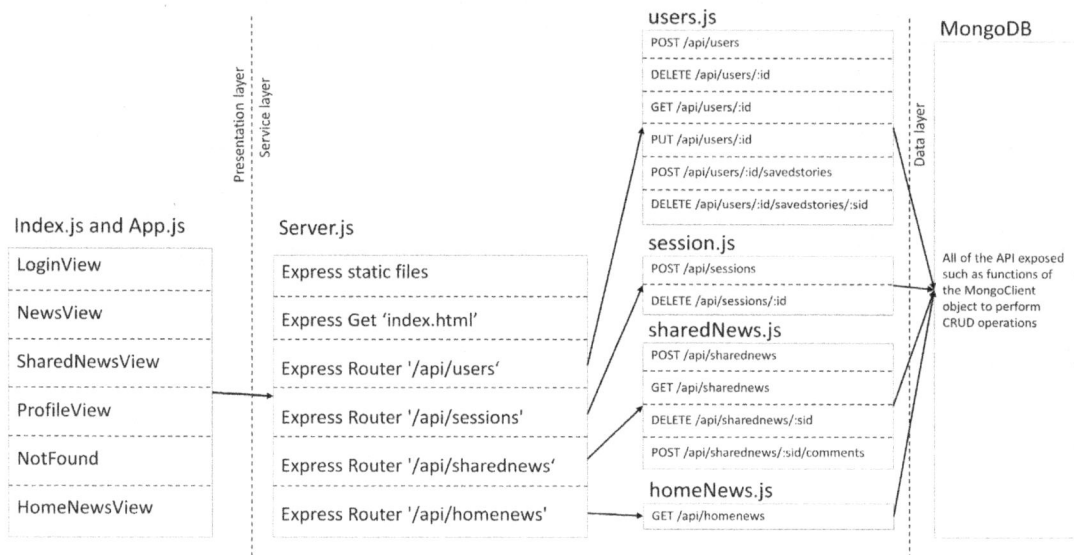

*Figure 59 - NewsWatcher application file diagram*

You can now look at the individual files used for each of the Express routes one at a time since each exists in its own file. These are the files where you find the servicing of the individual HTTP verbs (GET, POST, PUT, DELETE).

# 12.9 Session Resource Routing (routes/session.js)

The session route is used in the API to allow people to log in and out. Here are the specific routes for the session resource:

| Verb and Path | Result |
|---|---|
| POST /api/sessions | Create a login session token. |
| DELETE /api/sessions/:id | Delete a login session token. |

The post operation takes a user's email and password in the request body and basically logs a user into their account. A token is sent back in the response body. The client calling code can take the JWT and then keep passing it back on subsequent calls to identify that person. The token can be stored in client-side storage and used as needed.

# Chapter 12: NewsWatcher App Development

The `post` verb handler needs to first query for the user document to see if the person has a registered account. The user must exist, or they cannot be logged in. If there is a match for the email, then the stored password hash is validated with a hash of the password coming in.

If the password is validated, then a token is created and passed back in the JSON payload. The tokens can be used forever, but it would be easy to add a timestamp to set it to expire. With that added in, you may do something such as require a new login every six months.

As previously discussed, a bit of verification must happen with the token. You check it to make sure it is originating from where the token was originally assigned from. The IP address and the header setting for `user-agent` are kept with the token as additional verification.

There is no database storage of the token in the NewsWatcher application, so the delete route does not really need to do much. There is just a simple check to verify that the person logging out is the same as the one contained in the token.

The code uses the joi module to validate the incoming request body object parameters. Here is the code for session.js:

```
// session.js: A Node.js Module for session login and logout
"use strict";
var bcrypt = require('bcryptjs'); // For password hash comparing
var jwt = require('jwt-simple'); // For token authentication
var joi = require('joi'); // For data validation
var authHelper = require('./authHelper');
var express = require('express');
var router = express.Router();

// Create a security token for log in and use on subsequent calls
router.post('/', function postSession(req, res, next) {
  // Password must be 7 to 15 characters and other restrictions
  var schema = joi.object({
    email: joi.string().email().min(7).max(50).required(),
    password: joi.string().regex(/^(?=.*[0-9])(?=.*[!@#$%^&*])[a-zA-Z0-
9!@#$%^&*]{7,15}$/).required()
  });

  schema.validateAsync(req.body).then(value => {
      req.db.collection.findOne({type: 'USER_TYPE', email: req.body.email},
      function (err, user) {
        if (err) {
          err.status = 400;
          return next(err);
        }
        if (!user) {
          let err = new Error('User was not found.');
          err.status = 404;
          return next(err);
```

195

```
      }

      bcrypt.compare(req.body.password, user.passwordHash,
      function comparePassword(err, match) {
        if (match) {
          try {
            var token = jwt.encode({authorized: true, sessionIP: req.ip,
                                    sessionUA: req.headers['user-agent'],
                                    userId: user._id.toHexString(),
                                    displayName: user.displayName },
                                    process.env.JWT_SECRET);
            res.status(201).json({displayName: user.displayName,
                                  userId: user._id.toHexString(),
                                  token: token, msg: 'Authorized' });
          } catch (err) {
            err.status = 400;
            return next(err);
          }
        } else {
          let err = new Error('Wrong password');
          err.status = 401;
          return next(err);
        }
      });
    });
  })
  .catch(error => {
    let err = new Error(`Invalid field: password 7 to 15 (one number, one
special character): ${error}`);
    err.status = 400;
    return next(err);
  });
});

// Delete the token as a user logs out
router.delete('/:id', authHelper.checkAuth, function (req, res, next) {
  // Verify the passed in id is the same as that in the auth token
  if (req.params.id != req.auth.userId) {
    let err = new Error('Invalid request for logout');
    err.status = 401;
    return next(err);
  }
  res.status(200).json({ msg: 'Logged out' });
});
module.exports = router;
```

Notice the use of the middleware function `authHelper.checkAuth` in the delete route. This is something we will define next. This is what allows us to inject an authorization check before proceeding to the final function that does the work. If the authorization fails, then the function for the end route handling will not be called.

# 12.10 Authorization Token Module (routes/authHelper.js)

You have seen how a user logs in and a token is generated. You now need to create some middleware that will be inserted and run for verifying the token before proceeding on a route that needs to be secured. This code takes the passed-in token and makes sure it is valid. As you know, middleware can be inserted into any route. That is what will be happening here, where we use this middleware injection on routes that need to have a JWT validated.

Most of the routing modules will make use of the authHelper module to verify that a valid token is passed before performing any other action. This happens because each HTTPS/Rest call would have an x-auth header token value filled in. You could also use a header such as "Authorization: Bearer <token>" if you want to mimic standards such as OIDC.

You will see the use of the checkAuth function in many of the route handlers. The code here will simply verify that there is an x-auth header and, if there is, will decode it with the secret, and then set the decoded object in a request property named auth for further usage by anything in the processing chain. If the token is missing, has been tampered with, or does not contain what it is supposed to contain, an error is returned.

When you look at the password handling code of session.js, you see where all the information is placed in the token. It is up to you to decide what to put in there. With this token, we get the embedded information about the user being signed in. Do not store the user password (or other sensitive data) in the token. Here is the middleware that verifies that a token is valid:

```
// authHelper.js: Injected middleware that validates the request
"use strict";
var jwt = require('jwt-simple');
// Express middleware code to run and look at the header of the request
module.exports.checkAuth = function (req, res, next) {
  if (req.headers['x-auth']) {
    try {
      req.auth = jwt.decode(req.headers['x-auth'], process.env.JWT_SECRET);
      if (req.auth && req.auth.authorized && req.auth.userId) {
        return next();
      } else {
        let err = new Error('User is not logged in.');
        err.status = 401;
        return next(err);
      }
    } catch (err) {
      err.status = 401;
      return next(err);
    }
```

```
  } else {
    let err = new Error('User is not logged in.');
    err.status = 401;
    return next(err);
  }
};
```

If the JWT is validated, we simply call `next()` and the code proceed to the next processing step of express, which would be the actual route endpoint for a particular verb. Otherwise, you see if there is an error, we call `next(err)` and the execution ends up in our error handler in server.js and then returning back to the error response.

If you provide a checkbox in the UI for the user to stay logged in, then the user can give permission so you can store the token on the device and not have to log them in each time. If a user wants to get their token from the returned login request, they can. If they pass it on to anyone else, their account can possibly be compromised. There could be an additional test for the IP address to be the same.

# 12.11 User Resource Routing (routes/users.js)

The user resource represents information for a logged in user. A user document retrieved by their id will contain information such as their profile that has their news filters and the news stories that matched. For a given user, you can also make Rest calls to save a story or delete a saved story. Here are the specific routes for the user resource:

| Verb and Path | Result |
|---|---|
| POST<br>/api/users | Create a user with the passed-in JSON of the HTTP body. |
| DELETE<br>/api/users/:id | Delete a single specified user. |
| GET<br>/api/users/:id | Return the JSON of a single specified user. |
| PUT<br>/api/users/:id | Replace a user with the passed-in JSON of the HTTP body. |
| POST<br>/api/users/:id/savedstories | Save a story for user, content of which is in the JSON body. |
| DELETE<br>/api/users/:id/savedstories/:sid | Delete a story that the user had previously saved. |

Start by looking at the require statements at the very top of users.js. Some of the modules required are ones that you have seen before.

```
var express = require('express');
var bcrypt = require('bcryptjs');
var async = require('async');
var joi = require('joi'); // For data validation
var authHelper = require('./authHelper');
var ObjectId = require('mongodb').ObjectId;

var router = express.Router();
const { LambdaClient, InvokeCommand } = require("@aws-sdk/client-
lambda");
try {var LambdaClientInst = new LambdaClient({ region: 'us-east-1' });}
catch (e) {console.log(e);}

...code cut out here...

module.exports = router;
```

You may have noticed some code setting up the usage of the AWS SDK to invoke a Lambda call. We will explain more on this later. Let's look at the verb handling functions one by one.

**POST /api/users**
The post verb takes a JSON payload and creates a new user account as a document in the MongoDB database collection. This happens when a user account is first created. A password is passed in as part of the JSON body and you create a hash of it to store in the MongoDB database document for that user.

This call will fail if there already exists a document in the collection with that email value already. Relying on an email address to identify a user account is one way to keep user accounts unique and identifiable. The mongodb findOne() function is what is used to see if a user account already exists.

The code goes ahead and creates all the properties needed in the document. Default values are used that make sense. There is even a sample news filter set up for the user.

We can now go through the code to process the Post verb. Right at the start is the validation of the passed-in JSON body. We want to make sure that there are no extra properties and validate that the allowed properties conform to some known types and have safe values. The joi NPM module is used to perform the validations. Here is the code for the post verb handler:

```
router.post('/', function postUser(req, res, next) {
  var schema = joi.object({
    displayName: joi.string().alphanum().min(3).max(50).required(),
    email: joi.string().email().min(7).max(50).required(),
    password: joi.string().regex(/^(?=.*[0-9])(?=.*[!@#$%^&*])[a-zA-Z0-
9!@#$%^&*]{7,15}$/).required()
  });
  schema.validateAsync(req.body).then(value => {
```

```
req.db.collection.findOne({type: 'USER_TYPE', email: req.body.email},
function (err, doc) {
  if (err) {
    err.status = 400;
    return next(err);
  }

  if (doc) {
    let err = new Error('Email account already registered');
    err.status = 403;
    return next(err);
  }

  var xferUser = {
    type: 'USER_TYPE',
    displayName: req.body.displayName,
    email: req.body.email,
    passwordHash: null,
    date: Date.now(),
    completed: false,
    settings: {
      requireWIFI: true,
      enableAlerts: false
    },
    newsFilters: [{
      name: 'Technology Companies',
      keyWords: ['Apple', 'Microsoft', 'IBM', 'Amazon', 'Google'],
      enableAlert: false,
      alertFrequency: 0,
      enableAutoDelete: false,
      deleteTime: 0,
      timeOfLastScan: 0,
      newsStories: []
    }],
    savedStories: []
  };

  bcrypt.hash(req.body.password, 10, function getHash(err, hash) {
    if (err) {
      err.status = 400;
      return next(err);
    }

    xferUser.passwordHash = hash;
    req.db.collection.insertOne(xferUser,
    function createUser(err, result) {
      if (err) {
        err.status = 400;
        return next(err);
      }

      xferUser._id = result.insertedId;
```

```
            // Fire the Lambda to update the filtered news stories
            if (process.env.USE_LOCAL_LAMBDA === 'TRUE') {
              storiesRefreshLambda.handler
                  { params_call_type: "refreshForUserFilter_call",
                    doc: xferUser }, {},
                (error, data) => {
                  if (error) {console.log("Lambda: INVOKE ERROR!", error);}
                });
            } else {
              let params = {
                FunctionName: 'NYTStoriesMgmt', /* required */
                InvocationType: "Event", // "Event" for asynchronous
                LogType: "Tail",
                Payload: JSON.stringify(
                          { params_call_type: "refreshForUserFilter_call",
                            doc: xferUser })};
              const command = new InvokeCommand(params);
              LambdaClientInst.send(command, function (error, data) {
                  if (error) {console.log("Lambda: INVOKE ERROR!", error);}
                });
            }
            res.status(201).json(xferUser);
          });
        });
      });
    })
    .catch(error => {
      let err = new Error(`Invalid field: display name 3 to 50 alpanumeric,
valid email, password 7 to 15 (one number, one special character):
${error}`);
      err.status = 400;
      return next(err);
    });
});
```

Users are sign up for an account by providing a username, email, and a password. We don't store the actual password but will instead store an encrypted hashed value of the password. Even if anyone were to get a hold of that encrypted hashed value for a user, they would not be able to log in with it, and it is extremely difficult to decrypt and figure out the password from that.

Once a person is registered as a user, then the /api/session/ path will be used to accept their email and password to get their session going each time they want to log in and use the NewsWatcher application.

As explained in the chapter on authentication and authorization, you will be sending a token in the header of each HTTP/Rest request. When a user logs in, he or she will get a token that is subsequently used to identify the user for all further interactions.

# PART II: The Service Layer (Node.js)

Notice the call to `LambdaClientInst.send()`. This is running the Node AWS Lambda code to offload the filter processing for a newly added user. This call will take the new account and do the initial matching of stories for the filter. I have not shown you the code for the AWS Lambda process yet, but you can see that the concept is simple. Note that we can have a flag set to run the Lambda in our local development machine if we would like. That makes it easier to debug the code locally. To invoke the Lambda in code we create the `params` object and give the name of the Lambda function to be invoked, state that it is asynchronous and give the payload that gets passed into the handler for our own specific use. The code does an immediate return and does not wait for the Lambda to complete.

## DELETE /api/users/:id

With this path, you see the use of a passed in id. It comes in through the mechanism of Express. Specifying the path like this will have Express create a property of `req.params.id`. What you need to do is to verify that the request for a deletion of a user is actually the id that exists in the token. This way a user cannot delete an account that does not belong to them.

Look at session.js again and you see where the mongodb `_id` property of the retrieved document is captured. This is what is going to be passed back in the Rest request URL path portion to identify a user.

The middleware function `authHelper.checkAuth` is injected to do the verification that a valid token exists for the request. That middleware-injected function will return an error if the token is not acceptable and then the route function will never get called.

If everything proceeds correctly, the route function executes, and the document is removed from your collection, and the user account is gone. `findOneAndDelete()` is used so there will be one and only one document with that `_id`.

The `ObjectId()` helper function from the mongodb module to take the string and get it into the proper form needed. Here is the code for the user deletion handler:

```
router.delete('/:id', authHelper.checkAuth, function (req, res, next) {
  if (req.params.id != req.auth.userId) {
    let err = new Error('Invalid request for account deletion');
    err.status = 401;
    return next(err);
  }

  req.db.collection.findOneAndDelete
  { type: 'USER_TYPE', _id: ObjectId(req.auth.userId) },
  function (err, result) {
    if (err) {
      err.status = 409;
      return next(err);
```

202

```
    } else if (result.acknowledged != true && result.ok != 1) {
      let err = new Error('Account deletion failure');
      err.status = 409;
      return next(err);
    }

    res.status(200).json({ msg: "User Deleted" });
  });
});
```

### GET /api/users/:id

The Get verb route handler retrieves a single user by their id. The React app will have already called to get a session token and then will have access to the id of the user to pass it into this API call to retrieve the user document. Since you actually have the _id (object id), you can retrieve the document faster than if you had to query for it some other way. There is always an index created for the _id property to make this call performant.

The retrieval is done and populated with a transfer object. Notice that we are also altering the HTTP header for the response. That is necessary to stop caching from happening. Otherwise, when you get to the UI presentation code and are trying to retrieve a user, you may not get the most up-to-date one. Here is the code for the route handler to get a single user by their id:

```
router.get('/:id', authHelper.checkAuth, function (req, res, next) {
  if (req.params.id != req.auth.userId) {
    let err = new Error('Invalid request for account fetch');
    err.status = 401;
    return next(err);
  }

  req.db.collection.findOne
  { type: 'USER_TYPE', _id: ObjectId(req.auth.userId) },
  function (err, doc) {
    if (err) {err.status = 400; return next(err);}

    var xferProfile = {
      email: doc.email,
      displayName: doc.displayName,
      date: doc.date,
      settings: doc.settings,
      newsFilters: doc.newsFilters,
      savedStories: doc.savedStories
    };
    res.header("Cache-Control", "no-cache, no-store, must-revalidate");
    res.header("Pragma", "no-cache");
    res.header("Expires", 0);
    res.status(200).json(xferProfile);
  });
});
```

## PUT /api/users/:id

A `put` is used to update a user, such as in the case where a user has altered their own news filters. The code is very similar to what you needed for the initial post of the user, except that you now need to worry about a conflict happening upon a database write operation. Luckily, as mentioned before, the MongoDB driver handles the concurrency issues for us and when we use the `findOneAndUpdate()` function it will simultaneously to the find and update in one transaction for us. There is no need to show all the code for the verb handing here.

You can look at the code and notice the code that limits the news filter size. You will put code in the UI to limit that as well but be aware that limits can be hacked in the browser or by someone sending a bogus `put` request. You must guard against potential tampering as you will otherwise have a crash or at least a failure of the MongoDB update.

Here is the one interesting part of the code that we discuss. This is the call to MongoDB to find the user document and update it in one single operation.

```
req.db.collection.findOneAndUpdate(
{ type: 'USER_TYPE', _id: ObjectId(req.auth.userId) },
{ $set: {settings: { requireWIFI: req.body.requireWIFI,
                     enableAlerts: req.body.enableAlerts },
         newsFilters: req.body.newsFilters } },
{ returnDocument: 'after' },
function (err, result) {...});
```

The `$set` operation is used to update only individual properties and not the entire document. Also note that we use the `returnDocument` option so that we get the document after the operation completed. That way, we can pass on the document over to the Lambda call to look at the filters set up by the user and do the processing to stage their selected news stories for them.

## POST /api/users/:id/savedstories

In the user document, there is an array used for saving stories that a user wants to keep around. This route will take the JSON of the passed-in request body as the story to save. The id in the route is the id of the user who is requesting the saving of the story. This route is actually not being used by the UI at present.

A few checks need to happen before saving a story. You need to verify that the story is not already inserted. There is also a limit on the number of stories that can be saved, so you need to check the upper limit against what is saved.

Stories have an id associated with them to be able to identify them in cases where you don't want duplicates saved or shared. You will later see the code that creates that id. Here is the code for posting a story to be saved:

```
router.post('/:id/savedstories', authHelper.checkAuth, function (req, res,
next) {
  if (req.params.id != req.auth.userId) {
    let err = new Error('Invalid request for saving story');
    err.status = 401;
    return next(err);
  }

  // Validate the body
  var schema = joi.object({
    contentSnippet: joi.string().max(300).required(),
    date: joi.date().required(),
    hours: joi.string().max(20),
    imageUrl: joi.string().max(300).required(),
    keep: joi.boolean().required(),
    link: joi.string().max(300).required(),
    source: joi.string().max(50).required(),
    storyID: joi.string().max(100).required(),
    title: joi.string().max(200).required()
  });
  schema.validateAsync(req.body)
      req.db.collection.findOneAndUpdate({ type: 'USER_TYPE', _id:
ObjectId(req.auth.userId) },
        { $addToSet: { savedStories: req.body } },
        { returnDocument: "before" },
        function (err, result) {
          if (result && result.value == null) {
            let err = new Error('Over the save limit, or story already
saved');
            err.status = 403;
            return next(err);
          } else if (err) {
            err.status = 409;
            return next(err);
          } else if (result.acknowledged != true && result.ok != 1) {
            let err = new Error('Story save failure');
            err.status = 409;
            return next(err);
          }

          res.status(200).json(result.value);
        });

    })
    .catch(error => {
      error.status = 400;
      return next(error);
    });
});
```

*Note: The GitHub code has a line commented out that uses the $where operations. However this is a MongoDB Atlas features that does not work with the free tier and must be omitted.*

**DELETE /api/users/:id/savedstories/:sid**
This is similar to the other previous functions above and accomplishes the verification of the existing story, before being able to delete it. The $pull operator is used with the array property of the document to delete the entry. Here is the code for deleting a saved story:

```
router.delete('/:id/savedstories/:sid', authHelper.checkAuth, function
(req, res, next) {
  if (req.params.id != req.auth.userId) {
    let err = new Error('Invalid request for deletion of saved story');
    err.status = 401;
    return next(err);
  }


  req.db.collection.findOneAndUpdate({ type: 'USER_TYPE', _id:
ObjectId(req.auth.userId) },
    { $pull: { savedStories: { storyID: req.params.sid } } },
    { returnDocument: "before" },
    function (err, result) {
      if (err) {
        err.status = 400;
        return next(err);
      } else if (result.acknowledged != true && result.ok != 1) {
        let err = new Error('Story delete failure');
        err.status = 409;
        return next(err);
      }

      res.status(200).json(result.value);
    });
});
```

# 12.12 Home News Routing (routes/homeNews.js)

Home news stories are those stories that are visible when the application UI is first seen by the user. Only one single route is needed for the homeNews resource. A user does not need to be logged in to see these stories.

| Verb and Path | Result |
|---|---|
| GET /api/homenews | Get all the top news stories. |

At the top and bottom of the file is the usual code to expose the module as shown here:

```
var express = require('express');
var router = express.Router();
var cachedDoc = null;
var timeStamp = process.hrtime();

...this part left out...

module.exports = router;
```

Here is the route handler for setting the news stories for the home page.

**GET /api/homenews**

Retrieving all top news stories is done by directly getting the array that holds them from our one single document for that purpose. They are the same for all users. We have an interesting twist to this code. A caching mechanism is set up since the news stories list in the global document are not updated very often.

```
router.get('/', function (req, res, next) {
  const diff = process.hrtime(timeStamp);
  if (cachedDoc && process.env.USE_CACHE && diff[0] < 1800) {
    res.status(200).json(cachedDoc.homeNewsStories);
  } else {
    req.db.collection.findOne({ _id: process.env.GLOBAL_STORIES_ID
}, { homeNewsStories: 1 }, function (err, doc) {
      if (err)
        return next(err);
      cachedDoc = doc;
      timeStamp = process.hrtime();
      res.status(200).json(doc.homeNewsStories);
    });
  }
});
```

The array of news stories is sent back in the response and then React can bind it on the client-side in the JavaScript React rendering.

# 12.13 Shared News Routing (routes/sharedNews.js)

Shared news stories are those that are seen by all users. People can save, view, and comment on news stories. Here are the specific routes for the sharedNews resource:

| Verb and Path | Result |
|---|---|
| POST<br>/api/sharednews | Share a news story as contained in the JSON body. |
| GET<br>/api/sharednews | Get all of the shared news stories. |
| DELETE<br>/api/sharednews/:sid | Delete a news story that has been shared. |
| POST<br>/api/sharednews/:sid/comments | Add a comment to a specified shared news story. |

At the top and bottom of the file is the usual code as shown here:

```
"use strict";
var express = require('express');
var joi = require('joi'); // For data validation
var authHelper = require('./authHelper');

var router = express.Router();

...this part left out...

module.exports = router;
```

Here are each of the route path handlers.

**POST /api/sharednews**
This code is very similar to the code you already saw for saving a story in user.js. The only difference is that the story now is being copied into a different document type ('SHAREDSTORY_TYPE') where all NewsWatcher application users can view stories and comment on them.

A limit is set for the number of possible shared stories as well as a test to make sure the story has not been shared already. If all looks good, the document is created. Here is a small selection of the code that shows the use of the countDocument() MongoDB client call that can be used to first find how many stories are shared and then find out if a particular story was already shared.

```
req.db.collection.countDocuments({ type: 'SHAREDSTORY_TYPE' },
function (err, count) {
  if (err) {
    err.status = 400;
    return next(err);
  }

  if (count > process.env.MAX_SHARED_STORIES) {
    let err = new Error('Shared story limit reached');
    err.status = 403;
    return next(err);
  }

  // Make sure the story was not already shared
  req.db.collection.countDocuments({ type: 'SHAREDSTORY_TYPE',
                                 _id: req.body.storyID },
  function (err, count) {
    if (err) {
      err.status = 400;
      return next(err);
    }
    if (count > 0) {
      let err = new Error('Story was already shared.');
      err.status = 403;
      return next(err);
    }
    ...
```

**GET /api/sharednews**

Retrieving all shared stories is done by directly getting the documents of type SHAREDSTORY_TYPE as follows:

```
router.get('/', authHelper.checkAuth, function (req, res, next) {
  req.db.collection.find({ type: 'SHAREDSTORY_TYPE' }).toArray(function
(err, docs) {
    if (err) {
      err.status = 400;
      return next(err);
    }

    res.status(200).json(docs);
  });
});
```

Since you know there will not be more than 30 saved stories, it is fine to have an array returned and not use a cursor to iterate through the results. The array type works great, as you can send that back in the response and then React can bind it on the client-side.

**DELETE /api/sharednews/:sid**

Individual shared stories can be deleted. You actually will not be calling this from the presentation layer, but need it just for testing purposes to clean up after yourself. It can either be commented out or have some checks done to only allow an admin account to call it. The code is as follows:

```
router.delete('/:sid', authHelper.checkAuth, function (req, res, next) {
  req.db.collection.findOneAndDelete({ type: 'SHAREDSTORY_TYPE', _id:
req.params.sid }, function (err, result) {
    if (err) {
      err.status = 400;
      return next(err);
    } else if (result.acknowledged != true && result.ok != 1) {
      let err = new Error('Shared story deletion failure');
      err.status = 409;
      return next(err);
    }

    res.status(200).json({ msg: "Shared story Deleted" });
  });
});
```

**POST /api/sharednews/:sid/comments**

To add a comment, you need the id of the story and the JSON body for the comment. Since you have a partially normalized design here with separate documents for each story, you will not have as much write contention. You still will have concurrent access issues for each individual story as multiple comment additions possibly may conflict. This is why the findOneAndUpdate() call is used as it will handle these for you.

Notice that this can fail if 30 comments already have been added. The query criteria used has three different parts. The first two narrow down exactly what is being searched for. Then the $where operator is used, and the actual JavaScript object is accessed to check the array length. The shared story comment is added as follows:

```
router.post('/:sid/Comments', authHelper.checkAuth, function (req, res,
next) {
  var schema = joi.object({
    comment: joi.string().max(250).required()
  });
  schema.validateAsync(req.body)
    .then(value => { // eslint-disable-line no-unused-vars
      var xferComment = {
        displayName: req.auth.displayName,
        userId: req.auth.userId,
        dateTime: Date.now(),
        comment: req.body.comment.substring(0, 250)
      };
```

```
        req.db.collection.findOne({ _id: req.params.sid }, { comments: 1 },
function (err, doc) {
        if (err) {
          err.status = 403;
          return next(err);
        } else if (doc.comments.length > 30) {
          let err = new Error('Comment limit reached');
          err.status = 403;
          return next(err);
        }
        // Not allowed at free tier!!!req.db.collection.findOneAndUpdate({
type: 'SHAREDSTORY_TYPE', _id: req.params.sid, $where:
'this.comments.length<29' },
        req.db.collection.findOneAndUpdate({ _id: req.params.sid },
          { $push: { comments: xferComment } },
          function (err, result) {
            if (result && result.value == null) {
              let err = new Error('Comment insert failed');
              err.status = 403;
              return next(err);
            } else if (err) {
              console.log("+++POSSIBLE COMMENT CONTENTION ERROR?+++ err:",
err);
              err.status = 409;
              return next(err);
            } else if (result.acknowledged != true && result.ok != 1) {
              console.log("+++POSSIBLE COMMENT CONTENTION ERROR?+++
result:", result);
              let err = new Error('Comment save failure');
              err.status = 409;
              return next(err);
            }

            res.status(201).json({ msg: "Comment added" });
          });
        });
    })
    .catch(error => {
      error.status = 400;
      return next(error);
    });
});
```

# 12.14 AWS Lambda Node Processing (NYTStoriesRefresh.js)

In a Node.js backend server, you never want to have compute-intensive synchronous code run there for any reason. If you do, it could overwhelm the V8 JavaScript processing thread and your web service could become unresponsive to subsequent requests coming in. There are reasonable solutions to this, such as forking off other processes from your main process and having code execute there. If you have multiple CPUs, this could take a limited number of these requests for separate processing that each utilize a CPU.

However, there is a much more attractive alternative. It is better to use AWS Lambda or EKC/ECS for this type of processing. If you have tasks that need to run more than 15 minutes, you could investigate using EKC/ECS to run those tasks in a way that is offloaded from your node.js service. We will look at the usage of AWS Lambda for our implementation in this book.

To correctly architect your application, consider what needs to be moved off to secondary processing. In the case of the NewsWatcher application, we can identify a few pieces of code that need to be sent off to be run by AWS Lambda.

The code for the Lambda function is found in a file named NYTStoriesRefresh.js and is found in a folder named Lambda. This Lambda code is run for three different purposes:

1. A main timer that goes off every few hours to download stories from the New York Times news feed. This also updates all user news feeds per their individual filters.
2. Deletion of any shared stories that are more than a few days old to make room for news stories.
3. Processing to update a single user news feed per their filter changes in real time.

For the first two cases, these are run off a timer. We will use AWS Event Bridge to fire these. The third case gets fired in our actual route processing code in the Elastic Beanstalk Node code.

*Note: You can set the number of concurrent instances of your Lambda you want running. If you have reached your limit, subsequent invocations will queue up and retry as per configuration in AWS. To be more efficient and better manage the invocations, you can use SQS as the queuing mechanism and have that feed into Lambda and the Lambda concurrency will work with SQS to manage the throttling.*

**AWS Lambda and Node.js**
You can write a Lambda function in any number of coding languages. One of the supported frameworks is Node JavaScript code. This is what we will choose. For our code to run, we need a package.json file, the JavaScript code, and we need to do an "npm istall" command.

To do a deployment of you Lambda, we will upload a zip file that contains the JavaScript file(s), the package.json and the node_modules directory. This is different from Elastic Beanstalk where you do not need to provide the node_modules directory.

We can look at the package.json file for our Lambda code. There will not be as many dependencies here as our usage is simple. We just need the MongoDB driver, the async and the bcryptjs modules from npm. We have a script in the scripts section to make it easier to remember what needs to be zipped up for AWS deploy. Here is the package.json file for the Lambda that we will be using.

```
{
  "name": "NYTLambda",
  "version": "0.0.1",
  "private": true,
  "description": "Lambda to interface with NYT API",
  "main": "NYTStoriesRefresh.js",
  "author": {
    "name": "BUSHMAN",
    "email": "jsdevstack@outlook.com"
  },
  "scripts": {
    "zipLambda": "zip -r lambda.zip NYTStoriesRefresh.js package.json
node_modules"
  },
  "dependencies": {
    "async": "^3.2.3",
    "bcryptjs": "^2.4.3",
    "mongodb": "^4.3.1"
  }
}
```

**The AWS Lambda execution**
We can now look at the code in the NYTStoriesRefresh.js file and look at the top part of that code. The first lines will set up what is required for our module usage. As already mentioned, we require the usage of only a few modules.

After that you can see some code to create the connection to the MongoDB database. One thing to note is that the database connection can be used across invocations of the Lambda. This depends on if AWS keeps a hot copy of the Lambda around from a previous invocation. Thus, we have code to detect if the connection has already been established.

# PART II: The Service Layer (Node.js)

Let's say you have just deployed an AWS Lambda function. You make your first call to run it. This will be a brand-new load of the Lambda, and so all the usual loading and running of the Node.js application code will happen. The function will run to completion, and let's say it took 10 seconds. Now you invoke that Lambda function again. This time only the handler function is called, and the processing skips all the rest of the code, such as the require statements at the top.

Remember, you can set how many invocations can happen in parallel with your function. Let's say you have that set at 10. When you invoke the Lambda for the first time, it gets loaded and runs to completion. Now with two more immediate invokes of the Lambda, one will get the loaded Lambda and the second will get a fresh instance and need to do all its initialization again. Here is the top part of the code for you to review and learn from.

```javascript
var bcrypt = require('bcryptjs');
var https = require("https");
var async = require('async');
var ObjectId = require('mongodb').ObjectId;
var MongoClient = require('mongodb').MongoClient;

const NEWYORKTIMES_CATEGORIES = ["home", "world", "business",
                                 "technology"];
if (process.env.NODE_ENV !== 'production') {
  require('dotenv').config();
}

var globalNewsDoc;
var globalNewsDocFetchCnt = 0;
let cachedDb = null;

function connectToDatabase(uri, callback) {
  if (cachedDb && cachedDb.db && cachedDb.collection) {
    return callback(null);
  }

  MongoClient.connect(uri, { useNewUrlParser: true, useUnifiedTopology:
true, minPoolSize: 10, maxPoolSize: 100 }, function (err, client) {
    if (err == undefined || err == null) {
      cachedDb = {};
      cachedDb.client = client;
      cachedDb.db = client.db('newswatcherdb');
      cachedDb.collection = cachedDb.db.collection('newswatcher');
      console.log('=> Connected and returning');
      return callback(null);
    } else {
      console.log(`Failed to connected to MongoDB server err:${err}`);
      return callback(err);
    }
  });
}
```

This is all great and allows you to set up your Lambda with the ability to cache a connection to the MongoDB database and save yourself a second or two of initialization time.

*Note: If you want to reset your Lambda to be fresh and not load an existing instance, you can redeploy your code and that will force a clean load next time. Otherwise, you must wait for some AWS internal timeout for when it eventually onloads a Lambda.*

### The AWS Lambda Handler Function

AWS Lambda function file requires the export of a `handler` function. This is where an invoke call with execute and give us the ability to run some code.

```
module.exports.handler = (event, context, LambdaCallback) => {
  if (event.params_call_type === "deleteStaleSharedStories_call") {
    console.log('deleteStaleSharedStories_call');
    return deleteStaleSharedStories(context, LambdaCallback);
  } else if (event.params_call_type === "refreshForUserFilter_call") {
    console.log('refreshForUserFilter_call');
    return refreshStoriesForUser(event.doc, context, LambdaCallback);
  } else if (event.params_call_type === "refreshNYTStories_call") {
    console.log('refreshNYTStories_call');
    return refreshNYTStories(context, LambdaCallback);
  } else {
    console.log("Unknown fire of NYT times story handling");
    context.callbackWaitsForEmptyEventLoop = false;
    return LambdaCallback(null, {StatusCode: 200, body: 'Invalid input!' })
  }
};
```

We are making use of the `handler` function in three ways, and thus have created an object to be passed in to determine which code to run. The event parameter of the handler function has the payload object we pass in. The object has a property named `params_call_type`, which is used to branch with. Then we have a function to handle each of the three cases. We have an error case of any unrecognized type. There is code for where a user has updated their news filter in that code you can see another property of the event that holds the user document.

You will need to call the AWS `handler` callback function to terminate a Lambda. If you don't, your lambda will likely run until the timeout config for the Lambda and could add up to extra unnecessary charges. The following two lines terminate a lambda. You will see them scattered throughout the code in key places. If you want to signal that an error happened, then you pass an Error object as the first parameter to `LambdaCallback()`.

```
context.callbackWaitsForEmptyEventLoop = false;
LambdaCallback(null, {StatusCode: 200, body: 'Invalid input!' })
```

## PART II: The Service Layer (Node.js)

### Timer event to populate the master news list

Every few hours, we want to run the Lambda and fetch all news stories from the NYT news service provider. This batch processing can take some time before it finishes. This is a perfect use of a Lambda function because it is long-running, but not longer than 15 minutes. It is significant enough that it is better to separate out the processing to avoid usage in the main Node Elastic Beanstalk process. This is true even if we are horizontally scaled.

We use AWS EventBridge to define a timer and direct it to invoke our Lambda. We set up the Lambda call configuration so that the `event.params_call_type` will equal `"refreshNYTStories_call"`. Inside the processing code we send off an HTTP request to the NYT news feed API service. The results of the processing are placed in the master document `newsStories` array.

Once all the news stories are in place, the code will sequence through all the User documents, and for each, perform matching against the news filters. This is somewhat tricky, and uses the `async.doWhilst()` functionality. This way, the code can manage underlying async calls. The code will keep running as long as there is processing to do. Here is the code:

```
function refreshNYTStories(context, LambdaCallback) {
  async.timesSeries(NEWYORKTIMES_CATEGORIES.length, function (n, next) {
    var body = '';
    setTimeout(function () {
      https.get({
        host: 'api.nytimes.com',
        path: '/svc/topstories/v2/' + NEWYORKTIMES_CATEGORIES[n] +
'.json?api-key=' + process.env.NEWYORKTIMES_API_KEY
      }, function (res) {
        res.on('data', function (d) {
          body += d;
        });
        res.on('end', function () {
          next(null, body);
        });
      }).on('error', function (err) {
        next(err, body);
      });
    }, 500);
  }, function (err, results) {
    if (err) {
      context.callbackWaitsForEmptyEventLoop = false;
      return LambdaCallback(err);
    }
    // Do the replacement of the news stories in the single master Document
    connectToDatabase(process.env.MONGODB_CONNECT_URL, function (err) {
      if (err) {
        context.callbackWaitsForEmptyEventLoop = false;
        return LambdaCallback(null, true);
      }
```

216

```
cachedDb.collection.findOne({ _id: process.env.GLOBAL_STORIES_ID },
function (err, gStoriesDoc)
{
  if (!err) {
    gStoriesDoc.newsStories = [];
    gStoriesDoc.homeNewsStories = [];
    var allNews = [];
    for (var i = 0; i < results.length; i++) {
      try {
        var news = JSON.parse(results[i]);
      } catch (e) {
        console.error(e);
        return;
      }
      for (var j = 0; j < news.results.length; j++) {
        if (news.results[j].multimedia && // Validate good data
          news.results[j].multimedia.length > 0 &&
          news.results[j].title !== '' &&
          news.results[j].url !== '') {
          const hours = toHours(news.results[j].updated_date);
          let hoursString;
          if (hours === 0 || hours < 2) {
            hoursString = "1 hour ago";
          } else {
            hoursString = hours + " hours ago";
          }
          var xferNewsStory = {
            imageUrl: news.results[j].multimedia[0].url,
            link: news.results[j].url,
            title: news.results[j].title,
            contentSnippet: news.results[j].abstract,
            source: news.results[j].section,
            hours: hours,
            hoursString: hoursString
          };
          allNews.push(xferNewsStory);
          if (i === 0) { // Only home stories
            gStoriesDoc.homeNewsStories.push(xferNewsStory);
          }
        }
      }
    }
    async.eachSeries(allNews, function (story, innercallback) {
      bcrypt.hash(story.link, 10, function getHash(err, hash) {
        if (err) {
          return innercallback(err);
        }
        story.storyID = hash.replace(/\+/g, '-').replace(/\//g,
'_').replace(/=+$/, '');
        if (gStoriesDoc.newsStories.findIndex(function (o) {
          if (o.storyID == story.storyID || o.title == story.title)
            return true;
          else
```

```
            return false;
          }) == -1) {
            gStoriesDoc.newsStories.push(story);
          }
          innercallback();
        });
      }, function (err) {
        if (!err) {
          gStoriesDoc.homeNewsStories.sort((a, b) => {
            return a.hours - b.hours;
          });
          gStoriesDoc.newsStories.sort((a, b) => {
            return a.hours - b.hours;
          });
          globalNewsDoc = gStoriesDoc;
          globalNewsDocFetchCnt = 999999;
          cachedDb.collection.findOneAndUpdate(
          { _id: globalNewsDoc._id },
          { $set: { newsStories: globalNewsDoc.newsStories,
                  homeNewsStories: globalNewsDoc.homeNewsStories } },
          function (err, result)
          {
            if (!err) {
              let cursor=cachedDb.collection.find({type: 'USER_TYPE'});
              let keepProcessing = true;
              async.whilst(
                function test(cb) {cb(null, keepProcessing == true); },
                function iter(callback) {
                  cursor.next(function (err, doc) {
                    if (!err && doc) {
                      refreshStories(doc, function (err) {
                        callback(null, 1);
                      });
                    } else {
                      keepProcessing = false;
                      callback(null, 1);
                    }
                  });
                },
                function (err, n) {
                  context.callbackWaitsForEmptyEventLoop = false;
                  return LambdaCallback(null, true);
                }
              );
            }
          });
        }
      });
    });
  });
}
```

218

You will see the use of the async module that loops several times and has a 500-millisecond delay between each batch news request from NYT. This is because there is a restriction on the usage so you can't call it more than five times a second or risk disabling your IP address from accessing NYT's API.

Each news story from the feed needs an id so we can identify it and make sure there are no duplicate stories. While a global unique identifier (GUID) can be generated, the problem with a GUID is the same story may appear again in the next batch of news and cause problems if you thought it was a first-time news story. A hash of the link turns out to be the best way to uniquely identify a story.

***Note:*** *Anytime you use async functions, be ultra-careful of how they operate their async and sync capabilities. Make sure you correctly call the required callbacks in the right place. Error handling can be tricky here.*

The code makes use of the MongoDB cursor capability to sequence through all users. This is not efficient, as it processes one user at a time. A better approach would be to use the async module to process batches of users in parallel. Of course, if there were a million users, it would be necessary to have that part of the processing spread across another lambda that would process batches of users. Even better, would be not to process any users, and wait until a user logs in to kick off a lambda to refresh their new stories. Here is the refreshStories function for the User document updating.

```
function refreshStories(doc, callback) {
  for (var filterIdx = 0; filterIdx < doc.newsFilters.length;filterIdx++) {
    doc.newsFilters[filterIdx].newsStories = [];

    for (var i = 0; i < globalNewsDoc.newsStories.length; i++) {
      globalNewsDoc.newsStories[i].keep = false;
    }

    // If there are keyWords, then filter by them
    if ("keyWords" in doc.newsFilters[filterIdx] &&
        doc.newsFilters[filterIdx].keyWords[0] != "") {
      var storiesMatched = 0;
      for (let i = 0; i < doc.newsFilters[filterIdx].keyWords.length;i++) {
        for (var j = 0; j < globalNewsDoc.newsStories.length; j++) {
          if (globalNewsDoc.newsStories[j].keep == false) {
            var s1 = globalNewsDoc.newsStories[j].title.toLowerCase()
                    .split(/\s+|\./);
            var s2 = globalNewsDoc.newsStories[j].contentSnippet
                    .toLowerCase().split(/\s+|\./);
            var keyword = doc.newsFilters[filterIdx].keyWords[i]
                      .toLowerCase();
            const all = [...s1, ...s2];
            if (all.includes(keyword)) {
              globalNewsDoc.newsStories[j].keep = true;
```

```
        storiesMatched++;
      }
    }
    if (storiesMatched == process.env.MAX_FILTER_STORIES)
      break;
  }
  if (storiesMatched == process.env.MAX_FILTER_STORIES)
    break;
}

for (var k = 0; k < globalNewsDoc.newsStories.length; k++) {
  if (globalNewsDoc.newsStories[k].keep == true) {
    doc.newsFilters[filterIdx].newsStories.push
    (globalNewsDoc.newsStories[k]);
  }
}
}
}

// Do the replacement of the news stories
cachedDb.collection.findOneAndUpdate(
{ _id: ObjectId(doc._id) },
{ $set: { "newsFilters": doc.newsFilters } },
function (err, result)
{
  return callback(err);
});
}
```

This code is looking at the keywords of a user filter that they have chosen and looking to see if they appear in the title or content snippet of the story. The basic algorithm loops through each filter of a user. For each filter, the code checks if there are any stories in the master news list that match the keywords. There is a limit to the number of matching stories allowed. When the update of the user document happens, the $set operator is used to only update a single property of the document.

**Refresh of a user's filters**
We can now look at the function that runs when a single user has updated their filters. Most of this code should be familiar as it uses the same code to connect to the database. As mentioned before, it is possible that a previous Lambda instance will be reused. We can then not only reuse the connection to the database but also the global news stories retrieved from the NYT. This is done by setting up a global variable globalNewsDoc to hold the results and not fetch them again if not needed.

```
function refreshStoriesForUser(doc, context, LambdaCallback) {
  if (!globalNewsDoc || globalNewsDocFetchCnt++ > 100) {
    globalNewsDocFetchCnt = 0;
    connectToDatabase(process.env.MONGODB_CONNECT_URL, function (err) {
      if (err) {
```

```
      context.callbackWaitsForEmptyEventLoop = false;
      return LambdaCallback(null, true);
    }
    cachedDb.collection.findOne({ _id: process.env.GLOBAL_STORIES_ID },
    function (err, gDoc)
    {
      if (err) {
        context.callbackWaitsForEmptyEventLoop = false;
        return LambdaCallback(null, true);
      } else {
        globalNewsDoc = gDoc;
        refreshStories(doc, function (err) {
          context.callbackWaitsForEmptyEventLoop = false;
          return LambdaCallback(null, true);
        });
      }
    });
  });
  } else {
    refreshStories(doc, function (err) {
      context.callbackWaitsForEmptyEventLoop = false;
      return LambdaCallback(null, true);
    });
  }
}
```

## Deleting old stories

We need code that will delete shared news stories after a certain amount of time. This becomes another timer that periodically goes off to do this processing. The code is as follows:

```
function deleteStaleSharedStories(context, LambdaCallback) {
  connectToDatabase(process.env.MONGODB_CONNECT_URL, function (err) {
    if (err) {
      context.callbackWaitsForEmptyEventLoop = false;
      return LambdaCallback(null, true);
    }

    cachedDb.collection.find({ type: 'SHAREDSTORY_TYPE' })
    .toArray(function (err, docs) {
      if (err) {
        context.callbackWaitsForEmptyEventLoop = false;
        return LambdaCallback(null, true);
      }
      async.eachSeries(docs, function (story, innercallback) {
        var d1 = story.comments[0].dateTime;
        var d2 = Date.now();
        var diff = Math.floor((d2 - d1) / 3600000);
        if (diff > 72) {
          cachedDb.collection.findOneAndDelete(
          { type: 'SHAREDSTORY_TYPE', _id: story._id },
          function (err, result)
          {
```

```
            innercallback(err);
        });
      } else {
        innercallback();
      }
    }, function (err) {
      if (err) {
        context.callbackWaitsForEmptyEventLoop = false;
        return LambdaCallback(null, true);
      } else {
        context.callbackWaitsForEmptyEventLoop = false;
        return LambdaCallback(null, true);
      }
    });
  });
});
}
```

## Setting up AWS Lambda

With the code ready to go, we can deploy it to AWS to be run. To get started, navigate in the AWS portal to the Lambda service. You will find a button to click to add a new Lambda function. You will give it a name and select Node as the runtime. The UI looks as follows:

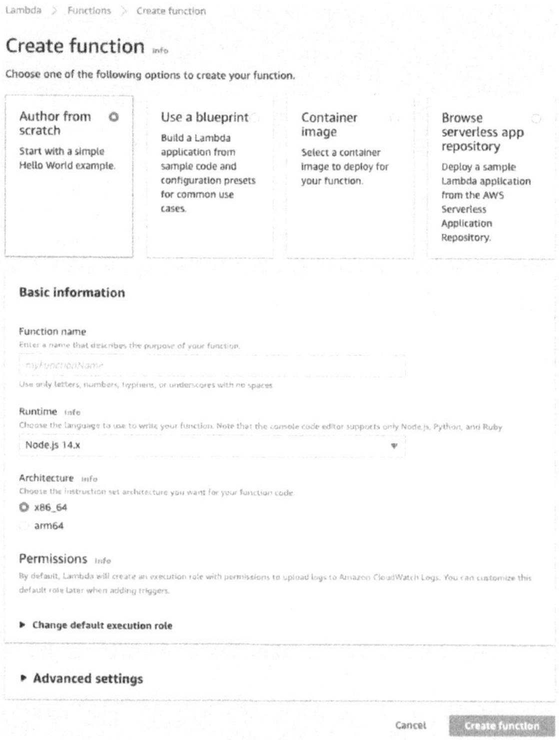

*Figure 60 - Creation of a Lambda function*

Then you need to go into the configuration for the Lambda and configure a few more settings. There will be a default setting for where the handler is found, however, in our case we have a set name of the file we want to use, so we need to change that to be "NYTStoriesRefresh.handler".

## Edit runtime settings

**Runtime settings** Info

Runtime
Choose the language to use to write your function. Note that the console code editor supports only Node.js, Python, and Ruby.

Node.js 14.x ▼

Handler Info

NYTStoriesRefresh.handler

Architecture Info
Choose the instruction set architecture you want for your function code.

◉ x86_64
○ arm64

Cancel  **Save**

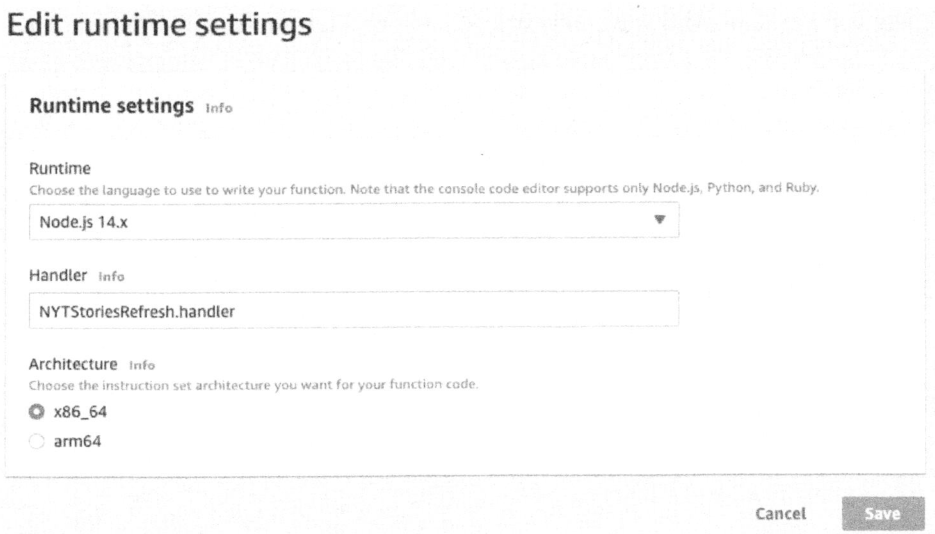

*Figure 61 - Setting the handler location*

By default, when you have signed up for an AWS account, you will have limits in place for various services you could be using. One of those if the number of concurrent Lambda invokes that can be running. That is set up as 1000. As with many of these limits, you can submit a request to AWS to increase this limit. For the concurrency setting, you can determine that as many invokes of this function can be made up to the allowed unreserved amount. You can also reserve an amount for a lambda you have created. This would be because you have other lambdas to run that also need priority. This is out only Lambda, so we can leave it as is.

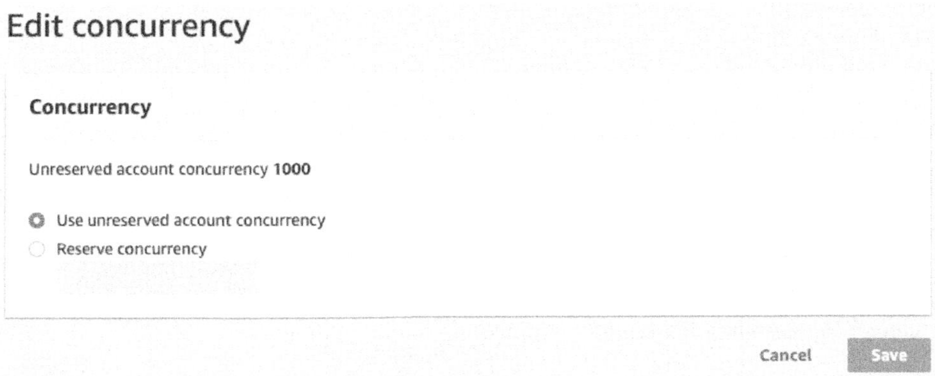

## Edit concurrency

**Concurrency**

Unreserved account concurrency **1000**

◉ Use unreserved account concurrency
○ Reserve concurrency

Cancel  **Save**

*Figure 62 - Concurrency setting*

When you create a Lambda, it has some basic permissions in its default role. To add the ability to access other AWS services (SQS, DynamoDB etc.), you need to add that access to the role. You can see in the figure below the role. We also set the timeout here. In our Lambda we have code to terminate the invocation when processing is completed, so the setting at 15 minutes is not an issue and we will not be running for that length of time. You would want to set a smaller timeout if you really did want to restrict the tie and terminate for some reason.

The setting of the amount of memory is critical and a bit of trial and error. The more memory, the more the cost per the time it runs. Also, the more memory you give the more CPU power you are allocated, including how many CPUs are used. Once you run a lambda, you can look at the log output in the Lambda service and see how much memory was used. Our Lambda uses less than 100 MB. So, you might think we could save money and set the memory at 128 MB and be good. However, if you do, it takes almost 90 seconds to run. Bump up the memory to 1024 and the Lambda can complete in around 10 seconds. There is a calculator in AWS to help you determine the optimal setting and get an idea of the cost you will have.

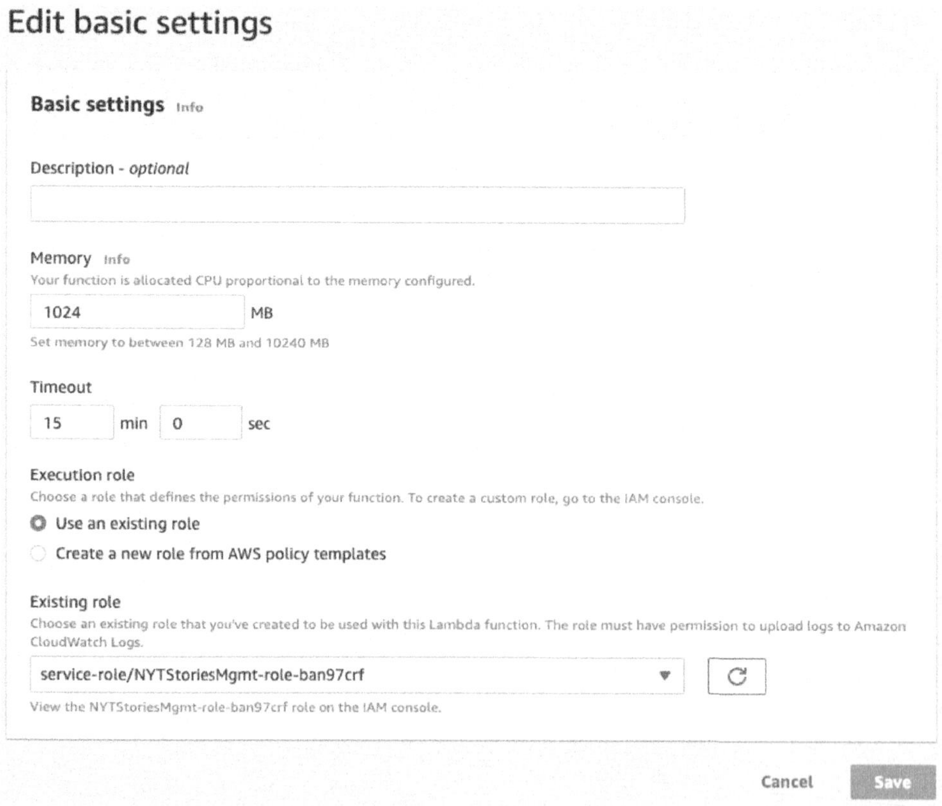

## Edit basic settings

**Basic settings** Info

Description - *optional*

Memory Info
Your function is allocated CPU proportional to the memory configured.

| 1024 | MB |

Set memory to between 128 MB and 10240 MB

Timeout

| 15 | min | 0 | sec |

Execution role
Choose a role that defines the permissions of your function. To create a custom role, go to the IAM console.

◉ Use an existing role

○ Create a new role from AWS policy templates

Existing role
Choose an existing role that you've created to be used with this Lambda function. The role must have permission to upload logs to Amazon CloudWatch Logs.

| service-role/NYTStoriesMgmt-role-ban97crf | ▼ | ↻ |

View the NYTStoriesMgmt-role-ban97crf role on the IAM console.

Cancel    Save

*Figure 63 – Basic settings like memory and timeout*

We have a setting in the .env file that can run the Lambda local on our machine, and it will use the environment setting there. In AWS, we need to set up environment variables so that is what you see in this UI. I have put in "<>" placeholders for places you need to add your own values.

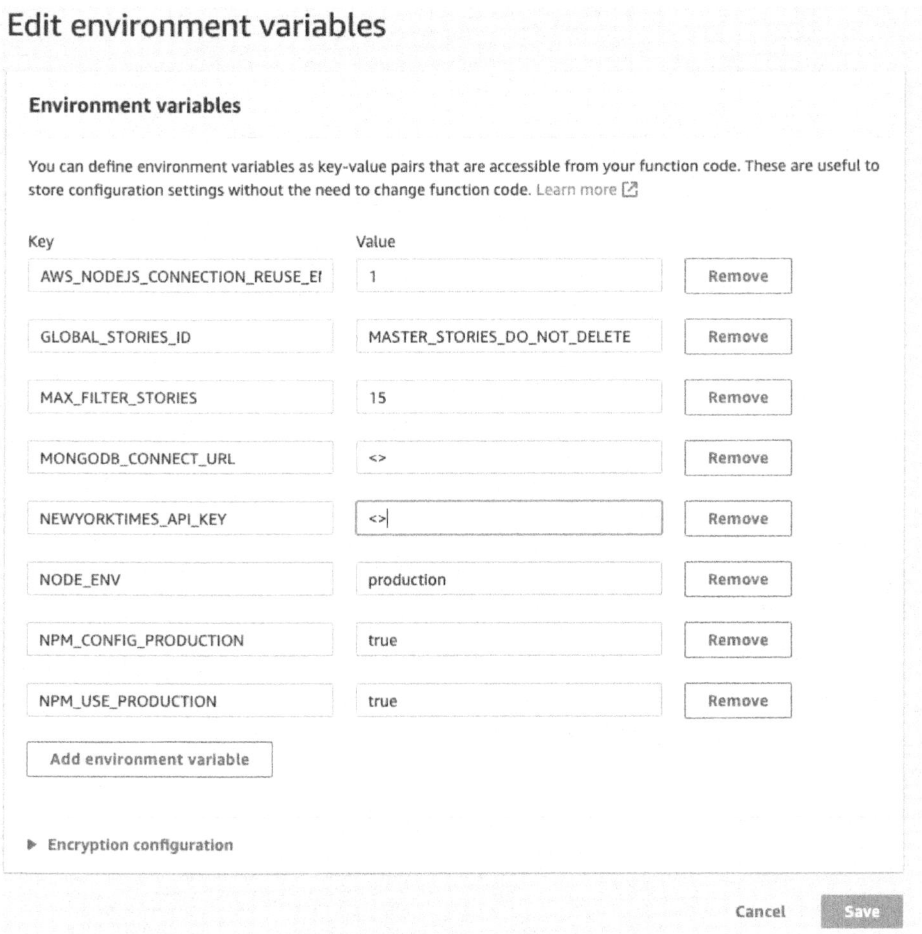

*Figure 64 - Environment variable setting*

# PART II: The Service Layer (Node.js)

Once you have uploaded you code through the zip file using the **Upload from** button, you can view the code and edit it there if you want to make quick changes and test them out.

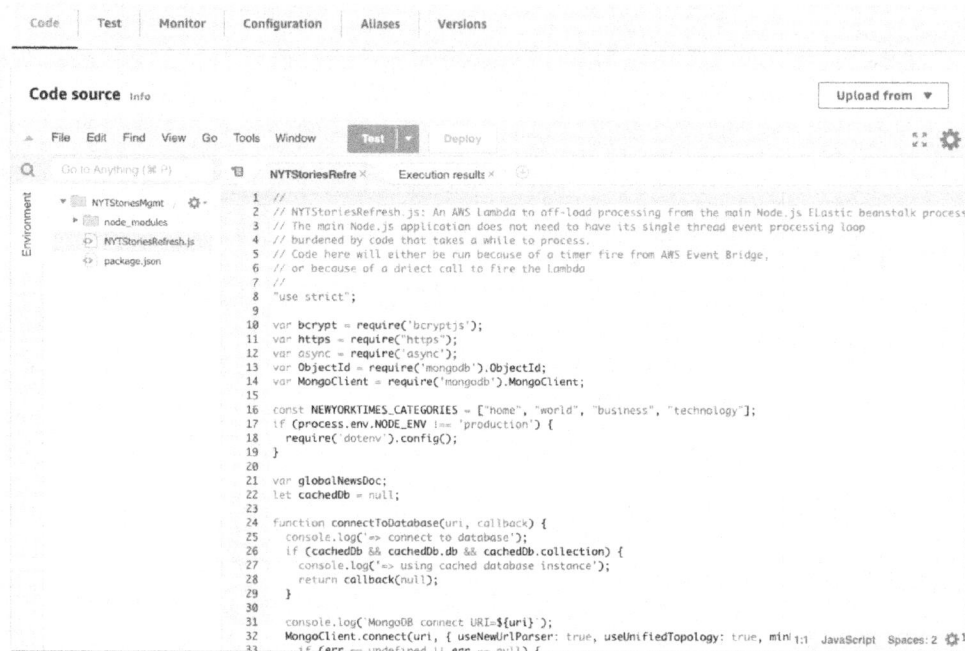

*Figure 65 - Code viewing and editing*

You can even invoke a Lambda right out of the AWS portal UI. That can be done with a button in the **Code** tab, or with the separate **Test** tab. For our test we have added the object to the event for the call type so that it will run the code to delete stale shared articles.

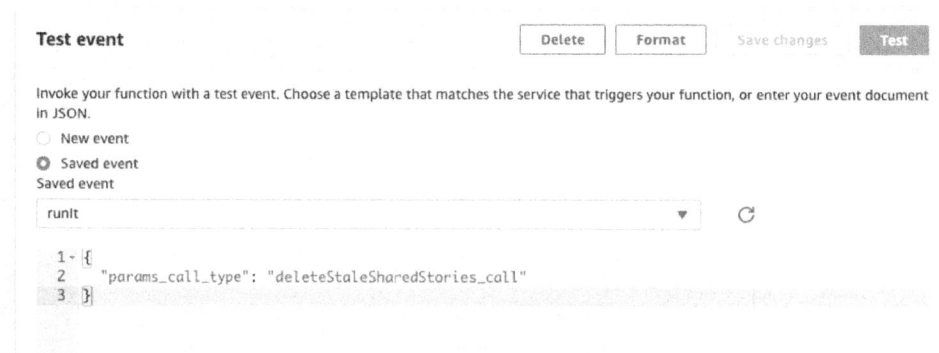

*Figure 66 - Testing a Lambda*

There is a **Monitor** tab you can go to for viewing the logs of each Lambda instance invoked. One of the links will pop open a new browser window and take you to the **CloudWatch** logs for that Lambda. All of your `console.log()` statements will appear in the logs, along with AWS provided logging, such as the final log entry that shows you how long it ran for and how much memory it used up.

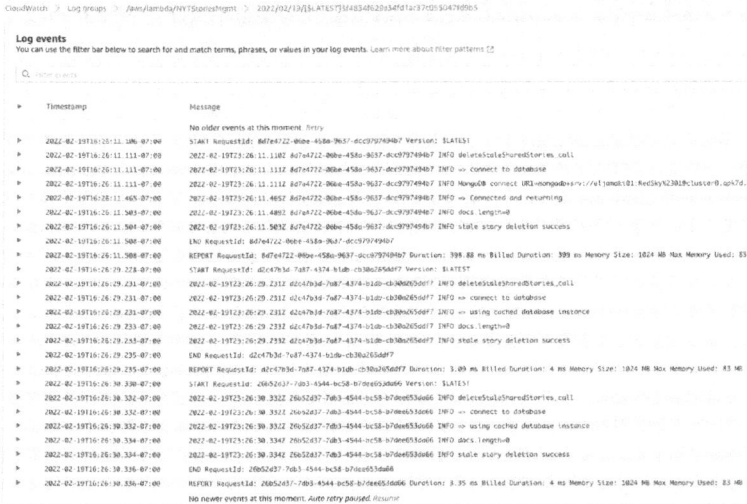

*Figure 67 - CloudWatch logs from the Lambda invoke*

***Note:*** *It is a good practice to edit the retention setting of the Lambda logs to be something like 7 days. It defaults to indefinitely and thus might end up costing you in storage charges that are unnecessary.*

There is a setting to determine the number of retries in case an invocation fails, and how long it will hang around for the retry.

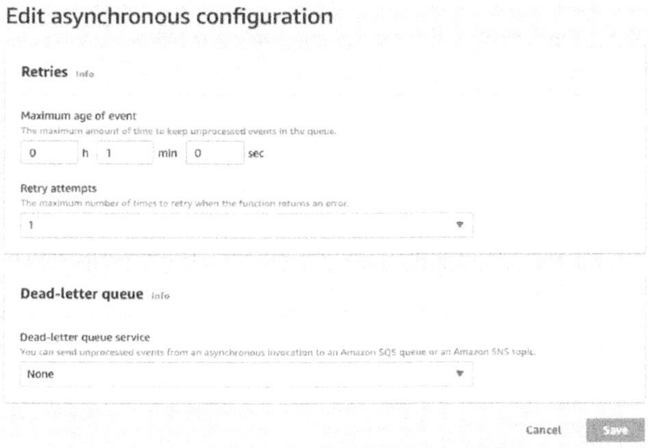

*Figure 68 - Async setting*

## Setting up AWS EventBridge

With the Lambda code in place and all configure and tested, we have it invoked from our Node.js Elastic beanstalk backend route code (that was already shown). We can also hook up the event firing on a timer using AWS EventBridge for invocations we want on a timer basis.

To get started, we go to the EventBridge service in the AWS portal and click on Rules and create two rules – one for the shared news story deletion and one for the master NYT news fetch refresh. You give the rules a name and set them up to be fired at a fixed rate schedule. We see that for this rule it is configured to run every four hours.

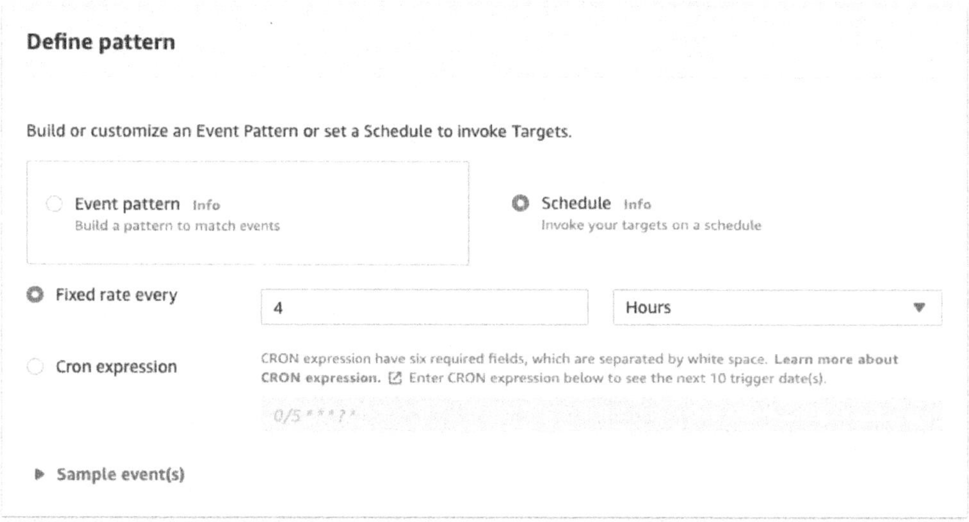

*Figure 69 - Scheduled timer interval*

***Note:*** *You can also have the rule set up to run due to some type of event external to AWS such as something in Zendesk, Shopify, MongoDB, Datadog and many more. You can also set it up for internal AWS events such as SQS, CloudWatch, Data Pipeline, DynamoDB and just about anything that is a service through AWS. The power here is that you can then define an event pattern to match for when the rule triggers*

The last thing to configure is the target to run when the timer goes off. There are many selections to choose from and we will select **Lambda function**. We then can see a list of our functions and select the one we are interested in and if needed, define some JSON that get sent into the Lambda handler. This is the same JSON we would have used in the testing from the Lambda **Test** tab.

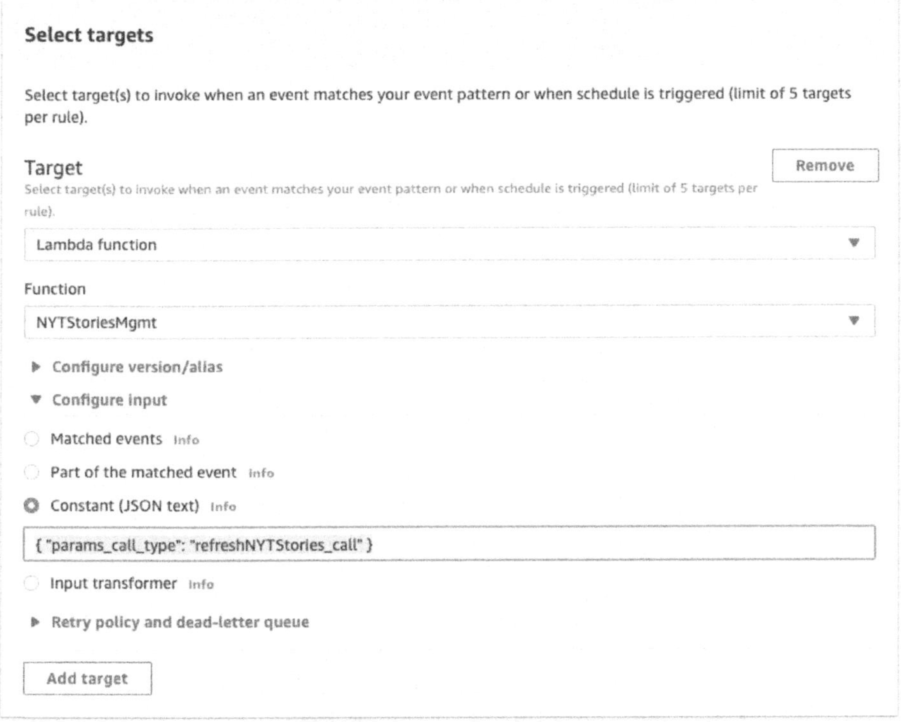

**Select targets**

Select target(s) to invoke when an event matches your event pattern or when schedule is triggered (limit of 5 targets per rule).

| Target | Remove |
|---|---|

Select target(s) to invoke when an event matches your event pattern or when schedule is triggered (limit of 5 targets per rule).

Lambda function ▼

Function

NYTStoriesMgmt ▼

▶ Configure version/alias

▼ Configure input

○ Matched events  Info

○ Part of the matched event  Info

◉ Constant (JSON text)  Info

{ "params_call_type": "refreshNYTStories_call" }

○ Input transformer  Info

▶ Retry policy and dead-letter queue

Add target

*Figure 70 - Setting the target to the Lambda*

# 12.15 Securing With HTTPS

At this point, we need to discuss an important security measure to be put into place. You can't just host your REST API endpoint without encrypting traffic back and forth. You fix this by only permitting HTTPS connections. That way all traffic is encrypted and signed so that it is tamper-resistant and harder to eavesdrop on.

Since the Node.js service is exposed through the Elastic Beanstalk app, you don't need to make any changes to your Node.js code to secure it as it is all done through AWS. If your Node.js instance is exposed directly to the internet and serving up traffic, then do some simple configuration on the Node.js side to install a certificate and key files, and then make a few modifications to the Node code.

Elastic Beanstalk keeps Node from being directly exposed. The Elastic Beanstalk load balancer acts as a reverse proxy and that is where you need to set up SSL. You will need to get your own domain name and install a certificate for your Elastic Beanstalk service.

*Note: When you launch your application through VS Code, it will not accept HTTPS locally on your machine. However, tests run against a production deployment must be altered to use HTTPS. Your test code needs to make the appropriate changes to the URL being tested.*

### Securing communications to MongoDB

For performance reasons, you will want to place your database in the same AWS datacenter as your Elastic Beanstalk Node app. As a bonus for doing this, the ability to secure the communication between your Node.js service and your database becomes easier. This is because everything is sent over the internal AWS data center network and never gets out over the public internet. As a further measure of security, you can also communicate over an SSL connection if you use a dedicated plan from MongoDB, Inc. With a dedicated plan, you also define custom firewall rules so that your database access is limited to specific IP address ranges and/or to specific AWS EC2 security groups.

# DNS and certificate setup

In the Introduction section of this book, you saw a physical topology diagram (see Figure 4) that showed a domain name being routed through a DNS server that had a certificate to enable HTTPS. We can now go about setting that up and show some of what you will encounter.

First, go to the Route 53 service management console. There will be a selection in the left-hand side menu for looking at **Registered domains**, go there and then click **Register Domain** to start the process of getting your own domain or you may need to click **Get started now** under the heading to do a domain registration, then click **Register Domain** and sequence through the steps. If you already have your own domain name and want to use it, you can transfer that, but that is not covered in this book.

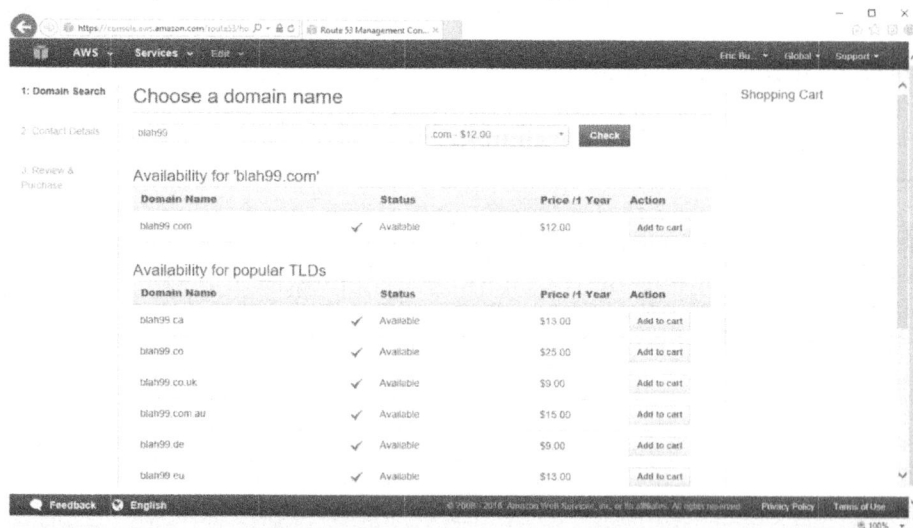

*Figure 71 - Route 53 domain availability*

You can type in different names until you find one that is available. You will incur an initial charge as well as a small recurring fee when you purchase a domain name. For this implementation of the NewsWatcher sample app, I settled on newswatcher2rweb.com since newswatcher.com was already taken.

It can take an hour or longer before everything is ready with your new domain name. Once it is ready, you can get a certificate set up that uses the AWS Certificate Manager Service. Navigate to **Certificate Manager** and click **Request a certificate**. Enter the domain name variations you want to be supported. You can use wildcards and have the flexibility you need. Here is the page where that happens:

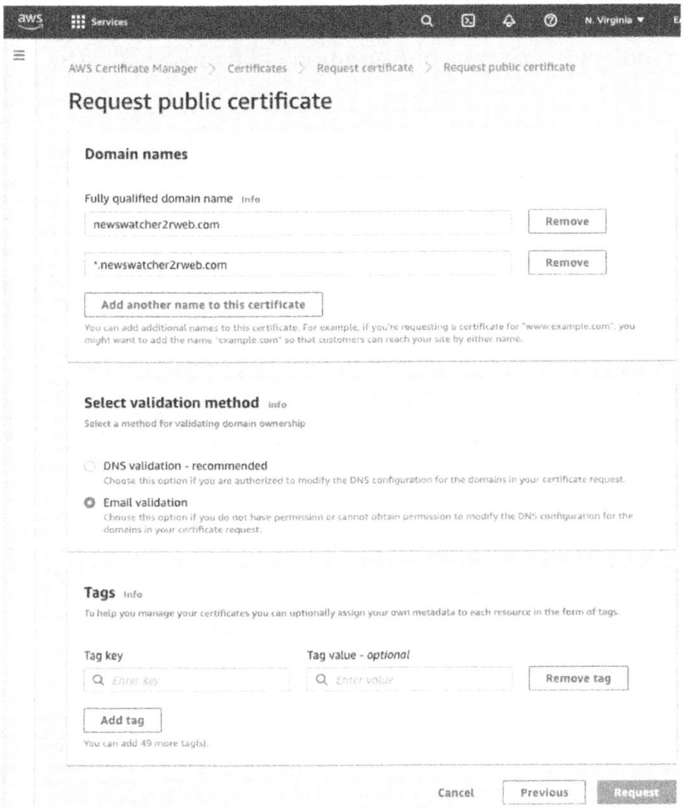

*Figure 72 - Add domain names to AWS Certificate Manager*

After the Add Domain Names screen, you have a few more to click through. At some point, you will also need to validate this action by replying to your confirmation to an email you will receive that verifies you are the owner of the domain name for the certificate that is being set up.

PART II: The Service Layer (Node.js)

A certificate can now be set on the Elastic Beanstalk load balancer. You do this through the AWS management portal. Here are the steps:

1. Open your AWS Elastic Beanstalk environment and on the left, find and select **Configuration**.
2. Click the **Edit** button for **Capacity**. And select the **Environment type** to be **Load balanced**. Save that.
3. Click the **Edit** button for **Load balancer**.
4. Click **Add listener** under the **Listeners** and set this up on port 443 with HTTPS and select your certificate from the dropdown for **SSL policy**.
5. Click **Apply** at the bottom of the page.

Now you can watch the status until it indicates that the configuration change is successful.

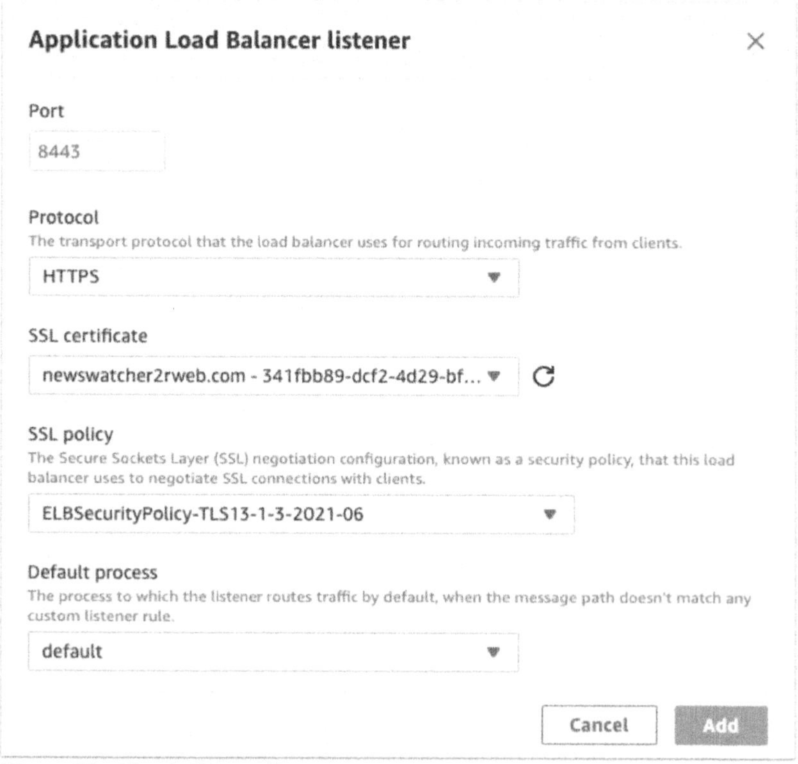

---

**Application Load Balancer listener**                                    ✕

Port

8443

Protocol
The transport protocol that the load balancer uses for routing incoming traffic from clients.

HTTPS ▼

SSL certificate

newswatcher2rweb.com - 341fbb89-dcf2-4d29-bf... ▼   ⟳

SSL policy
The Secure Sockets Layer (SSL) negotiation configuration, known as a security policy, that this load balancer uses to negotiate SSL connections with clients.

ELBSecurityPolicy-TLS13-1-3-2021-06 ▼

Default process
The process to which the listener routes traffic by default, when the message path doesn't match any custom listener rule.

default ▼

Cancel    Add

---

*Figure 73 - Elastic Beanstalk status*

Now you can set up the DNS routing to your Elastic Beanstalk load balancer. Go back into the Route 53 management UI in the AWS portal and click the **Hosted zones** link. Then click on your Domain name. Click **Create Record**. Make sure the Alias switch is on and you can set the two dropdowns and then select your environment.

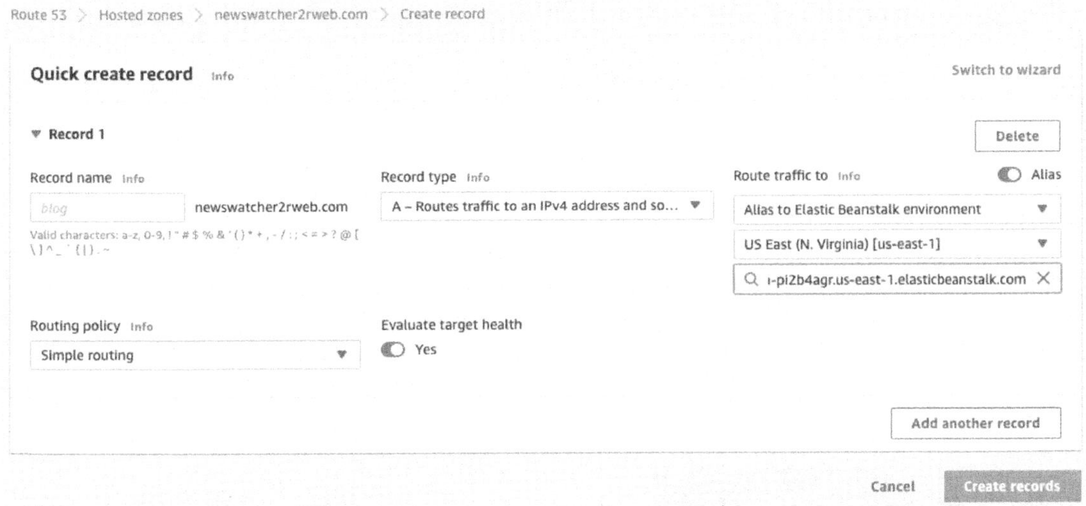

Route 53 > Hosted zones > newswatcher2rweb.com > Create record

**Quick create record** Info                                                    Switch to wizard

▼ Record 1                                                                      Delete

Record name  Info                    Record type  Info                          Route traffic to  Info          ⬤ Alias

[blog]              newswatcher2rweb.com    A – Routes traffic to an IPv4 address and so... ▼    Alias to Elastic Beanstalk environment   ▼

Valid characters: a-z, 0-9, ! " # $ % & ' ( ) * + , - / : ; < = > ? @ [         US East (N. Virginia) [us-east-1]    ▼
\ ] ^ _ ` { | } . ~

                                                                                🔍 i-pi2b4agr.us-east-1.elasticbeanstalk.com  ✕

Routing policy  Info                 Evaluate target health

Simple routing           ▼           ⬤ Yes

                                                                                Add another record

                                                        Cancel    **Create records**

*Figure 74 - AWS Certificate Management Create Record*

You can add a second record set with **Record name** "www". Be careful here and be sure to select the Elastic Beanstalk load balancer environment.

You are almost finished. You now need to turn off HTTP access at the load balancer. To do that:
1. Go to the EC2 service management console and click **Load Balancers**.
2. Click the **Listeners** tab, then check the box next to the HTTP : 80 listener and click **Delete**.

Back at the Elastic Beanstalk environment page, when you click on the URL at the top, you will find it no longer works. In your browser, when you change the URL to start with https:, you will be told the certificate does not match. Now type in the URL to the domain name you registered, and you will see everything up and working. HTTPS is now working. When you try **https**://www.newswatcher2rweb.com/ it will work, but the other **http**://www.newswatcher2rweb.com/ will not work unless you have it set up to forward to HTTPS.

# 12.16 Deployment

At this point, you have everything in place to start trying out your middle-tier web service API that is implemented as an HTTP/REST endpoint. You obviously would not build a service like this without testing it along the way. All the discussion involving the testing of a service is presented in the next chapter. For now, you can see it deployed at this point and then gain a deeper understanding of the testing that is necessary.

# PART II: The Service Layer (Node.js)

At this point, you can zip up your code and get ready to deploy it to AWS. Before doing so, you can test things out with some tests that will be described in Chapter 13. You will want to test on your local machine as well as deploy to some known staging cloud location and test there as well. You could also run the test code from your local machine to go against a staging site that is hosted in AWS.

Here are the folders and files I selected to have zipped up on my Windows machine. There is a right-click selection to create a zip file under **Send to** on a Windows machine. The build and public folders may not actually there yet, so leave those out. They folders that get added that will hold the React application. In the package.json file is a script to do the zip of the files you need – "zipForEB": "zip -r eb.zip build public routes package.json server.js .npmrc". You follow the same instructions previously detailed to get your application code up and running. You now at this point have secured traffic through HTTPS.

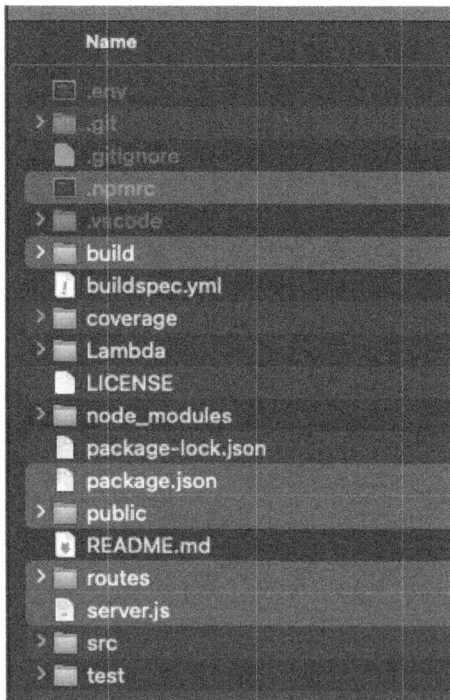

*Figure 75 - Selections to make a zip file*

***Note:*** *If you find errors upon deployment, you can click on the **logs** selection in the UI of the Elastic Beanstalk AWS console. From the **Request Logs** menu, select to download the full logs. In the folder you download, you will find a file named eb-activity.log. Scroll and see if there are any indications as to why the npm install command failed. An AWS Elastic Beanstalk installs sometimes runs out of memory when the install occurs. To work around this, switch to a larger EC2 type and then try your install again.*

# Chapter 13: Testing the NewsWatcher REST API

Now comes the exciting part where you will get to see the HTTP/REST Web API exercised. We will prove the service is up and running locally, then test the deployment to verify everything in production. Once the service is proven, we can start the final task of creating a React UI for NewsWatcher and be confident that the integration will go smoothly.

This chapter will present several practices you will want to follow for exercising your code to fully test it. It is a lot simpler to test and debug issues locally on your machine than in production.

# 13.1 Debugging During Testing

Let's first talk about debugging techniques. You will need to perfect those skills as you run tests and examine the execution of your code.

In some cases, logging to the Node console will provide you with enough clues to track down an issue. This means that you must log important things that are happening in the application. Beyond this, you will need a few tools to help you do your investigations.

One tool at your disposal is the VS Code debugger. Before deploying anything, you can run our code locally. You can use VS Code to debug the Node.js project code.

If you want to debug your Node.js code, you open your project and launch the Node.js project by pressing F5. You can set up your breakpoints in advance or add ones as needed that you want to hit. Once your project is running, you run Jest test code and exercise your code through tests you have written. Then you can step through your code.

*Note: VS Code is very versatile and lets you run the Jest test code and at the same time the service. You can set up a Launch file to define the test configuration that the debugger dropdown will let you choose from.*

To set up debugging in VS code, click the debug icon. You will see a gear icon at the top of the window that you can click to create the launch.json file. This is the file that instructs VS Code how to proceed. You can set it up to have multiple configurations in the file. One can be for launching your node process with debugging capability. Another entry could be for attaching to an already running Node process. When you click the gear icon, select the Node selection, and create the file. Your file may look as follows:

# PART II: The Service Layer (Node.js)

```
{
  "version": "0.2.0",
  "configurations": [
    {
      "type": "node",
      "request": "launch",
      "name": "Launch Program",
      "program": "${workspaceFolder}/server.js"
    },
    {
      "type": "node",
      "request": "launch",
      "name": "Debug Jest api_endpoint Tests",
      "program": "${workspaceRoot}/node_modules/jest/bin/jest.js",
      "args": [
        "--verbose",
        "--runInBand",
        "--no-cache",
        "--testTimeout",
        "999999",
        "${workspaceFolder}/test/api_endpoint.test.js"
      ],
      "runtimeArgs": [
        "--nolazy"
      ],
      "outputCapture": "std",
      "internalConsoleOptions": "openOnSessionStart",
      "disableOptimisticBPs": true
    },
    {
      "name": "Debug Jest React App Tests",
      "type": "node",
      "request": "launch",
      "env": {
        "CI": "true"
      },
      "runtimeExecutable": "${workspaceRoot}/node_modules/.bin/react-scripts",
      "args": [
        "test",
        "--runInBand",
        "--no-cache",
        "--testTimeout",
        "999999"
      ],
      "cwd": "${workspaceRoot}",
      "protocol": "inspector",
      "outputCapture": "std",
      "internalConsoleOptions": "openOnSessionStart",
      "disableOptimisticBPs": true
    }
  ]
}
```

236

To the left of the gear icon is the start button (or you can press F5). The drop-down menu will show you the config section options that come from your launch.json file. The far right greater-than symbol in the box opens the output console window. It is a good idea to always have that open to view any statements or errors that may get displayed.

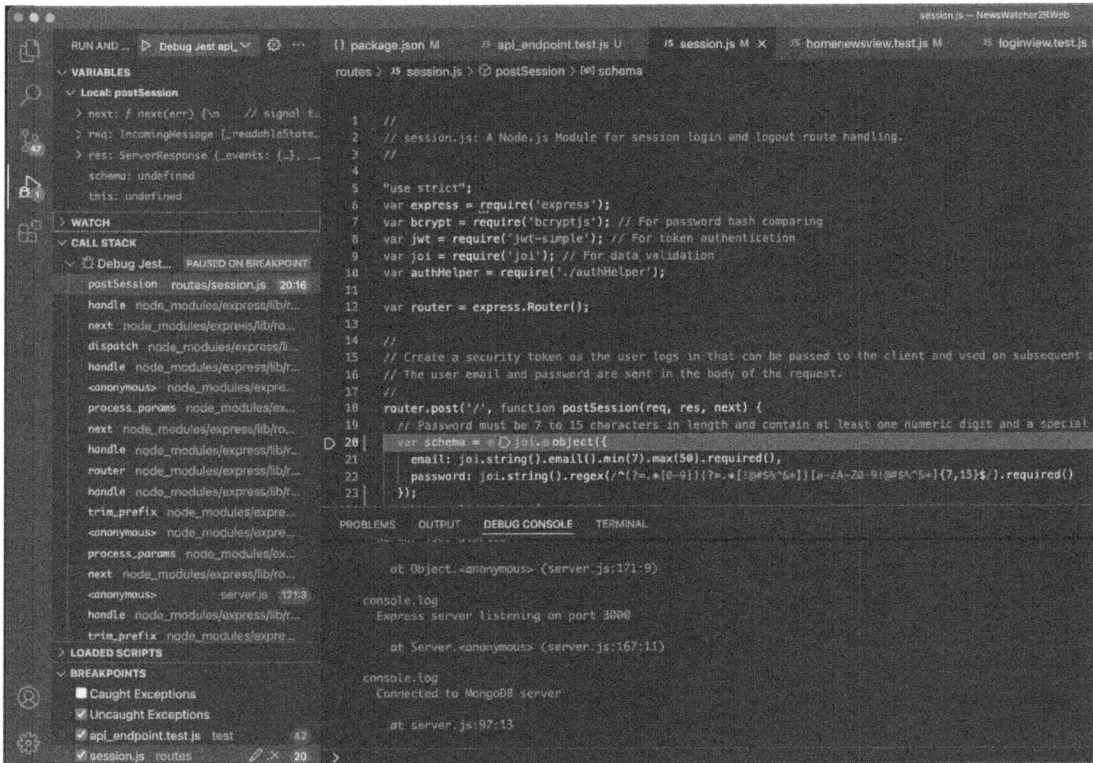

*Figure 76 - Visual Studio Code debugging*

You can open the launch.json file and look at it, but you don't need to make any adjustments to it, as the defaults are just what you need, unless you have it set to start node with another file. You can look up the documentation on the settings that can be used in the launch.json file on the Node website and VS Code website.

To place a breakpoint, open a JavaScript file and click out to the left of the margin or click on the line and press F9. Once you hit a breakpoint while running code, you get full access to inspect the call stack and variables.

# 13.2 Tools to Make an HTTP/REST Call

You will probably want to use a tool to test out calls to your HTTP API. That way, you can get everything verified before you throw a test harness into the mix. I have used Postman, Fiddler, and Curl.exe on my machine for testing individual HTTP/REST calls. With these tools, you can send HTTP requests to your service and view the returned responses.

In Postman, you can set up the verb to use and then configure the request headers and the JSON body content. You can do a POST and then look at the returned response. Let's look at the UI of Postman and see the basics of how to use it to call your API. This will then act to send an HTTP request to the route handler such as for registering a user. There will then be the `app.use('/api/users', users)` code in your server.js file routes to the users.js function of `router.post('/', function(req, res, next) {}`. This is the code that creates a new registration.

You will want to make a post verb call and pass in a JSON body that contains the display name, email, and password. You then expect a 201 return code to be given back to indicate a successful creation. Within the return response will be the returned document that you can inspect.

You can now start up your Node service with your project open and press F5 to debug or press Ctrl+F5 to run without debugging. The first thing you will notice is that you get a console app window that opens in the VS Code UI. That represents your Node process running and the executing of your server.js file. You will see all your console logging appear in this window. If you don't see it, click on the toolbar on the left to get to the debug area and you can open it from there.

Your Node.js server is running as a local process now. To make an HTTP request, you can connect to localhost with the port number of 3000 and interact with the REST API. You can open Postman and try this out. You will do a post to the /api/users path and pass in a body that contains the email, displayName and password. The verb is a Post. Once it is run you can look at the body of the return to see if it was successful. As one of the headers, add one to be Content-Type of application/json. The following figure shows what this looks like after the sending of the request. You can see that we filled out the body for the request and we received back a 201 status and the body as shown

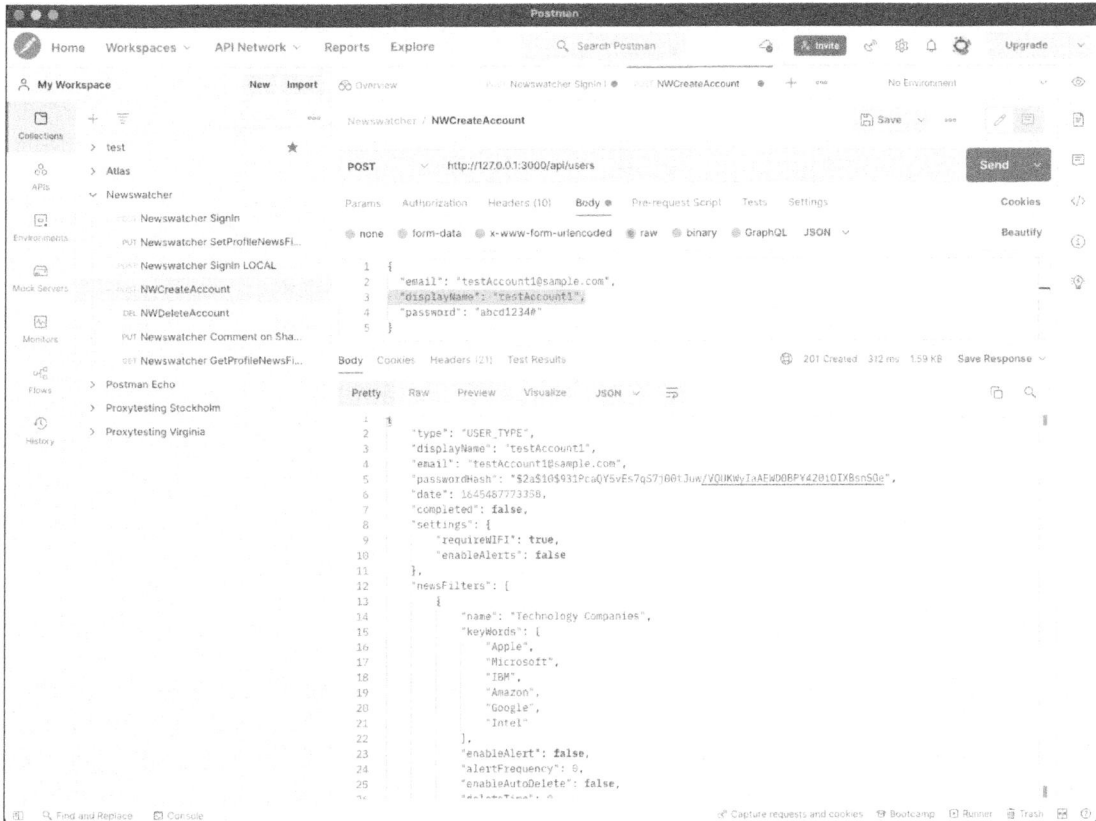

*Figure 77 – Postman usage*

You have now successfully seen your API exercised. Open the MongoDB Atlas portal and you also can see that there is a new document created. In the portal, you will see something like the following:

```
_id: ObjectId("6214269df0fe5b3b66b6509e")
type: "USER_TYPE"
displayName: "testAccount1"
email: "testAccount1@sample.com"
passwordHash: "$2a$10$931PcaQY5vEs7qS7j00tJuw/VQUKWyIaAEWD0BPY420iOIXBsnSOe"
date: 1645487773358
completed: false
> settings: Object
> newsFilters: Array
> savedStories: Array
```

*Figure 78 - Atlas management portal with a new document*

PART II: The Service Layer (Node.js)

You can now enter every single request through Postman to prove them all. Next, try to log in and get a token back. This would look as follows:

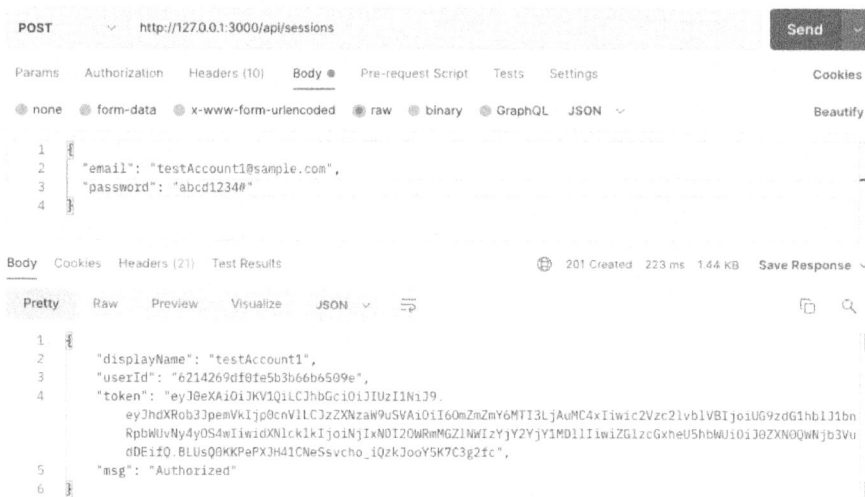

*Figure 79 - Sign in to get a JWT back*

Look at the result and you will see a token in the response if you open it up. You can use this token in subsequent requests. You can next make a call to retrieve the user document and see if there are any news stories that have matched the filter. This is great progress, and you can continue to try all your API calls and debug each one if needed. As you get each one working, you can create a test case for each in your Jest test code that will be covered next.

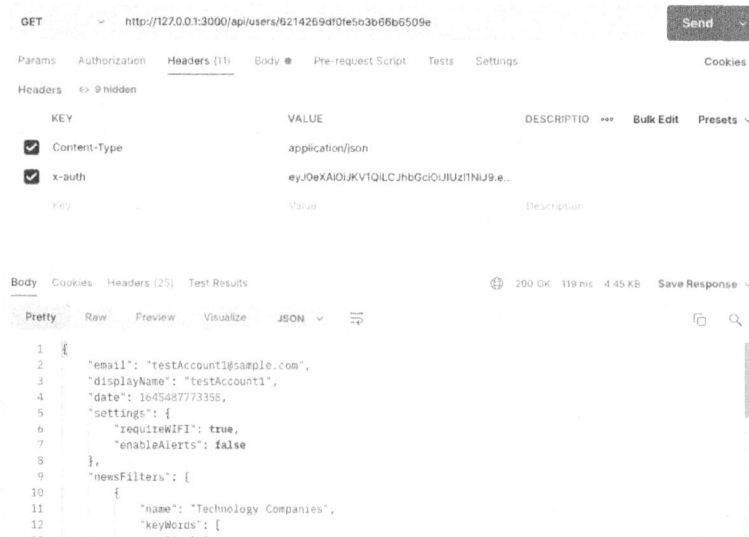

*Figure 80 - User Get request*

240

# 13.3 Functional Test Suite with Jest

We can now take the next step of automating the HTTP API testing using Jest. This will become part of your functional test pass. You can start by implementing some of the same operations that were already tried when you used Postman. To start with, you need to add a new folder to your VS Code project to hold the tests. You can also set up the added Node modules you need for the tests.

*Note: NPM lets you download different types of modules. Like modules I've shown you. You download them and then use a* require *statement to utilize them as code modules. Others like Jest are not code modules, but are command-line tools. What you do is write a JavaScript file that Jest will interpret and run. You then use a command-line to execute Jest, and it does its work. Jest is installed as a local part of your project. This means the executable will be referenced from that location. Jest can also be installed globally if you prefer.*
Add the jest and supertest modules to your dev dependencies of package.json.

```
npm install --save-dev jest
npm install --save-dev supertest
```

The devDependencies section is reserved for non-production modules that you will not need to deploy to a production build. They are only needed to run your test code locally.

We can now go through some of the files in the test folder to see how the NewsWatcher application can be tested. This first file will be used to hit the API endpoint to exercise the routes.

## Writing Jest tests (api_endpoint.test.js)

You will make use of the supertest module and the assert module. The necessary require() calls can be set up for those. The server object also needs to be retrieved and given time to be initialized. There are special blocks of code that Jest will run before and after all tests. We need a timer to delay the start of the test because the connection to MongoDB is asynchronous and take some time at startup. At shutdown we can gracefully close the DB connection and the app.. The rest of the code makes use of Jest test blocks to run tests with.

The request object is what you set up from the supertest module. With that, you make the HTTP post verb call and register a user.

You will use the describe keyword for any major test block and use the individual test keywords for each test inside of that. With just one test to run, it would look as follows:

241

```
var assert = require('assert');
var app = require('../server.js');
var request = require('supertest')(app);

describe('User cycle operations', function () {
  // Wait until the database is up and connected to.
  beforeAll(function (done) {
    setTimeout(function () {
      done();
    }, 5000);
  });

  // Shut everything down gracefully
  afterAll(function (done) {
    app.db.client.close();
    app.close();
    done();
  });

  test("should create a new registered User", function (done) {
    request.post("/api/users")
    .send({
      email: 'bush@sample.com',
      displayName: 'Bushman',
      password: 'abc123*'
    })
    .end(function (err, res) {
      assert.equal(res.status, 201);
      assert.equal(res.body.displayName, "Bushman", "Correct name");
      done();
    });
  });
});
```

I will go through the previous code to make sure you understand it. To start with, you have the usual Node.js require statements at the top. For the supertest module, you specify that you are going to hit the passed in web server endpoint of your Node project. The strings you set as parameters to describe() and test() are printed out for you as part of the test run. You should use text that helps you remember what you are testing.

With this code, you have a test that will verify that you can register a user. The test block takes a string to describe what you are testing and then a function to run. There is a done() function that you call to signal to Jest that it can move on to the next test block. These tests are each run sequentially. If you don't call done(), the test will eventually time out.

To use supertest, you specify a verb operation to use. This one is using post. You string together the function calls send() and end(). Each one will get called in sequence. The send function does the HTTP/REST request and has the body set.

The `end()` function can get the response and validate the return code and values from the returned body. You use the assert module for validations.

There are some cases where you have a second test that relies on the results of the first test. This can be tricky if there is delayed processing of the first test code. You can either stack one test inside the other or use a JavaScript `setTimeout()` call to delay your second test run by a bit and then have it run.

For example, when a user changes their news filters, that operation will return immediately but kick off the asyncronouse backend Lambda function. This means that the test code would move on to the next test. The NewsWatcher application works by invoking the Lambda function to update the news stories. If you want to test a change to a profile filter string, you will realize that it will take a second to do that in the background, so you put in a delay before the next code can run.

The following code shows the use of a delay of three seconds before running a test:

```
test("should allow access if logged in ", function (done) {
    setTimeout(function () {
        request.get("/api/users/" + userId)
        .set('x-auth', token)
        .end(function (err, res) {
            assert.equal(res.status, 200);
            savedDoc = res.body.newsFilters[0].newsStories[0];
            console.log(JSON.stringify(savedDoc, null, 4));
            done();
        });
    }, 3000);
});
```

Here is a more complete set of tests that registers a user and then makes sure the user can log in and then deletes the account to clean things up. The first test right at the start is to verify that a person cannot log in if they don't first register an account.

There may be negative tests you can put in place to verify your error handling code. In this code below, I make use of local valuables inside the describe block to pass these between tests such as the token variable. For example, you need to capture the token at sign in time and keep using it on subsequent REST calls that you are testing.

```
var assert = require('assert');
var app = require('../server.js');
var request = require('supertest')(app);
```

```
describe('User cycle operations', function () {
   var token;
   var userId;
   var savedDoc;

   // Wait until the database is up and connected to.
   beforeAll(function (done) {
      setTimeout(function () {
         done();
      }, 3000);
   });

   // Shut everything down gracefully
   afterAll(function (done) {
      app.db.client.close();
      app.close();
      done();
   });
   test("should deny unregistered user a login attempt", function (done) {
      request.post("/api/sessions").send({
         email: 'bush@sample.com',
         password: 'abc123*'
      })
       .end(function (err, res) {
         assert.equal(res.status, 500);
         done();
      });
   });

   test("should create a new registered User", function (done) {
      request.post("/api/users")
       .send({
         email: 'bush@sample.com',
         displayName: 'Bushman',
         password: 'abc123*'
      })
         .end(function (err, res) {
         assert.equal(res.status, 201);
         assert.equal(res.body.displayName, "Bushman", "Name of user should
be as set");
         done();
      });
   });

   test("should allow registered user to login", function (done) {
      request.post("/api/sessions")
       .send({
         email: 'bush@sample.com',
         password: 'abc123*'
      })
       .end(function (err, res) {
         //<Session&Cookie code>cookies = res.headers['set-cookie'];
         token = res.body.token;
```

244

```
        userId = res.body.userId;
        assert.equal(res.status, 201);
        assert.equal(res.body.msg, "Authorized", "Message should be
AUthorized");
        done();
    });
  });

  test("should delete a registered User", function (done) {
    request.del("/api/users/" + userId)
      .set('x-auth', token)
      .end(function (err, res) {
      assert.equal(res.status, 200);
      done();
    });
  });
});
```

You run your Jest test suite from a command prompt. On Windows machines, you open a **Node.js command prompt**. You can also use GitHub Desktop and right-click your repository and select **Open in Git shell**. Once you are at the command prompt, navigate to the location of your project if needed. This is not necessary if you open a prompt from GitHub desktop.

You don't need to start up the node.js application because the Jest code will do that. You can run Jest from the local project folder in the Git shell window as follows:

```
./node_modules/ jest/bin/jest.js --runInBand --testTimeout 999999 --
collectCoverage true test/api_endpoint.test.js
```

You may need to play around with the timeout argument. It is possible to get false failures because of Jest timing-out and moving on too quickly to the next `test` block. The package.json script section in the sample application has a script you can run as follows:

```
npm run test-API
```

You should realize that, even if you are running against your local Node.js service, you are still hitting the real AWS hosted MongoDB database. You may need to open the Atlas portal and delete unwanted documents that you created through your tests. The tests included with the NewsWatcher sample application are written to clean up after themselves. Of course, you could install a local copy of MongoDB on your machine and connect against that as well.

To run against the deployed Node.js application in AWS, you can change the supertest usage to go against the production URL, such as https://www.newswatcher2rweb.com. You can see in the GitHub project that I have a few commented out lines for different ways to run the tests.

The output for the complete functional test suite looks as shown in the following figure.

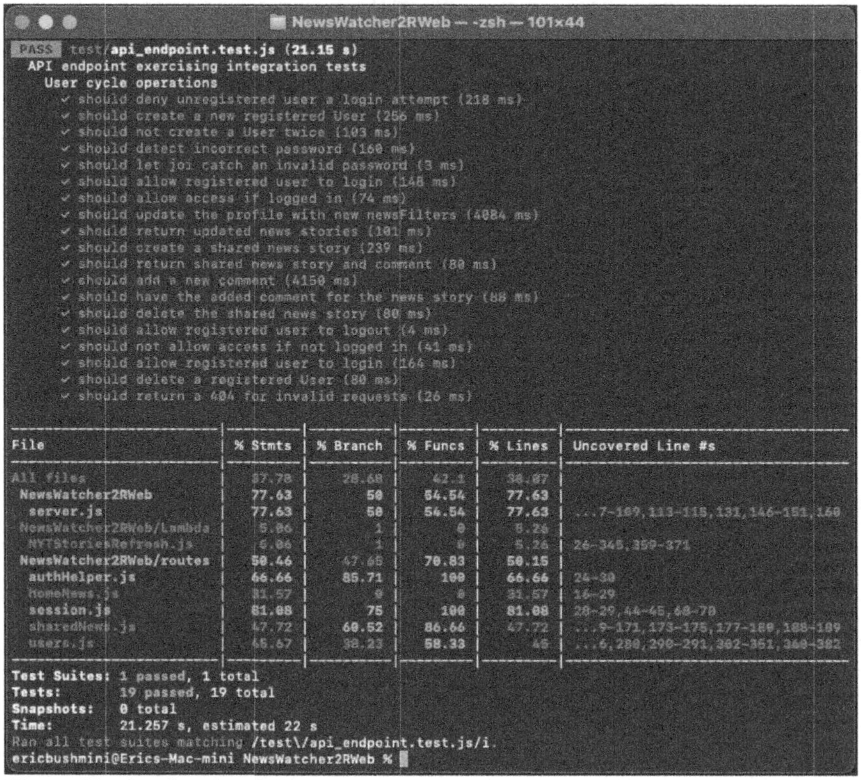

```
NewsWatcher2RWeb — -zsh — 101x44
PASS test/api_endpoint.test.js (21.15 s)
  API endpoint exercising integration tests
    User cycle operations
      ✓ should deny unregistered user a login attempt (218 ms)
      ✓ should create a new registered User (256 ms)
      ✓ should not create a User twice (103 ms)
      ✓ should detect incorrect password (160 ms)
      ✓ should let joi catch an invalid password (3 ms)
      ✓ should allow registered user to login (148 ms)
      ✓ should allow access if logged in (74 ms)
      ✓ should update the profile with new newsFilters (4084 ms)
      ✓ should return updated news stories (101 ms)
      ✓ should create a shared news story (239 ms)
      ✓ should return shared news story and comment (80 ms)
      ✓ should add a new comment (4150 ms)
      ✓ should have the added comment for the news story (88 ms)
      ✓ should delete the shared news story (80 ms)
      ✓ should allow registered user to logout (4 ms)
      ✓ should not allow access if not logged in (41 ms)
      ✓ should allow registered user to login (164 ms)
      ✓ should delete a registered User (80 ms)
      ✓ should return a 404 for invalid requests (26 ms)

---------------------------|---------|----------|---------|---------|------------------------------------
File                       | % Stmts | % Branch | % Funcs | % Lines | Uncovered Line #s
---------------------------|---------|----------|---------|---------|------------------------------------
All files                  |   37.78 |    28.68 |    42.1 |   38.87 |
 NewsWatcher2RWeb          |   77.63 |       50 |   54.54 |   77.63 |
  server.js                |   77.63 |       50 |   54.54 |   77.63 | ...7-109,113-116,131,146-151,160
 NewsWatcher2RWeb/Lambda   |    5.06 |        1 |       0 |    5.26 |
  NYTStoriesRefresh.js     |    5.06 |        1 |       0 |    5.26 | 26-348,359-371
 NewsWatcher2RWeb/routes   |   50.46 |    47.68 |   70.83 |   50.15 |
  authHelper.js            |   66.66 |    85.71 |     100 |   66.66 | 24-30
  homeNews.js              |   31.57 |        0 |       0 |   31.57 | 16-29
  session.js               |   81.08 |       75 |     100 |   81.08 | 28-29,44-45,68-70
  sharedNews.js            |   47.72 |    60.52 |   86.66 |   47.72 | ...9-171,173-175,177-180,188-189
  users.js                 |   45.67 |    30.23 |   58.33 |      46 | ...6,280,290-291,302-351,360-382
---------------------------|---------|----------|---------|---------|------------------------------------
Test Suites: 1 passed, 1 total
Tests:       19 passed, 19 total
Snapshots:   0 total
Time:        21.257 s, estimated 22 s
Ran all test suites matching /test\/api_endpoint.test.js/i.
ericbushmini@Erics-Mac-mini NewsWatcher2RWeb %
```

*Figure 81 - Jest functional test output*

# 13.4 Performance and Load Testing

Writing an application that can serve a single user is not a big challenge. The real challenge comes when multiple people are hitting the web service REST API all at the same time.

To write the NewsWatcher sample app Node API endpoint and get it to work for a single user was just two weeks of work for me. To get it to scale and handle the simulated load of many users required much longer than that to work out all the issues. You certainly don't want to wait until your big production rollout to find that your code falls flat on its face when more than one person uses it.

How are you going to accomplish testing your code at scale? The only way to accomplish this is with a test suite that can provide usage in parallel and simulate multiple users.

# Chapter 13: Testing the NewsWatcher Rest API

There are UI testing tools that can record your usage of a website and replay it. They can be replayed more than once at the same time to simulate lots of interaction happening. This might work well for some sites that serve static content and have no concept of people logging in and exercising some unique workflow in the backend service layer.

Load testing enables you to prove the scaling of your application. You also use this to measure your Service Level Agreement (SLA) values under a constant load. You can experiment by increasing the load until you find the breaking point. This will tell you the absolute peak values you can run under. To do this, you need to alter the URL to be that of the production or cloud staging environment.

Running the load testing suite can also help you test out your Elastic Beanstalk (or other hosting such as AWS ECS/EKS/Fargate) scaling strategies and topology. It will be useful in verifying the performance of MongoDB that you have worked hard to optimize. I happen to have written what I consider to be a very useful platform as a service to do scenario-based load testing.

The platform is called ApiZapi (https://apizapi.com) and has zero infrastructure onboarding for new customers because it is written using AWS Lambdas. It is set up as a serverless offering where customers only get charged for the time that tests are running. It is for API testing at high loads, and a great tool for setting up testing of an API in a scenario-based way. You can sign up for free usage and try it out. The top key points of ApiZapi are as follows:

1. Scenario-based configuration in an easy-to-use unique UI. Usable by non-developers, but powerful enough for a developer to want to use. No code or script writing is required.
2. All test runs are done with AWS serverless technology using AWS Lambda functions. Current scale limited of 100K Virtual Users, but can go higher, even to a Million! (not currently enabled, but could be if you really need it and ask me about it nicely)
3. AWS "Cloud Native" usage such as SQS, SES, Elastic Beanstalk, Route 53, EventBridge, CloudWatch, Lambda, S3, ALB, EC2 and CodeBuild. Complete multi-tenant SaaS offering.
4. Fully functional code base of about 28K lines of code. Interesting code includes:
   a. The code to calculate the fanout and starting up at the right time in the ramp up steps of VUs
   b. Use of the feature rich and performant Node.js HTTP built in Client. This in turn is built using the low level Libuv port that is written in C code to efficiently manage low level machine operating system TCP/HTTP connections with all available CPUs and report back for servicing on the single thread of Node with concurrent processing.
   c. All based on the MERN stack with JavaScript throughout all the tiers.

# 13.5 Running Lint

You can install the npm packages "eslint" and "eslint-plugin-react". These allow you to run validations that scan the code and look for syntax errors, uninitialized variables, and even specific styling that you want. I would state that it is more of a requirement than an option to have this in place.

If you are writing in a language such as C++ that is compiled up front, you get some of that checking at compile time, but you still want to enforce style. When you write code with JavaScript, you don't catch some subtle errors that could be caught until each line executes. Of course, you could also opt to write code using TypeScript, which would give you type checking.

With the use of a npm module to create a React app, you will have few added lines in the package.json that hooks up the use of eslint in your project to work with the React code. Then you will not need to install eslint or eslint-plugin-react on your own.

```
"eslintConfig": {
  "extends": [
    "react-app",
    "react-app/jest"
  ]
},
```

Then you can run from the command prompt the complete command or add a script to the scripts section of the package.json file as follows:

```
"lint": "eslint --ignore-path .gitignore ."
```

The command to run eslint then becomes:

```
npm run lint
```

This gives you some default checks and you can add an additional "overrides" section if you want to customize beyond that.

# Chapter 14: DevOps Service Layer Tips

It is a fabulous accomplishment to get the code all tested and deployed to production. Don't get too comfortable though — it is quite another matter to manage the operations of a full-stack application. This chapter will present some key skills that will make your life easier when it comes to running a 24x7 operations for your service layer.

Chances are that you will experience some type of catastrophic failure before too long. First off, you absolutely want to do everything upfront to put preventative measures in place. As the old saying goes, "An ounce of prevention is worth a pound of cure."

You also need to put a plan in place for handling a crisis when it happens. You want to be able to have all the information at your fingertips to make it possible to recover in the shortest amount of time. Some general techniques will be presented here, but you will have to come up with your own specific strategies that fit your own environment.

Let me make a brief comment about continuous integration and continuous delivery (CI/CD). If you are working on any kind of substantial project that will be going on for a while or has multiple people contributing, you need to implement CI/CD. Manually performing the tasks of building, testing, and deploying code gets old fast. Always remember that manually doing these things is prone to human error. If your DevOps process is not automated with full integration testing, then it is not complete.

Another wise saying states, "You have to slow down to speed up." You can interpret that to mean that a little investment upfront pays huge dividends over and over. This ability to centrally coordinate CI/CD greatly helps groups with an agile process iterate more rapidly. Productivity goes up because of this automation and team downtime is reduced because integration bugs are not spread across the rest of the team. It is much more expensive in time and money to catch bugs in production, so there is a huge savings here.

Also, remember the saying to "Get right and stay right." This means you have your pipeline up and running and are adding tests for all new code so that you never face a mountain of unexpected bugs coming in. Establish this practice from the start and you can stabilize early and keep it that way.

# 14.1 Console Logging

Writing messages to a console output or log file is an age-old practice that may be useful if you can avoid being overwhelmed by too much logging and then be able to interpret the information to solve problems.

Node has the Console module that is useful to write trace output to. This is available in your application already, so you don't need to use a `require` statement. Here are some useful methods you can use that are found on the console object:

| | |
|---|---|
| `log()` `info()` `error()` `warn()` | Basic method for outputting text. It comes in several different forms all with the same function signature. <br><br> `console.log("we made it here %d", someVarNumber);` |
| `dir()` | It is useful to view an object you might have in your code. There are options available for this; for example, to recurse further than the default depth of 2 levels. <br><br> `console.dir(someObject);` |
| `time()` and `timeEnd()` | To log elapsed time, you use these two methods. You will get the elapsed time when you do the following: <br><br> `console.time("start");` <br> `// some code operations that you want to time…` <br> `console.timeEnd("start");` |
| `trace()` | For showing a stack trace from the point in your code where this is called. <br><br> `console.trace("someLabel");` |
| `assert()` | This is the standard assertion usage commonly available. <br><br> `console.assert(valid, "Hi");` |

You can set up logging to go into a file that then can be looked at. Sometimes logging will point you in the general direction and then you can use the debugger to further diagnose an issue. As another option, you can send every log message to a special collection in MongoDB. You can set up MongoDB documents to have a time to live (TTL) before they are automatically deleted or set the collection as capped.

# 14.2 CPU Profiling

The V8 engine can provide you with CPU usage reports. The simplest way to do this is to launch your Node process with the profile flag set and, after running your test code, you stop the process, and you will then have a file available to view. You will launch Node as follows on your local machine:

```
node --prof server.js
```

If you then run some code, such as your load testing suite, you can get some idea of where your code might be spending most of its time. Once you have run your test code for a bit, you stop the Node process and a file named something like "isolate-000001D213C4D490-v8.log" will be saved.

```
node --prof-process isolate-000001D213C4D490-v8.log > processed.txt
```

Open the txt file in an editor and you can inspect the output. You will see reports that look as follows:

```
Statistical profiling result from isolate-0x118008000-4600-v8.log, (6254 ticks, 38 unaccounted, 0
excluded).

 [Shared libraries]:
   ticks  total  nonlib   name
     37    0.6%           /usr/lib/system/libsystem_malloc.dylib
     24    0.4%           /usr/lib/system/libsystem_pthread.dylib
     18    0.3%           /usr/lib/system/libsystem_c.dylib
     15    0.2%           /usr/lib/libc++.1.dylib
      8    0.1%           /usr/lib/system/libsystem_platform.dylib
      5    0.1%           /usr/lib/system/libsystem_kernel.dylib
      2    0.0%           /usr/lib/libc++abi.dylib

 [JavaScript]:
   ticks  total  nonlib   name
     59    0.9%    1.0%   LazyCompile: *_encipher /Users/ericbushmini/Documents/GitHub/
NewsWatcher2RWeb/node_modules/bcryptjs/dist/bcrypt.js:959:23
     50    0.8%    0.8%   LazyCompile: *listOnTimeout internal/timers.js:505:25
     22    0.4%    0.4%   LazyCompile: *ensureMinPoolSize /Users/ericbushmini/Documents/GitHub/
NewsWatcher2RWeb/node_modules/mongodb/lib/cmap/connection_pool.js:334:27
     15    0.2%    0.2%   RegExp: ^((}(?:[0-9]|[1-9][0-9]|1[0-9][0-9]|2[0-4][0-9]|25[0-5])|(?::(?:
[0-9a-fA-F]{1,4})){1,7}|:)))(%[0-9a-zA-Z-.:]{1,})?$
     10    0.2%    0.2%   LazyCompile: *getValidatedString /Users/ericbushmini/Documents/GitHub/
NewsWatcher2RWeb/node_modules/bson/lib/parser/deserializer.js:641:28
      8    0.1%    0.1%   LazyCompile: *Module._load internal/modules/cjs/loader.js:725:24
      7    0.1%    0.1%   LazyCompile: *processTimers internal/timers.js:485:25
      6    0.1%    0.1%   LazyCompile: *remove internal/linkedlist.js:15:16
      6    0.1%    0.1%   LazyCompile: *_key /Users/ericbushmini/Documents/GitHub/NewsWatcher2RWeb/
node_modules/bcryptjs/dist/bcrypt.js:1096:18
      5    0.1%    0.1%   RegExp: ; *([!#$%&'*+.^_`|~0-9A-Za-z-]+) *= *("(?:[\u000b\u0020\u0021\u0023-
\u005b\u005d-\u007e\u0080-\u00ff]|\\[\u000b\u0020-\u00ff])*"|[!#$%&'*+.^_`|~0-9A-Za-z-]+) *
```

*Figure 82 - Sample statistical profiling result*

PART II: The Service Layer (Node.js)

You can also utilize the V8-profiler using a Node module to start it up in your code if you set up route handlers and want to profile things in production. The V8 engine, which is one of the components that Node.js is built on, lets you run an analysis of production running code. Profiling data can be sent to an external file that you can later open and view the results with the F12 Chrome development tool (under More Tools, open JavaScript Profiler). Here is the code to accomplish a profile run:

```
var fs = require('fs');
var profiler = require('v8-profiler');
profiler.startProfiling('1', true);

...do some processing...

var profileResult = profiler.stopProfiling();

profileResult.export()
  .pipe(fs.createWriteStream('profile.json'))
  .on('finish', function() {
    profileResult.delete();
  });
```

An easier technique to profile code is starting node from the command line as "node -inspect server.js". Then open Chrome and go to "chrome://inspect/#devices" and in that page click on "inspect". This will open a DevTools app.

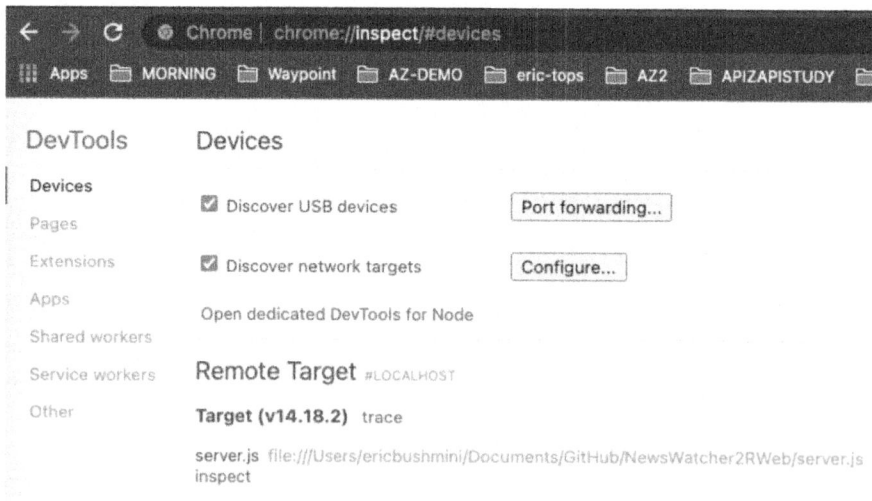

Figure 83 - The DevTools application

252

If you then open the React app or otherwise exercise the HTTP API, you can then go to the **Profiler** tab of DevTools and can see what is taking most of the processing time.

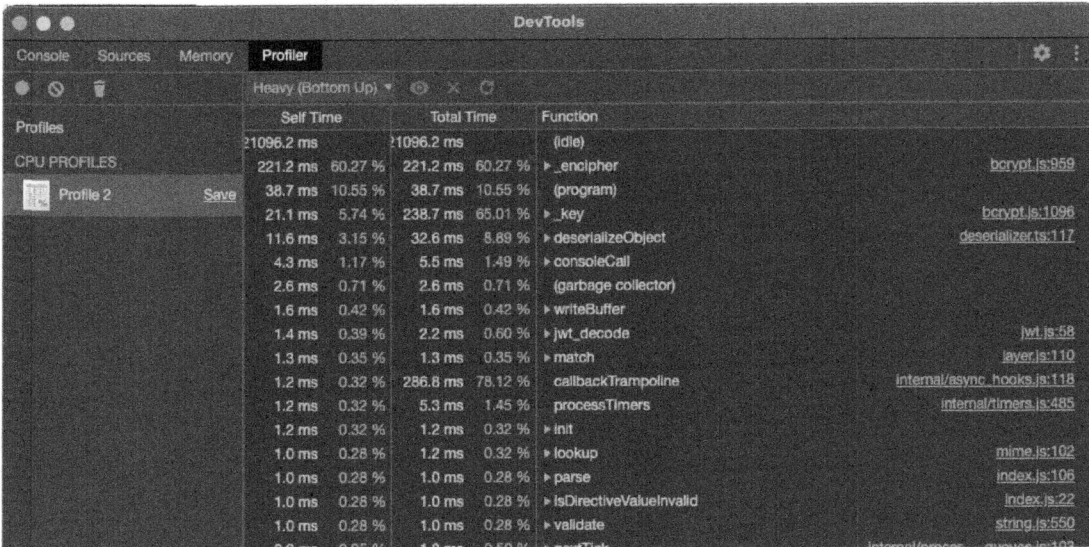

*Figure 84 - Chrome developer tool V8 profile report*

With profiling turned on while a load test was run, I was able to produce a report that showed a few issues I could address. One of them turned out to be the use of `bcrypt` for password hashing. As seen in the profile analysis, you see a bottom-up view of the calling tree. If you expand until you see your functions, you can then understand where the call is being made from. Here, you see that the `bcrypt.hashSync()` function call took 19% of the time.

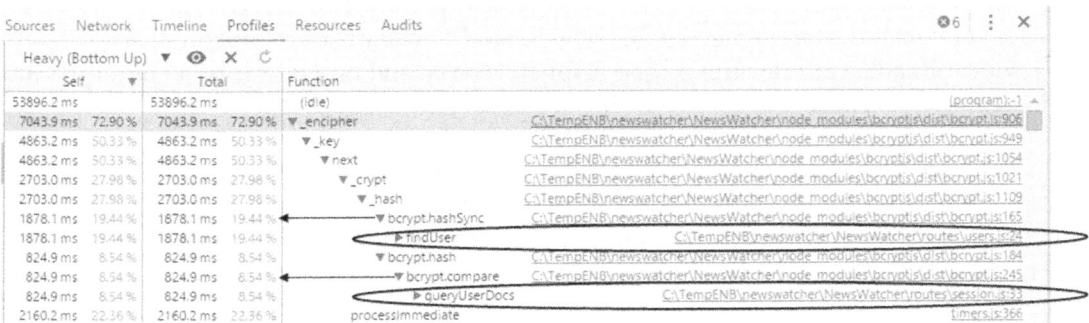

*Figure 85 - Profile drill-down*

Here is the code for what you see listed for the first issue:

```
findUserByEmail(req.db, req.body.email, function findUser(err, doc) {
    ...
        passwordHash: bcrypt.hashSync(req.body.password, 10),
```

It was obvious, given the name of the function, that I was not using the async version of the hash-generation function. Go figure, I should have known that `hashSync()` was not a good idea to call, as it is a synchronous blocking operation. As you know, you want to minimize CPU usage on the main Node thread. Checking the documentation, I found the async version and made the change to use it:

```
bcrypt.hash('bacon', 8, function(err, hash) {
});
```

*Note: It is helpful to give names to all anonymous functions, otherwise you will just see a lot of "anonymous" functions and it is harder to pinpoint which functions are in the profile listing.*

# 14.3 Memory Leak Detection

In a managed language framework, you don't directly allocate memory and subsequently free it up. You can do a `new` and `delete` of an object, but you still don't have control over that memory, such as the actual reclaiming of it or doing things like having memory pointers into it. Instead, a garbage collector (GC) keeps track of memory references, and the GC decides when to run and when memory can be recycled.

The truth is that with garbage collection running, you can still run into memory leaks that will eventually cause your application to run slowly, to completely freeze, or to crash.

You can obviously create a memory growth problem if you have something as simple as an array that you continually push data into and never free up. If you hold on to an object reference permanently after you no longer need it, the GC will not ever reclaim it. A thorough code review of callbacks, closures, constructor functions, and arrays can be a starting point to finding memory leaks.

Ultimately, your brain might not be able to trace through all the intricacies of your code and you will need to take memory snapshots that can be compared across time. Node.js applications are always built with several, if not dozens, of downloaded modules. You must be suspicious of those as well since they might contain memory leaks.

You can watch the OS reporting of memory for your Node.js process over time and see what kind of graph you have. If you see an ever-increasing amount of memory being taken up, you can then dive in and investigate. It is not unheard of for people to resort to restarting their Node.js processes every day just to circumvent any memory leak problems. You may, in reality, not have a leak at all if you can see the GC kick in and do its job over time. Compare memory snapshots over a 24-hour time span.

You can use the V8-profiler module that was previously used for CPU profiling to take memory snapshots. You can likewise have the output files viewed in the Chrome debugger.

```
var fs = require('fs');
var profiler = require('v8-profiler');
var snapshot = profiler.takeSnapshot();

snapshot.export()
  .pipe(fs.createWriteStream('snapshot.json'))
  .on('finish', snapshot.delete);
```

*Note: Remember to never run your CPU profiling at the same time as you take heap snapshots. The overhead memory usage for the CPU profiler will inundate you.*

It is probably easier to use the DevTools application to collect and view the memory usage. There is a **Memory** tab in the DevTools to click on and take snapshots and see the difference over time and determine if you have any memory leaks.

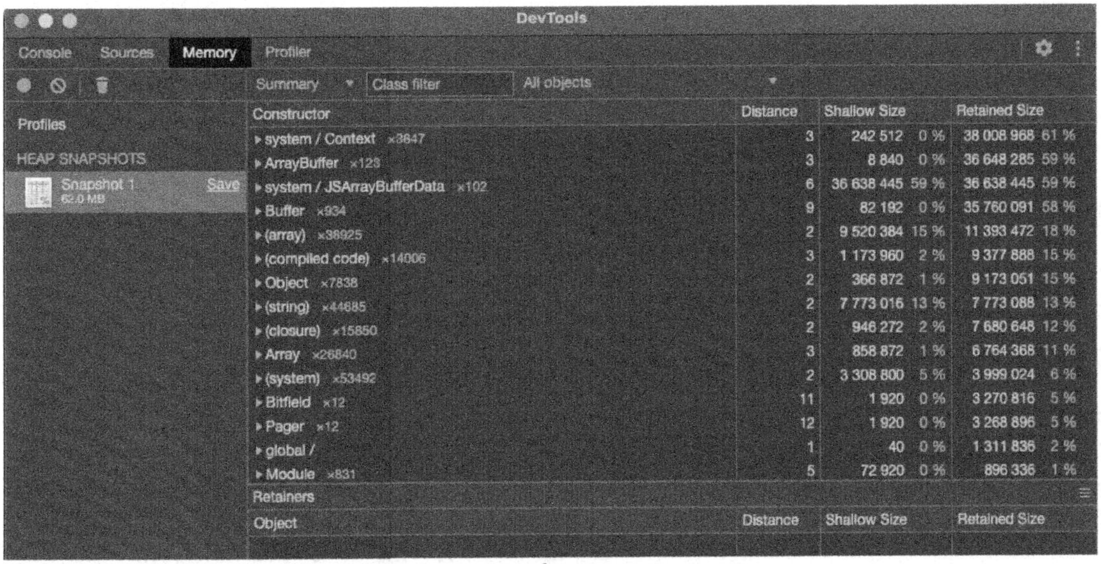

*Figure 86 - Chrome developer tool with memory snapshot*

The **Distance** column shows you how many steps removed from the root object the memory reference is. You can usually assume that the object with the shortest distance is the one causing a memory leak. The **Shallow Size** is just what this one usage is taking. **Retained Size** gives you all the space that would be freed up that this object is referencing and thus holding on to. This is only true if those objects are also no longer referenced by anything else.

There are different views you can try out, such as Containment, which helps you also view low-level memory internals.

# 14.4 CI/CD

Do you work on a project where the tasks of building, testing, and deploying code are all done manually? Why is that? From my experience, the reason teams don't automate manual tasks is typically because they either lack incentive or they lack the knowledge. Lacking incentive is rather a poor excuse. There is plenty of incentive if you honestly look at the return on investment of implementing a CI/CD process. If your DevOps process is not automated with full integration testing, then it is not complete.

I once heard a speaker at a conference say to developers, "It ain't done until it is automated." He was saying that you need to automate everything from the check-in, testing, and integration through building and deploying. In this book, I will say the 'D' in CD stands for delivery. This means that code is served up and ready to be deployed. We will, however, take the process all the way through to delivery and can say we implemented a CI/CD+ solution.

As already stated, "You have to slow down to speed up." In the introduction to this chapter, I incorporated that statement which I've often heard to emphasize how a little investment upfront in time and/or costs pays off in multiple ways. To summarize would be to say - the ability to centrally coordinate CI/CD is key to an agile environment, increases productivity, reduces, or eliminates tea downtime, and streamlines production. That's incentive enough.

Now let's assume that it really comes down to knowledge. After reading this chapter, you will have gained enough knowledge to make changes to automate manual tasks and to implement CI/CD.

## Many tools and ways to accomplish CI/CD

Implementing CI/CD can, of course, be a complex undertaking. No one article or book can tell you everything you need to know because there are so many tools out there. No one solution will work for all projects. All projects are unique in their code structure and frameworks used. We will only scratch the surface and only do so for a Node.js project in a very narrow niche. We also narrow our focus down by relying on a PaaS solution.

If you choose not to go the route of setting up a complete CI/CD tool, you can still automate all of this on your own and launch things manually. For example, you can implement a series of Gulp tasks that will do a lot of what a CI/CD system does.

***Note:*** *You can do everything in your DevOps environment through the command line. Once again, I have chosen to take the route of using UIs where available.*

# Developer tested code

As a preparatory step, a developer will have prepared some code and privately tested any new feature before submitting it to the CI/CD process. The testing should be as thorough as possible in the context of the rest of the integrated system. If that is not possible, then it can be tested as an isolated unit, perhaps with stubbed out or mocked functionality injected. The testing should be runnable by a developer by typing "npm test", which will run lint and then all tests for that part of the code.

Once the developer has had their own private verification completed, the developer can do a code check-in that will then be staged to go into the pipeline for consideration. If you are using Git and GitHub, the code will be pushed to the branch you designate for the CI/CD to be triggered from. You will not necessarily push it into the master branch. The master branch should only be used for the code that is running in production. You can push to master and tag that code, yet not deploy it to production. Then you can do final testing and still have manual acknowledgment to push that to production.

CI/CD is all about increasing your iteration speed and the quality of everything written. Of course, you must provide high-quality comprehensive test suites to achieve this. Once your code is ready to commit, then the pipeline workflow of CI/CD takes place. So, what does the cycle look like? Let's go through each of the steps. Here is a simple diagram to show you the pieces that will make up a simple Node.js CI/CD system:

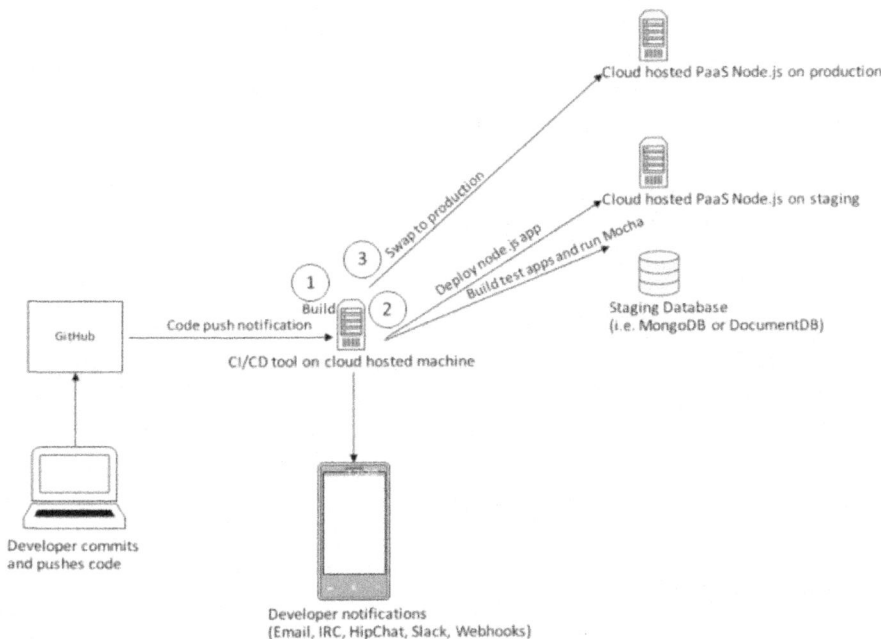

*Figure 87 - Cycle of CI/CD workflow*

# AWS tools

A lot of companies make tools to help with doing CI/CD. In the past, I have loved using the open source tool Jenkins and it is a good choice. There are even companies hosting Jenkins in a PaaS environment if you don't want to install and manage it yourself. GitHub Actions should also be considered.

This book will teach you about AWS CodePipeline as an overall orchestration tool to mage the build, test, and deployment steps. The build and test are done with AWS CodeBuild. Deployment can be done with CodeDeploy to deploy to ECS for a Microservices architecture. In our case, we use Elastic Beanstalk. Since there is a direct integration to that in CodePipeline, so CodeDeploy will not be used.

# Step One: Install

Once the code is committed into the source code repository, then the code needs to be staged in the overall application. CI/CD tools listen to GitHub and will kick off their processing pipeline immediately upon seeing a code push. A CI/CD system can see the push and do a local clone of the branch, to have a freshly integrated code version in the CI/CD system.

All code dependencies such as NPM modules need to be updated in this clone and that is part of what the CI/CD will kick off. This is done with a npm install command. Since we want to get a production environment up and running, we will have to do an extra npm install to get the DevDependencies since those don't get installed if the environment variable NODE_ENV is equal to "production". I will explain this again as we go through the steps.

# Step Two: Build

Once all the dependencies are installed, we can do any build and test steps. In our case, the Node.js code is set to go, but there will be a build step required for our React code so things such as Lint can be run. If Lint passes, then the Jest tests can be run.

Comprehensive testing is central to getting CI/CD working. This step could run a thorough suite of tests at all levels, such as Unit, integration, feature, load, performance, and UI automation testing. A perfect pass is expected, and you will get generated reports to show that all went well and produce code coverage reporting and have memory leak detection.

# Step Three: Post-Build and deploy

This is the final step that can generate the files necessary to deploy and do the deployment if you wish. Since we installed development dependencies such as Lint and Jest, we can back those out and just have the dependencies needed for the production environment. The deployment can be automatic, or an email can be sent to a specific person to do manual acknowledgment to approve it. If your team is in favor of Testing in Production (TIP), you can deploy to a reserved hidden portion of your production environment.

You should have a testing environment that is as close to production as possible. One thing to remember is to also have in place a dedicated database for testing purposes hosted in PaaS that is always available. You do not need to continually deploy there if it is simply the location of the data, and nothing needs to be preconfigured. You can run some cleanup script as part of the CI/CD step here if that is necessary and some pre-population script if you require certain documents to be in place. PaaS offerings for MongoDB and others are available as document-based databases.

If the testing fails, you will be notified, and the CI/CD process will not proceed further. You will want failures automatically entered in an issue tracking system to officially track and resolve issues.

# NewsWatcher CI/CD with AWS CodePipeline

Here is a quick introduction on how to set up a CI/CD process with AWS CodePipeline. To start with, make sure you already are signed into GitHub. Then in the AWS portal, select **CodePipeline** from the **Services** list. You can find the button to create a new pipeline. One of the steps will ask you for a "source provider" and you will select GitHub. Then you can click to connect to GitHub and basically through Oauth2 establish the permission to continue to connect and interact with your GitHub account to pull from the chosen repository. Here is what I have in the buildspec.yml file:

```
phases:
  install:
    commands:
      - echo Installing Node Modules...
      - npm install
      - npm install --only=dev
  build:
    commands:
      - echo Build started on `date`
      - echo Building React Web application
      - npm run build-react
      - echo Performing Test
      - npm test
  post_build:
    commands:
      - echo Final build, without devDependencies
      - rm -rf node_modules/
      - npm install
artifacts:
  files:
    - '**/*'
```

PART II: The Service Layer (Node.js)

Once you are ready to go, you can merge and push a code change up to GitHub and watch in the AWS console with CodePipeline as each step progresses until you see the code pushed out to production. Here is a screenshot showing the full process after it is run. If you happen to have an error in the CodeBuild step, then you can open that up and see the details and see how to fix the error.

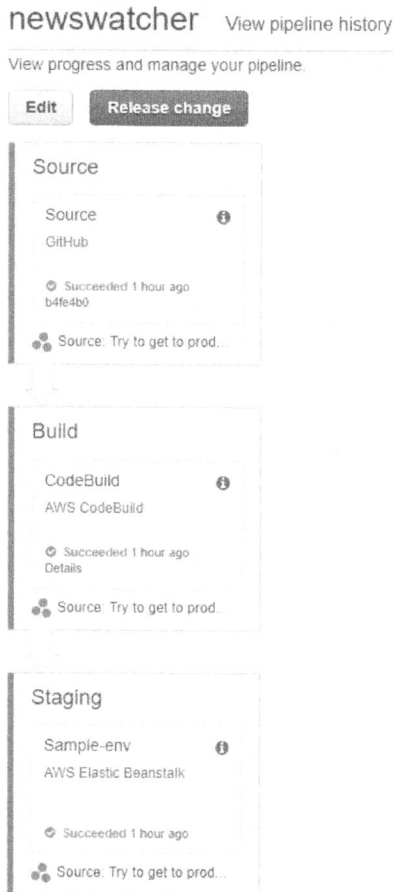

Figure 88 – AWS CodePipeline usage

You can also include the build, test, and deployment for a Lambda function. You would need to add lines like the following:

```
cd Lambda
npm install --production
npm run zipLambda
aws lambda update-function-code --function-name NYTStoriesMgmt --zip-file
fileb://lambda.zip
```

# 14.5 Monitoring and Alerting

The AWS Elastic Beanstalk management portal has a Monitoring page with which you can view key machine performance metrics. You want to open the portal and look at what is available to be monitored. Not only can you look at trending charts, but you can also set up alerts that send you emails or even text messages in the case of thresholds being crossed. Here is the Monitoring page:

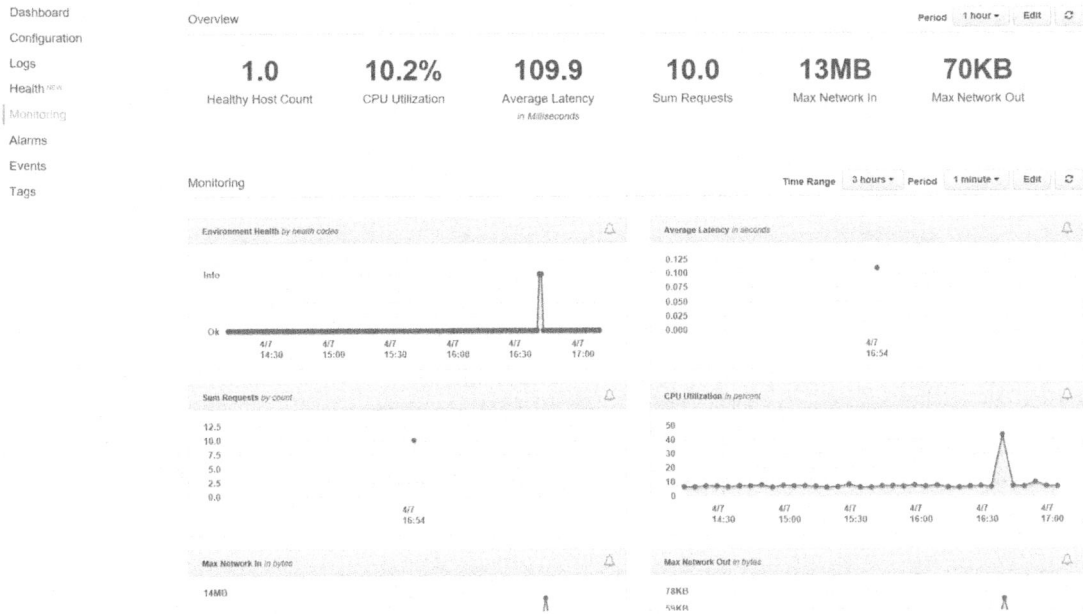

*Figure 89 - Elastic Beanstalk monitoring page*

You can click **Edit** and select what you want to be graphed out. You can also click the alarm bell icon and create an alarm that notifies you if you crossed some threshold. Alternatively, you can open the CloudWatch management console in AWS and explore similar capabilities for metric viewing, logs, events, and alarms. The following image shows the editing of settings for an alarm,

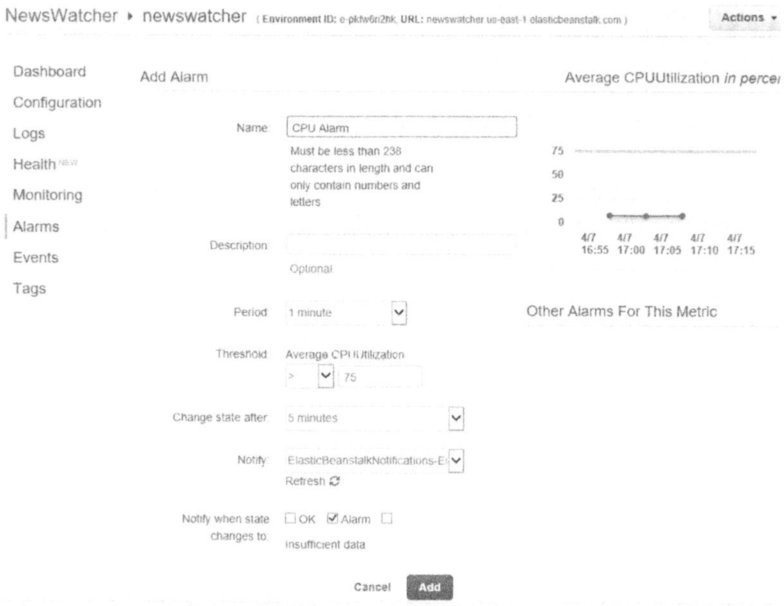

# PART II: The Service Layer (Node.js)

*Figure 90 - AWS Alarm setting*

Be advised that the Node.js process can crash and, if that happens, it needs to be restarted. In an IaaS environment, you will be responsible to provide some mechanism to detect this and restart your Node.js process. NPM packages such as "forever" and PM2 do this. AWS Elastic Beanstalk uses Nginx and the restarting is handled automatically for you. This is another example of how PaaS can make your life easier when it is really done right.

If your application is not available, reliable, and performant, then your customers and your business will suffer. The goal of monitoring and alerting is to maintain application availability, reliability, and performance. You do this by implementing Application Performance Monitoring (APM). APM allows you to discover problems before anyone else does and then enables you to achieve resolutions in the least amount of time.

To begin implementing an APM strategy, you need to instrument your code and surface events, logs, and metrics. Then you use an APM tool to chart out performance metrics and set up alarms. You need metrics so that you are not flying blind. A pilot can fly a plane in the dark because of instrumentation and telemetry. The last thing you want to have to do is remote into individual servers and start poking around to search for a "needle in a haystack." Instrumentation and monitoring are the only way you will be able to scale and survive.

Check out a tool such as New Relic for a complete suite of capabilities from log monitoring to machine resource monitoring to transaction monitoring, to client-side monitoring. This literally only takes a few minutes to get set up. The transaction monitoring gives you a deep trough through all the stacks of code for any given transaction.

262

# Transaction monitoring with AWS X–Ray

AWS offers X-Ray to instrument your code and have the telemetry viewed in a portal that allows you inspect each of the traces that your application is processing. A trace consists of information for what is occurring in your service layer. For example, you can see a trace for every HTTP route endpoint and each of the verbs that are being handled.

To get X-Ray tracing to work with your Elastic Beanstalk application, you need to turn on a setting. Go into the Elastic Beanstalk console and then into the Configuration settings for the Software Configuration and check the box to enable the X-Ray daemon. This will run the agent on the EC2 machines that collect the data and forward it to the location it can be collected for viewing and alerting. This will enable X-Ray to get at machine resources.

*Note: NewRelic is a tool that has been around for some time and is a very comprehensive solution for gathering machine resource usage and transaction instrumentation. Like AWS X-Ray, you can add code to your Node.js project to instrument the sending of telemetry. You can customize what is sent back and add custom metrics and events. You can also add a script to your HTML client to instrument all usage from the client-side. NewRelic has a rich capability to set thresholds and alert on them.*

Costs are associated with using X-Ray, but you must process a lot of tracing to incur any cost. Look over the AWS documentation for an explanation of costs. You can control what X-Ray captures on a route-by-route basis. You can completely ignore a route or set it up to only capture a certain percentage of traffic.

X-Ray understands how to collect tracing information for transactions that are handled by your Express usage in Node through the aws-xray-sdk npm module.

*Note: To download the module, you need to be running your machine console as admin. Then you can run "npm install aws-xray-sdk –save".*

Here is code to use AWS X-Ray in Node JavaScript:

```
var app = express();
var AWSXRay = require('aws-xray-sdk');
app.use(AWSXRay.express.openSegment('NewsWatcher'));

// All your API routes
app.use('/api/users', users);
app.use('/api/sessions', session);

app.use(AWSXRay.express.closeSegment());
```

PART II: The Service Layer (Node.js)

Then you can see the traces when you go to X-Ray in the AWS console. You can also see all of the AWS resources in your service map. On the map, you can click on a trace and see its details. Here is what the trace view looks like.

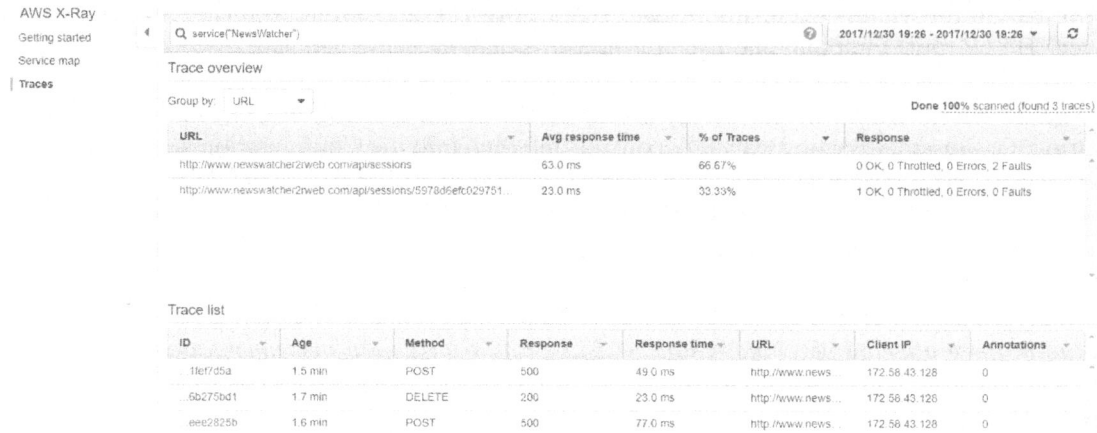

*Figure 91 – X-Ray trace*

If your code has an unhandled exception, then you can see the stack trace in the details of the trace. You can also add extra information to any trace as follows:

```
var segment = AWSXRay.getSegment();
segment.addMetadata("errorHandler", err.toString());
```

If you are using the AWS SDK in your code, you can add some code to make it aware of those calls, and then the usage will be instrumented.

It is possible to put in code to measure any part of a transaction trace, such as when you want a further breakdown of some processing code to see its timing in the overall trace time. This can be synchronous or asynchronous code.

When there is an incoming request to your Node.js service, such as a POST to an API endpoint, a single trace id is assigned to that request. No matter what happens up to the very end of that request, every sub-part of work that goes on shares that same id. That means that all the work in your code calls to other backend HTTP services. Any use of other AWS resources is all tied together and viewable under that single id in the X-Ray console.

264

# PART III: The Presentation Layer (React/HTML)

Part Three of this book will teach you about the presentation layer of a three-tier architecture. In doing so, the sample application will be extended to where the UI is functional and interacting with the services layer. Before reading this chapter, you should obtain a basic understanding of HTML.

In this part of the book, I present the technology of React as a framework. You will see that it fits nicely into the overall architecture, to bind data from the service layer web service into a UI. The UI will be rendered as an HTML website and be made "responsive" so that it also looks good on a mobile device.

We already build the middle service tier code and will need to make use of that for the serving of the React JavaScript file and associated static resources. Thus, the Node.js application performs the dual role of servicing not only the HTTP/REST API, but also serves up your React HTML and JavaScript files, etc., to a browser as a SPA application.

*Note: It is not the intention of this book to be a comprehensive guide to UX design or SPA web design. I will only touch on some concepts of these topics and then stick to a narrow technology presentation. Many books are devoted entirely to React and React Native, so you can see that I only cover the basics of how each works to get you going.*

# Chapter 15: Fundamentals

We can now go over the fundamental concepts of the presentation tier in our three-tier architecture. To start out, we will see what capabilities are essential and find a list of questions to consider when doing your code design. We can then get into the specifics of React as a JavaScript UI framework and learn how it will be used to implement the NewsWatcher sample application. One of the reasons I picked React as a technology, is because it fulfills the needs of this top layer of the application architecture and does so with the JavaScript language.

# 15.1 Definition of the Presentation Layer

Any application that requires user interaction needs a presentation layer. Humans need a presentation layer to view data and allow them to also input data. For example, an online bookstore offers a list of books for users to browse through. Each book may contain a photo of the book and data associated with it, such as the title, author, description, publication date, and cost. Input gathered from the user are such things as book orders, reviews, and customer service questions.

## MV* and SPA designs

You need to employ the techniques of abstraction and componentization in all the layers of an architecture, and this is also true with coding in the presentation layer.

One of the benefits of choosing a framework for your presentation layer is that most frameworks are set up to employ some type of MV* pattern that lends itself to an organized set of components that make up your code.

Become acquainted with MV* design patterns and with what a SPA design entails. Several excellent books are available on MV* and Single Page Applications along with plenty of online material to study. Many things about SPA designs make them a great choice today.

*Note: If you are concerned about Search Engine Optimization (SEO), be aware that you may need to switch some of your page rendering to server-side and not be fully rendered as a client-side SPA. This is sometimes required for search engines to index your site. React can be set up to give you the balance you need for this. I've devoted a short chapter on how to do this.*

# Presentation layer planning

Many decisions go into creating a presentation layer. The first step involves planning what type of interactions your users may want to take. You will want to make an initial sketch of the UI of your application early on before coding anything. You can do this concurrently as you design the service layer. It is wise to pursue a simultaneous bottom-up and top-down approach.

Knowing what operations and workflows you need is another important early step in fleshing out a presentation layer. The following questions will help you determine the design of a presentation layer:

- Have you done any of the following — sketching, prototyping, storyboards, surveys, contextual inquiry, stakeholder interviews, A/B testing, wireframes, sitemaps, personas, scenarios? How about human interaction, usability, and accessibility studies?
- What are your data security and privacy requirements?
- How do people sign in and become authorized?
- Are there multiple steps that are progressively revealed one after another?
- Is there a need for a customizable UI?
- What are your globalization and localization requirements?
- What devices are you targeting, such as desktop and mobile platforms?
- How can you keep data presentation to a minimum to avoid overwhelming the user?
- What form is best for presenting data?
- What business need can be met or accomplished?
- What are the accessibility requirements to accommodate disabilities?
- What are the navigation levels of the UI? Did you map out how the navigation works?
- Do you need a user feedback mechanism?
- Do you have offline requirements?
- How is state stored in the client? Is it centralized across an all the view hierarchy?
- How will the UI be deployed and updated?
- How will users enter data? What tests are needed to validate the data?
- Can you map out the multi-step data entry forms and show the branching conditions?

Carefully document your answers to these questions. Before you roll anything out into your production environment, have experts review everything.

React is a great choice as a UI framework and fulfills all the needs of the presentation layer of an application architecture. It uses JavaScript, so it fits in perfectly with the overall development stack we have been pursuing thus far. You now can learn the specifics of React and see a practical implementation of how it is used in the NewsWatcher sample application.

# 15.2 Introducing React

React is a framework that gives you the ability to dynamically render HTML content in a browser web page. The interesting twist is that you write JavaScript alongside your HTML markup. The JavaScript code set up logic to determine what is rendered and can do HTTP/REST requests, data binding and navigation, and much more. The overall capabilities of React provide the mechanisms that enable you to build a SPA, a traditional server-side rendered website, or a hybrid.

Other frameworks have achieved some popularity, such as Knockout.js, Ember.js, Angular, Vue.js, and Backbone.js. However, React is my choice in this book as it has many things going for it in its design, is widely adopted and is officially sponsored as an open-source project by a large viable company (Facebook). You can be up and running on the desktop, mobile web, and even produce native mobile applications in a short amount of time.

*Note: You see me referring to React as a framework. I do so in the larger sense. React, along with your chosen additions for other functionality, provide a framework that you can build your application presentation layer with. That is the point of a framework; it is extensible. On its own, React is not as prescriptive and feature-capable as some frameworks, such as Angular. It is up to you to combine React with other libraries of your choice to provide all the features you need.*

Because of its ability to bind and affect DOM elements, you will no longer need to utilize libraries such as jQuery. React frees you from some of the laborious code you used to have to write to do DOM manipulation.

React makes HTML dynamic so that data values flow back and forth for you in your code. To use React, you write component files that render to the DOM. React itself does not provide control elements or styling for you, as that is left to the HTML markup and other libraries such as material design, Bootstrap, and many more that are available.

The React library is consumed by placing a script element inside your HTML file that pulls in React as JavaScript to execute it. You also can locally bundle React through Webpack through tooling that is easy to install and get started with.

There are also third-party UI design studios tools, where you can drag and drop elements, play around with them, and have React code generated and edited. This is handy to be able to visualize a component in an isolated way to save you time. Otherwise, you need to see the component in an overall application. Look into tools such as Storybook and React Styleguidist.

*Note: The Node.js Express framework has the concept of serving up templates of "HTML-like" files and binding data to them on the server-side so that they arrive all filled out on the client-side. Express supports many template formats such as Jade, EJS, mustache, and handlebars. Instead of using those, I give the client-side React to request data through a REST API with JSON and then doing the binding on the client-side. This alleviates the back-and-forth HTML page requests. This is like a native mobile application. You also can stick with HTML rather than learn a new template markup syntax, such as Jade.*

# 15.3 React with Only an HTML File

You could incorporate the React library through a script element in an HTML file. *This is not a good idea*, however I show this to help you understand what ultimately must happen to bring any code into an HTML page. React is just a library that runs as a script in a browser that exists in an HTML rendered page. You can place the following text in a file and open it with your browser and see that it works.

```
<!DOCTYPE html>
<html>
  <head>
    <meta charset="UTF-8" />
    <title>Hello World</title>
    <script
src="https://unpkg.com/react@18/umd/react.development.js"></script>
    <script src="https://unpkg.com/react-dom@18/umd/react-
dom.development.js"></script>
    <!-- Don't use this in production: -->
    <script
src="https://unpkg.com/@babel/standalone/babel.min.js"></script>
  </head>
  <body>
    <div id="root"></div>
    <script type="text/babel">
      const root = ReactDOM.createRoot(document.getElementById('root'));
      root.render(<h1>Hello, world!</h1>);
    </script>
  </body>
</html>
```

To get started with React development, you will want to install the necessary tooling on your development computer where you can build of your application to produce a bundled file that has everything in it from your React application to run. This means React is bundled as a .js file and served as a static file. Using React, you end up with a generated HTML file that looks something like the following:

```
<!doctype html>
<html lang="en">
<head>
  <meta charset="utf-8" />
  <link rel="icon" href="/favicon.ico" />
  <meta name="viewport" content="width=device-width,initial-
scale=1" />
  <meta name="theme-color" content="#000000" />
  <link rel="apple-touch-icon" href="/logo192.png" />
  <link rel="manifest" href="/manifest.json" />
  <title>React App</title>
  <script defer="defer" src="/static/js/main.4567589a.js"></script>
  <link href="/static/css/main.073c9b0a.css" rel="stylesheet">
</head>
<body><noscript>Enable JavaScript to run this app.</noscript>
  <div id="root"></div>
</body>
</html>
```

The JavaScript file that you see being pulled in with the script tag is generated in your build process on your machine and contains all the needed React code and your own code that is necessary to load and run your SPA. The next section will walk you through how to arrive at the file shown above in the simplest possible way.

# 15.4 Installation and App Creation

If you want to go through all the work of installing each of the modules and tools for doing React development on your own, you certainly can do that. To do so, you go to the official React website and follow their downloading instructions. You could do npm installs of modules such as "react" and "react-dom". Then you need to install tools for cross-compiling and bundling. You install tools from npm such as webpack, Babel, Autoprefixer, ESLint, Jest, and other tools. The following two terms describe what some of these tools accomplish.

**cross-compiling:** Allows you to use the latest JavaScript syntax and have it turned into JavaScript that can run in older browsers.

**bundling:** This is how to gather all the files together for much better performance and a controllable loading capability. Otherwise, you must reference the minified JavaScript files in your HTML and take the hit on the initial page load. Going with bundling also gets you prepared for being able to deploy as a native application on a mobile device as that is the only way to accomplish that.

Instead of installing the tooling on your own as described above, you can simply run a tool that populates a React project for you with some example code and all the tooling necessary.

# Using create-react-app

The task of getting the initial project files in place to have the traditional "Hello World" application running may seem a little daunting at first. The best approach is to use the `create-react-app` utility.

You can run this, and you will have an application ready to run. What we will do is create the React application in a separate folder from the Node.js application. Then simply copy over the necessary folders and files from the React project over into the Node project. This is the best approach to take for now. Run the following on the command line in a completely new directory. This will create your basic React application that you can pull from:

```
npx create-react-app my-app
```

Copy over the "public" and "src" directories. Look in the package.json and the .gitignore files and bring over what is missing in the ones in the Node.js files. While you are editing the .gitignore file, add ".chrome" as a line, as this will be needed later. For the package.json file that means just bringing over a few of the script commands. I renamed them so as not to conflict with anything I had for my Node.js project.

```
"start-react": "react-scripts start",
"build-react": "react-scripts build",
"test-react": "react-scripts test --env=jsdom"
```

You also need to copy the Node.js package.json lines. Then you can run npm install and have those ready to use. Here is what would be in the newly created React package.json.

```
"dependencies": {
  "@testing-library/jest-dom": "^5.16.4",
  "@testing-library/react": "^13.0.1",
  "@testing-library/user-event": "^13.5.0",
  "react": "^18.0.0",
  "react-dom": "^18.0.0",
  "react-scripts": "5.0.1",
  "web-vitals": "^2.1.4"
},
```

To run the React application, you can now run the command "npm run start-react" and that will be served up for you on your local machine. If you run that, you will see the following launched in the Chrome browser:

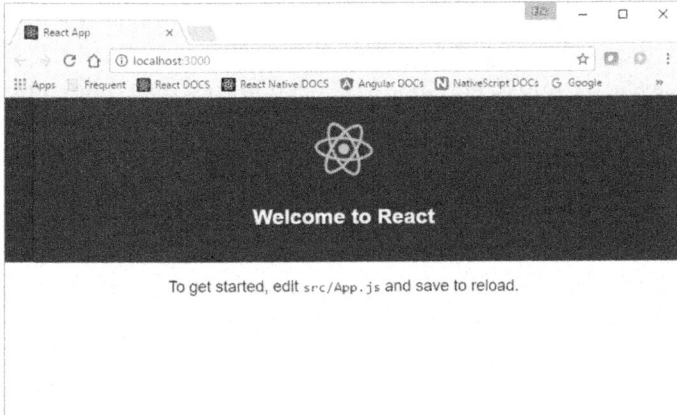

*Figure 92 - Starter project screen for React install*

**Note:** *Node.js has its own commands to get an initial application structure set up and configured on your development machine. While it can be confusing for what to do for your initial project creation, you have just seen that there is also a command line utility to create a React application. There is a lot more to a React install, however, especially if you use a framework setup such as the create-react-app utility. Thus, I had to essentially get each project set up independently and then bring one over to the other. There are other utilities out there such as Yeoman that can give you the complete scaffolding for everything in one single initialization. There is also a generator named react-fullstack that can be used as a starting point for your project. Of course, you can also clone anyone's GitHub application, such the one for this book, and have everything ready to use.*

# Serving up React from Node.js and Express

Before getting into the heavy coding with React, make sure that the initial React page is served up correctly from the Node.js service. The details of making your Node.js Express application serve up the React index.html and JavaScript code as a web page are extremely simple. Here are the bare minimum lines of code you need to do that.

```
const express = require('express');
const path = require('path');
const app = express();

app.get('/', function (req, res) {
  res.sendFile(path.join(__dirname, 'build', 'index.html'));
});

app.use(express.static(path.join(__dirname, 'build')));

app.listen(3000);
```

In the next chapter, you will see the details of the NewsWatcher application code for the presentation layer. You will see how these lines are incorporated into the server.js file. To run it with Node, you put in those lines shown above and then you need to first do a build and bundle of the React code before the Node service can be started up to serve it. To do the build and bundle of React, you run the following:

```
npm run build-react
```

You can then press F5 in Visual Studio Code and then your application is once again being served up, this time through Node, and you can go to http://127.0.0.1:3000/ in your browser to see it. You may want to inspect the build directory that is created and look at each of the files to see exactly what is being served and run on the client browser if you are curious.

# The index.html template and the generated one

If you recall, I showed some HTML that was all self-contained. It had everything needed to run a React backed web page and it had the scripts necessary to use the React libraries. However, with the project structure created with the create-react-app utility, this works completely differently. You still have the index.html file. That is used as the starting point for the one being generated. You will notice that it does not have any scripts pulled in. It looks as follows:

```
<!doctype html>
<html lang="en">
  <head>
    <meta charset="utf-8">
    <meta name="viewport" content="width=device-width, initial-
scale=1">
    <link rel="shortcut icon" href="%PUBLIC_URL%/favicon.ico">
    <title>React App</title>
  </head>
  <body>
    <div id="root"></div>
  </body>
</html>
```

This HTML file is serving as a point template file and has that same div with an id of "root". By template, I mean, it is taken at build time and altered and the final index.html file is created and placed in the build directory. Here is what that generated file looks like:

```
<!doctype html>
<html lang="en">
<head>
    <meta charset="utf-8" />
    <link rel="icon" href="/favicon.ico" />
```

```
    <meta name="viewport" content="width=device-width,initial-
scale=1" />
    <meta name="theme-color" content="#000000" />
    <meta name="description" content="Web site created using
create-react-app" />
    <link rel="apple-touch-icon" href="/logo192.png" />
    <link rel="manifest" href="/manifest.json" />
    <title>React App</title>
    <script defer="defer"
src="/static/js/main.45cd6978.js"></script>
    <link href="/static/css/main.073c9b0a.css" rel="stylesheet">
</head>

<body><noscript>You need to enable JavaScript to run this
app.</noscript>
    <div id="root"></div>
</body>
</html>
```

# The starting point for your code

We finally can get to the material you have been waiting for. You can now learn what the starting point is for where you place your own custom code for your React application. This code belongs in the index.js file found in the src folder. This is what is used at build time to be the starting point of your React app. Here is the file that the React build tools know about and use to bootstrap your application:

```
//index.js
import React from 'react';
import ReactDOM from 'react-dom/client';
import './index.css';
import App from './App';
import reportWebVitals from './reportWebVitals';

const root = ReactDOM.createRoot(document.getElementById('root'));
root.render(
  <React.StrictMode>
    <App />
  </React.StrictMode>
);
```

You see the same `root.render()` call that was in the initial self-contained HTML file. It is just split out and used at build time to be pulled in. What you see here is that the index.js file is importing the App.js file. In there you will find the definition of the component named App that is being rendered. React code is based on the concept of components that are assembled to build the UI. Eventually, it all leads down to HTML.

PART III: The Presentation Layer (React/HTML)

The index.js file is used by React to do the initial element insertion into the div that had the "root" id. This code uses the App.js file as the starting point for where you place all the JavaScript and HTML tags to do the rendering of what your UI will be. Here is what the initial contents are for the rendering as found in the App.js file when we created it with the `create-react-app` command:

```
// App.js
import logo from './logo.svg';
import './App.css';

function App() {
  return (
    <div className="App">
      <header className="App-header">
        <img src={logo} className="App-logo" alt="logo" />
        <p>
          Edit <code>src/App.js</code> and save to reload.
        </p>
        <a
          className="App-link"
          href="https://reactjs.org"
          target="_blank"
          rel="noopener noreferrer"
        >
          Learn React
        </a>
      </header>
    </div>
  );
}
export default App;
```

When you run "`npm run build-react`" from the command line, this builds and bundles everything for usage to be served up from your Node.js service. After the build command completes, you will find that you have a new folder named "build". It is in here that the files are placed that are fully capable of being used as pages to be loaded and seen in a browser. For example, you will now find a new index.html file in the build folder that was shown already.

The big difference now between this file and the one in your public folder is there is a style sheet and script file brought in that are used to run the React application. Therefore a Node.js server can be set up to serve up the build folder index.html file and how it can get at its own built files for all the React and provided JavaScript files to run.

272

# Debugging React code

You will want to know how to debug this new React code. Debugging the Node.js code is simple. You just place a breakpoint in your Node JavaScript code and you stop on it when that is executed and step through it.

For React code, it is easy to debug your code in the Chrome browser. You can get to the code and set breakpoints from the Chrome menu **More tools->Developer tools**. You can inspect the DOM and look at console output and much more.

You can run Node separately, either from the command line ("npm start") or VS Code (F5). Then you can go to http://127.0.0.1:3000/ or http://localhost:3000 in Chrome and debug the React code with the developer tools. You can debug server-side Node code in VS Code.

To set a breakpoint, just open the **Sources** tab and then, on the left side, find the folders with the code in it. Go to the static folder and then the js folder. If you have NewsWatcher running, open the views folder and select a file such as loginview.js and you can set breakpoints and step through the code. Find more about the Chrome Developer tools online.

In the Chrome browser, you are looking at your code before it is cross-compiled and bundled. Mapping files are created behind the scenes that allow the actual code that is being run to be mapped to the original lines for your ease in debugging it all. Here is what it looks like with the NewsWatcher application running and hitting a breakpoint in the code.

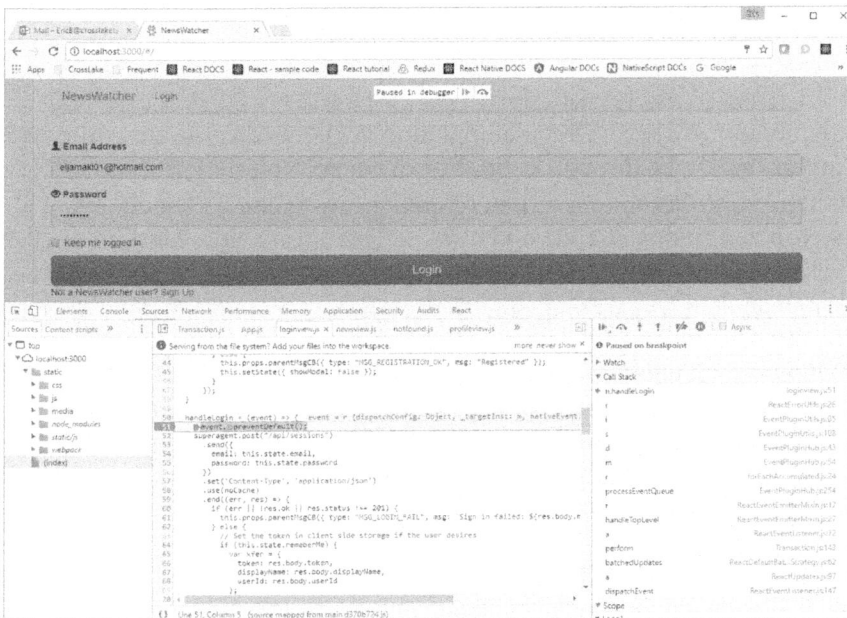

*Figure 93 - Chrome Developer tools*

# 15.5 The Basics of React Rendering

Perhaps you have heard that React does not use HTML files like a traditional Web Server might serve up. React uses code files to create components that get pieced together to create your UI. It is all done through standard DOM manipulation code at the lowest level. Don't be thinking you are getting away from HTML markup and CSS styling by any means. There will ultimately some of that required to create your application.

With all the React component you write, at the lowest level you will end up providing HTML markup that gets rendered to the DOM. You have all the standard set of HTML elements to pull from. You will use tags such as `<p>`, `<a>`, `<ul>`, `<button>`, `<input>`, `<iframe>`, `<img>`, etc. You can also add content, attributes, and styling. There are also plenty of online resources to pull in as components others have written. For example, there is no need for you to write your own sophisticated spreadsheet or charting component.

While React has you provide HTML, you do it in a JavaScript file (.js or .jsx) instead. The code is JavaScript that allows certain additional syntax. These files are cross-compiled into the actual JavaScript code that a browser can run. At the start of this chapter, you learned that there is a top-level index.html file that has a div element in it as follows (see the line in bold type):

```
<body>
    <div id="root"></div>
</body>
```

If you understand a little about browser DOM manipulation APIs, you know that you can create anything you want in the DOM through code. For example, you may insert a `text` element using JavaScript code. Furthermore, you can reference any existing element by an id you give it. In that way, you can find it in code and alter it. This means you can find a `div` and add a child HTML element to it. That child element then shows up in the rendered UI as well.

Here is some plain HTML code with no React usage at all. You can place this in an html file and open it in your browser to try it out. It uses some JavaScript code inside a script tag. The standard `document` object available in browser JavaScript is used to access your web page. It is part of an agreed-upon API standard available in any browser. The code creates a `text` element as a child of the `div` element.

Chapter 15: Fundamentals

```
<!DOCTYPE html>
<body>
<div id="root"></div>
</body>
<script>
  var textnode = document.createTextNode("Welcome to React");
  document.getElementById("root").appendChild(textnode);
</script>
</html>
```

React employs this exact mechanism to take anything you want to be rendered and presented on the page. To begin with, React is rendering to the main div you saw above with the id of 'root'. There is a starting piece of code in a file named index.js that is used as the page gets rendered. At the time the initial React application is created, it looks as follows:

```
// index.js
import React from 'react';
import ReactDOM from 'react-dom/client';
import './index.css';
import App from './App';
import reportWebVitals from './reportWebVitals';

const root = ReactDOM.createRoot(document.getElementById('root'));
root.render(
  <React.StrictMode>
    <App />
  </React.StrictMode>
);
```

This code eventually gets built and morphed into another JavaScript file that utilizes the React API to make calls that use the document object and do things like call appendChild(). This is the first piece of code that the React library uses to render something.

The call to root.render(...) is accepting what component you want to render. App is a component from another file that is imported. The previous call to ReactDOM.createRoot() is telling React where to render the UI. That is where we are referring to the 'root' div in the index.html file.

The last file to look at in the process of understanding how React does DOM rendering is the file that contains our custom defined component. In this case, the file is named App.js. This then is the pattern for the code that you will create over and over as you define the UI you want to be rendered. Here are the contents of that file:

```
// App.js
import logo from './logo.svg';
import './App.css';

function App() {
  return (
    <div className="App">
      <header className="App-header">
        <img src={logo} className="App-logo" alt="logo" />
        <p>
          Edit <code>src/App.js</code> and save to reload.
        </p>
        <a
          className="App-link"
          href="https://reactjs.org"
          target="_blank"
          rel="noopener noreferrer"
        >
          Learn React
        </a>
      </header>
    </div>
  );
}
export default App;
```

In summary - React is a JavaScript library that exposes an API. You will create a complete hierarchy of other components you define and make use of. You construct your UI from Components that the App Component will bring in. In the app.js file, the App component is exported so that it can be imported and used in the index.js file.

*Note: By convention, you start your custom components with an uppercase letter. This then helps you distinguish these from HTML elements that are always in lower case (e.g., div).*

# The returning of HTML markup

The Component class requires you to return some HTML markup. You can create your components as something that extends a React provided component named Component, or you can create what are called functional components. If you extend Component it will be more like a JavaScript class and will need a method named "render". Functional components just have a return in their function. In either case, React takes your return/render to get anything that needs to be rendered in the DOM.

This use of JavaScript mixed in with HTML markup is what is referred to as the JSX syntax. It allows you to mix HTML tags and JavaScript in one file. For example, you can have

276

conditional code that determines what elements should be rendered on a page. Look at the following example:

```
if (user.account.total > 1000) {
  return (<h1>You maintained the minimum required balance!</h1>);
}
return (<h1>You are below your minimum required balance</h1>);
```

Be aware that you can only return one single HTML element from the render function, which means there needs to be a single parent element such as a div or <></>. For example, you can't return two h1 elements unless they are wrapped in a div element. You can however return a single h1 element as shown above.

There are a few subtle differences in the HTML as it is used with React. These are as follows:

- The React DOM property naming convention is to use camel casing and has slightly altered some of the standard properties. For example, tabindex, becomes tabIndex and class becomes className.
- You use curly braces to embed JavaScript expressions in an attribute. For example, specifying the source image for an <img> element, you will have to do this: src={myObject.someImageURL}. You can use any object in your code this way.

To summarize - you can understand that a React application consists of provided Components that return HTML to be rendered. This then is taken by the React API, and using the document object, React renders that into the actual browser DOM.

*Note: The React API also includes calls for actual low-level creation of individual elements. For example, there is a method React.createElement() that takes arguments specifying the element tag, attributes, and content. This is what the JSX gets turned into in the build process where the Babel compiler is run. This book will stick with using Components in JSX file syntax.*

# More about mixing JavaScript with HTML markup

You briefly saw that in JSX syntax, you can have conditional code that uses control flow in JavaScript to decide what markup to return for rendering. This is really one of the great advantages of using React.

Imagine you have a backend service that you call to retrieve data from, such as from an HTTP/REST web service. You can take that data and then process it in the JSX code and produce whatever UI you prefer.

# PART III: The Presentation Layer (React/HTML)

The next example code will show the taking of some passed in data to show how code can make use of it. The data is a list of comments to be displayed. The code will first determine an appropriate piece of text to display and will then loop through the comments and display each using another lower-level component. The looping code makes use of the JavaScript map() function on the array object to loop through and generate the HTML.

The CommentListItem would be some custom component you would provide that would be reusable and know how to render an individual comment. I did include a paragraph element just to help clarify what is happening.

```
function Comments(props) {
  return (
    <>
      {props.comments.length > 0 &&
        <h2>There are {props.comments.length} comments</h2>
      }
      <ul>
        {props.comments.map((comment, idx) =>
          <CommentListItem key={idx} title={comment.title} />
          <p>{comment.title}</p>
        )}
      </ul>
    </>
  );
}
export default Comments;
```

You can see what is possible in code that combines JavaScript code with HTML tags being returned. JavaScript can be inserted in the middle of HTML markup by placing it inside curly braces.

*Note: You have the full power of pulling in JavaScript libraries such as "lodash" that can be used to help produce logic to output the markup you want rendered.*

You can set up blocks of code that render based on some test or a value, such as the one you see testing props.comments.length. Finally, you see the use of the map() function that lets you loop through and iterate to render a component. In this case, it is utilizing another component to create the "li" elements that will be placed inside the "ul" element to contain each list item.

*Note: If you ever have logic that needs to decide between showing some UI or hiding it, it is better to not use CSS to hide the tags. Instead, choose to not return anything at all. You simply return null from the code or have the HTML skipped all together.*

278

## Functional component or extending from React.Component

This book will show components being built that are functional components. However, you might need to read code of others that is using the Component class syntax. Here is some example code that shows both. All are consumed in the same way and have the props available to pass in. With the class usage, the props are of course part of the class instance.

```
// Code using a Component class
class Greeting1 extends React.Component {
  constructor(props) {
    super(props);
  }
  render() {
    return <h1>Greetings, {this.props.name}</h1>;
  }
}

// Code as a function
function Greeting2 (props) {
  return <h1>Greetings, {props.name}</h1>;
}

// Code as a function with arrow function syntax
const Greeting3 = ({name}) => {
  <h1>Greetings, {name}</h1>;
}

// Usage
const g1 = <Greeting1 name="Joe" />;
const g2 = <Greeting2 name="Mary" />;
const g3 = <Greeting3 name="Paul" />;
```

# 15.6 Components and Props

You have seen a simple component named App. This was used in the initially created React application. Each component is really existing to render some HTML. Any user-defined component acts as a wrapper around the complexity of some underlying UI logic and HTML. You create components to split functionality out in a way that can be self-contained and reusable.

Components can be very complex and can have properties passed into them to control different aspects of their rendering. For example, you could pass in an array of data that the component then renders.

Components can also maintain internal state, such as data about what is going on within the component. Passed in properties that change and internal state changes cause the UI to re-render for the portions affected. React detects changes, and updates will be batched to provide better performance.

The capability of having properties on a component is what React calls "props." A component can have multiple props on it. Their usage looks just like that of attributes when you use a component in other markup. For example, you could pass in some text as a prop as shown in the following example. We can alter the index.js file to look as follows:

```
root.render(
  <App name="Joe"/>,
  document.getElementById('root')
);
```

You can alter the App component as follows and then see it work. Here is the altered App.js file to use the passed in name. The text is taken and displayed along with the welcome text that specifies a person's name. The line in bold is the only modification needed.

```
// App.js
import logo from './logo.svg';
import './App.css';

function App(props) {
    return (
      <div className="App">
        <div className="App-header">
          <img src={logo} className="App-logo" alt="logo" />
          <h2>Welcome to React {props.name}</h2>
        </div>
        <p className="App-intro">
          To get started, edit <code>src/App.js</code> and save.
        </p>
      </div>
    );
}
export default App;
```

Every prop shows up in the component as properties on the props object and are available in the component class via props. The name property is accessed in the preceding example as props.name.

A property is only used as an input mechanism to a component. It is important to know that props are immutable (can't be changed) inside the component. For example, you could not have code that alters `props.name`. Props are meant to be passed in and consumed by the Component when it renders itself. The calling code is what sets the `props` and the parent code can change them as needed at any time. For example, the calling code is could have a timer that later changes the `name` prop, and then the UI will re-render at that point.

You will learn that there are ways in the parent usage of a component where it can be notified of some change of data by a child component. This is done by passing a callback function as a prop that the child component can use. This is totally up to you to define. There are predefined callbacks that can be used as standard HTML attribute notification mechanisms, such as `OnClick()`. These, however, are not surfaced back up to the parent, unless you take steps to provide the mechanism to feed it back up.

# 15.7 Components and Internal State

As previously mentioned, properties are passed into a component and consumed via `props`. React also provides a way to define internal variables that hold state. This is more than simply doing something like defining a const. Because React can't really track that a variable is changing and update the UI.

```
const name = 'Fred';
setTimeout(() => {name = 'John'}, 5000);
return (<div><h2>Welcome to React {name}</h2></div>);
```

What react has provided is what called the `useState` hook. Using this will allow React to know about state changes and it can refresh the UI to reflect any changes. Look at the following code and you will find the use the `this.state` object. Here is also where you will see the usage of a constructor for your Component class.

```
import { useState } from 'react';
function App() {
  const [name, setName] = useState('Fred');
  setTimeout(() => {setName('John')}, 5000);
  return (<div><h2>Welcome to React {name}</h2></div>);
}
```

You can set the initial value of the state variable. These can be any datatype supported by JavaScript, including objects. You use `setName()` or whatever you have called it to make changes. React will do a UI refresh when it sees state changes have occurred and may batch up several before it does a UI refresh.

With the usage of `setName()` you can access the current state before you make your change. In the previous example, we don't use that capability.

You should strive to moved state values up to the highest component that has it in common across any child components. A parent can have a state property of its own and pass that into the child component as a prop. When the parent state changes, the child will see the update and React will re-render the component if necessary.

# Model, View, and Controller

If you want to think in terms of how an MVC architecture works, you can recognize that React fits into this model by providing both the View and the Controller capability. The View is what renders DOM elements from your component return statement or Render function if it is a class. The Controller code is what exists in the rest of the logic that acts on behalf of the passed in props and managed internal state.

Your React component can do the fetching of the Model data. You have code fetch the data and pass it as props to a component or, in the component itself, you would fetch it and set the state. You can utilize libraries such as Flux/Redux to do the coordination of data fetching and the setting of state in the code of the component. Here is an image that might help you visualize how this all fits together in an MVC design:

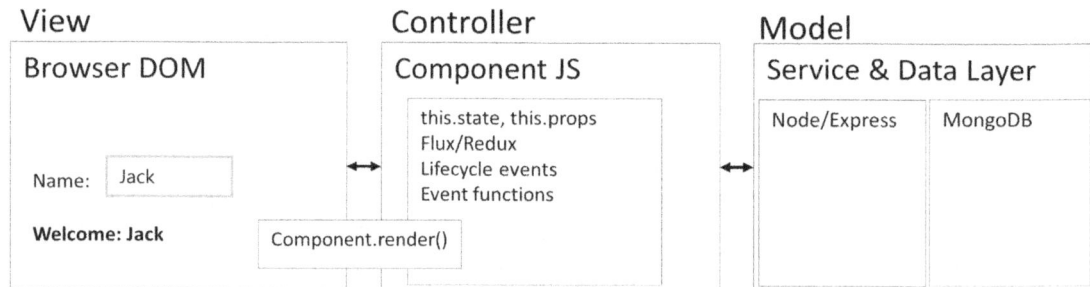

*Figure 94 - React MVC concepts*

***Note:*** *You can debate exactly where to keep the Controller and the View code. You can have it in one single component. It may be best, however, to separate the Controller code from the View code. To do this, you can have a data component that uses a data store and then the data component would pass props to a view-only presentation component. These two types of components are called Container and Presentation Components (also called "smart" and "dumb"). More will be said about this later in the section on Redux usage.*

# 15.8 Event Handlers

The HTML tags used in rendering can have event handlers given to them as appropriate. These are used for events such as a clicks or key presses. In React JSX files, this means you provide a JavaScript function in your React Component class and then use curly braces for specifying the event handler. Events are specified with camel case characters.

Here is an example that adds a handler for a click event on a button. Notice the use of the current value of count in the setCount call.

```
import { useState } from 'react';
function App() {
  const [count, setCount] = useState(0);

  const handleClick = (e) => {
    setCount(currState => (currState + 1));
  }

  return (
    <div>
      <h1>You have clicked the button {count} times</h1>
      <button onClick={handleClick}>
        Click it
      </button>
    </div>
  );
}
```

The 'e' function argument to the click handler is an event object that has a lot of properties and methods on it to be used as needed. For example, you may want to capture a certain key that was pressed and then prevent it from propagating. There are cases where you don't want a click or keypress to be processed any further. To prevent propagation of an event, you call e.preventDefault().

## Passing up Event Handling

You can have the parent of a child component be notified of events form the child by exposing a property that the child calls. If you have multiple events, you need to name each event property something different. In the code below, MyButton just the prop callback method in its onClick event. In this way, the App component can have the code that responds to the button click.

```
function App() {
  const [count, setCount] = useState(0);

  const handleClick = (e) => {
    setCount(currState => (currState + 1));
  }

  return (
    <div>
      <h1>You have clicked the button {count} times</h1>
      <MyButton myClickHandler={handleClick} />
    </div>
  );
}

function MyButton(props) {
  return (
    <div>
      <button onClick={props.myClickHandler}>
        Click it
      </button>
    </div>
  );
}
```

# 15.9 Component Containment

There is a special props property that is always available on a component that is used for passing in child components or child HTML elements directly. The name of the property is 'children'. This means that you can enclose children in the usage of a custom component tag, and then get access to them. Here is the previous example, but with the text of the button passed as part of the children.

```
function App() {
  const [count, setCount] = useState(0);

  const handleClick = (e) => {
    setCount(currState => (currState + 1));
  }

  return (
    <div>
      <h1>You have clicked the button {count} times</h1>
      <MyButton myClickHandler={handleClick}>
```

```
      <h1>Click it</h1>
    </MyButton>
  </div>
 );
}

function MyButton(props) {
 return (
   <div>
     <button onClick={props.myClickHandler}>
       {props.children}
     </button>
   </div>
 );
}
```

The previous example shows an HTML tag being enclosed as part of the children. You can also have other JSX components or string literals be children as well.

*Note: When the JSX is taken and rendered, any whitespace at the beginning and end of a line is removed, as well as blank lines. New lines that occur in the middle of string literals are condensed down to a single space character.*

You can always pass in any name for a property that contains multiple elements. This may become necessary if you have multiple sub-components being used in elements and are displaying them. For example, in the previous example the button text can be passed in as an additional property as follows:

```
function MyButton(props) {
  return (
    <div>
      <button onClick={props.myClickHandler}>
        {props.myButtonText}
      </button>
    </div>
  );
}
```

# 15.10 HTML Forms

React supports the creation of HTML forms via the standard form tag that exists in HTML. You can create a component that renders itself and provides the form tag and supports the event handling function for the submission. A simple example is as follow:

```
function App() {
  const [value, setValue] = useState('');

  const handleChange = (event) => {
    setValue(event.target.value);
  }

  const handleSubmit = (event) => {
    alert('The test submitted is: ' + value);
    event.preventDefault();
  }
  return (
    <form onSubmit={handleSubmit}>
      <h2>Text is {value}</h2>
      <label>
        Name: <input type="text" onChange={handleChange} />
      </label>
      <input type="submit" value="Submit" />
    </form>
  );
}
```

Not only does the above code have the submission handling, but you also see code to handle the binding for changes of an input element so that the changes are rendered. You can also get events on changes for a checkbox, list, select control, and others. The onChange attribute is used for those events.

# 15.11 Lifecycle of a Component

React sets up a series of stages that a component goes through as it gets rendered, eventually destroyed, and taken out of the DOM. Depending on if you are using React.Component extended classes or functional components, this will look different

## Class Components

For Components derived from React.Component, you need to learn about what are called lifecycle events. You can provide methods on your class that get called when the component enters each of these stages.

These methods are called "lifecycle hooks." For example, the componentDidMount() method is called after the component output has been rendered to the DOM. You then can know that your component is visible and take any action you need. You may want to create a timer that refreshes the state of the component every few seconds if data changes in the backend services layer may happen.

286

Another method available is `componentWillUnmount()`, which is called just before a component is taken out of the DOM.

## Functional Components

Functional components use what are called Hooks. You have already seen the hook used to set the local state of a component. There also two hooks you use in conjunction with Redux usage. Those will be explained later. The hook you can become familiar with, as far as rendering lifecycle, is the Effect hook. This serves the same purpose as the componentDidMount, componentDidUpdate, and componentWillUnmount in React classes.

The `useEffect()` in your code is called after changes to the DOM are flushed and in that code you get a chance to further affect the rendering. Inside the code, you have access to the props and state of the component. For example, the component will load and render and then call your `useEffect()` function you provide and then you could do something like do a fetch of data from the backend systems and push that into the local state for rendering. You can have multiple `useEffect()` calls.

# 15.12 Typechecking Your Props

You can make use of the `propTypes` property of your component to set the expected property types being used in your component. This way, you can catch errors in its usage. This is only seen when running in development mode in the debug output.

You can test for individual props such as an object, array, func, string, number, and many other types. You can also specify if a given prop is required or not. There is also a means of specifying default values for props that are not passed in by a parent. Here is an example that shows off these capabilities:

```
import PropTypes from 'prop-types';
function Comments(props) {
  return (<><h2>Count is {props.comments.length}</h2></>);
}
Comments.propTypes = {
  comments: PropTypes.array.isRequired
}
Comments.defaultProps = {
  title: 'Hi'
}
export default Comments;
```

# 15.13 Getting a Reference to a DOM Element

In some cases, you need to call the actual DOM manipulation functions on an element. To do that, you need to get a reference to be able to use it. React has a special syntax that you can employ to set the reference to the element. Here is what that looks like. You will only need to do this if you can't store a reference in your component class to an input element. In a Class Component, you would do the following.

```
<input type="text" ref={(input) => { this.textInput = input; }} />
```

Then in other code, you can make a call to use `this.textInput` to get or set its value. There is also a function provided in the React library named `findDOMNode` that can be used to find any element in code. I prefer to find ways around using either of these mechanisms as they really are there just for the odd case. Usually there is a way to find any component in context and get its value in simpler ways.

Functional components don't have instances so there is no `this` to use. There is a workaround, but hopefully nothing you will need to be concerned with. Refer to the React documentation.

# Chapter 16: Further Topics

That about covers the basics of using React. This chapter will cover further topics that you will want to know about to achieve a full robust application.

# 16.1 Using React Router

If you have a website comprised of many different pages to be viewed, you will need a way to navigate between them all. You will need to render some type of navigation bar with links to click on to accomplish this.

React does not come with any built-in capability for handling page view routing. You could create your own code to render some type of navigation bar across the top of your main landing page and then do the work to change to different page views inside of that.

The solution in this book utilizes a popular open source npm package named React Router. There are two different packages that were created by the same group of people — one is for websites and the other is for mobile app development. We will use the website version that is found here, https://www.npmjs.com/package/react-router-dom. In case you are interested, the native app version is named react-router-native.

React Router gives you the ability to create a dynamic navigation experience on your site. To get started, you can install the package into your project as follows:

```
npm install --save react-router-dom
```

*Note: React Router can be utilized in two ways - on the server-side to return pages that are pre-rendered with data or as a pure client-side SPA site. This book focuses on the client-side SPA usage and shows how views are populated from client-side fetches of data from the server HTTP/REST API created in Part Two of this book.*

At the top of your App.js file, you can put an import statement to pull in the needed components provided by the react-router-dom package. These are `HashRouter`, `Route`, and `Link`. Here is the code that can be placed into App.js for a simple usage with no CSS styling:

```
import { HashRouter, Routes, Route, NavLink } from 'react-router-dom'
import './App.css';
const Home = () => (
  <div><h2>Home</h2></div>
)
```

```
const About = () => (
  <div><h2>About</h2></div>
)

function App() {
  return (
    <HashRouter>
      <div>
        <NavLink to="/" >Home</NavLink>
        <span className="d-lg-none"> &sdot; </span>
        <NavLink to="/about" >About</NavLink>
        <hr />
        <Routes>
          <Route exact path="/" element={<Home />} />
          <Route path="/about" element={<About />} />
          <Route path="/404" element={<NotFound />} />
          <Route path="*" element={<Navigate replace to="/404"/>} />
        </Routes>
      </div>
    </HashRouter>
  );
}

export default App;
```

The `HashRouter` is the main component wrapper that creates the container for the UI to be rendered as navigation occurs. Different components are rendered as assigned by the `element` attribute of the Route. For a site where you want browser history kept, use `BrowserRouter` instead of `HashRouter`. In the above case, we are discussing a SPA, which is implemented with the hash routing.

The attribute `exact` is used to match only a single specific path. If you have multiple components being rendered at the same time, this is needed and will fix that problem.

The `NavLink` component used as an anchor tag may also be used, but adds the prop "to" for specifying the route to transition to. The `NavLink` component works hand-in-hand with the `Route` component that specifies what component to render for a given route. The "replace" attribute specifies that clicking the link will replace the current entry in the history stack instead of adding a new one. This is what we want for a SPA client-side site.

The last Route listed is one that lets all other paths be rendered with an error. If you end up there, a `NotFound` component states that this is a "404" because it was not expected. For example, this `NotFound` UI will render if you go to something like localhost:3000/#/blahblahblah.

Here is what the UI ends up looking like for this previous example code:

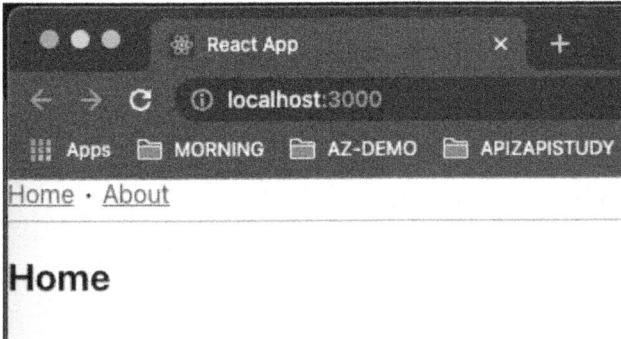

*Figure 95 - React Router usage*

# 16.2 Using Bootstrap with React

Styling a website to create a modern responsive site that looks good in a full-screen browser as well as on a small mobile phone is a challenge. You have much to consider when constructing your HTML in this case. This book is not a way for you to learn CSS or responsive Web Design techniques. What this book does instead is take a simple and effective approach by using Bootstrap. Bootstrap was created by Twitter and is a set of UI HTML templates and CSS styles that you can use to accomplish responsive website designs.

To make this even easier in a React application, there is a npm package that exports React Components to be easily used. You just need to run the following install to get everything available in the project:

```
npm install --save react-bootstrap bootstrap
```

The only other change to make is to pull in the needed CSS files so that they are exposed for usage in the application. Here is the altered index.js file that is found in the src directory.

```
// index.js
import React from 'react';
import ReactDOM from 'react-dom/client';
import App from './App';
import 'bootstrap/dist/css/bootstrap.css';
import './index.css';

const root = ReactDOM.createRoot(document.getElementById('root'));
root.render(
  <App/>,
  document.getElementById('root'));
```

PART III: The Presentation Layer (React/HTML)

Here is an altered App.js file that now shows the usage of a Bootstrap styled button:

```
// App.js
import { Button } from 'react-bootstrap';

function App() {
  return (
    <div>
      <h1>Welcome to React with Bootstrap</h1>
      <p><Button
        bsstyle="success"
        bssize="lg"
        href="https://react-bootstrap.github.io/"
        target="_blank">
        View React Bootstrap Docs
      </Button></p>
    </div>
  );
}
export default App;
```

Here is what the rendering of the example above will look like:

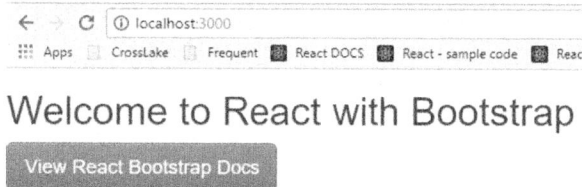

← → C ① localhost:3000

::: Apps   CrossLake   Frequent   React DOCS   React - sample code   Reac

# Welcome to React with Bootstrap

View React Bootstrap Docs

*Figure 96 - Bootstrap usage with React*

You can visit this React Bootstrap site to see all the available components you can use in your React application. The NewsWatcher application will make use of several of them.

***Note:*** *The other technology worthy of use for the same purpose is material-ui.*

# 16.3 Making HTTP/REST Requests

Every application you create will most likely need to make HTTP/REST requests to some backend service to retrieve data. React purposely does not have the built-in capability to do that because several libraries are already good at this. One option is to use the built-in standard of `fetch()`, as found in the JavaScript language and available in browser side code. Alternatively, other npm modules such as superagent or axios can be installed and used. This book utilizes `fetch()`.

Typically, you will need to fetch data in the hook that runs when you component is first rendered. That code will update some internal state properties and causes React to do any necessary DOM updates. Here is an example `useEffect()` hook that outputs some fetched data to the console:

```
useEffect(() => {
  fetch('/api/homenews', {
    method: 'GET',
    cache: 'default'
  })
    .then(r=>r.json().then(json=>({ok:r.ok,status:r.status,json })))
    .then(response => {
      if (!response.ok || response.status !== 200) {
        throw new Error(response.json.message);
      }
      console.log(response.json);
    })
    .catch(error => {
      console.log(`GET failed: ${ error.message }`);
    });
}, []);
```

This type of code will be used when the NewsWatcher application is put together. Basically, you specify the URI or relative path. As part of the second parameter, you give an object that contains the HTTP verb usage, such as put, post, get, or delete. You can also set any headers you need. Fetch can be used as promise-based code, and that is what is shown here. You can verify the HTTP status code returned, handle errors, and get at the body of the returned data.

# 16.4 State Management with Redux

You have now learned how to use props and local state in a React component. State and sometimes props should not always be kept as part of a component but should sometimes be kept in a central storage location that is accessible by all components, not matter their level in the hierarchy. If you find that the props or state of a component really needs to be shared with other components, then store those values centrally.

Redux is a library that helps you share state across all your React components. Its purpose is to provide a central repository for all components to interact with. Think of it as fulfilling the same function that state does in a component but shared across all components. With Redux, you can also accomplish data transformations and data routing in the process.

The state that you store can be some data you collected from user interactions or data that comes from a network request. Just about any type of data can be kept there.

Imagine retrieving data that needs to be shared across the hierarchy of components. For example, the UI in one part of the component hierarchy might have state that needs to be shared with any number of other components strewn all over the hierarchy. It will get confusing if you try to send that state up and across, and then down to another component using properties. That quickly becomes incomprehensible.

Redux solves these problems by being the central state repository for your complete hierarchy across all components. To use Redux, you first need to create the Redux store. You do this in your initialization code. When you create the store, you need to pass in the reducers to be used. A reducer is what gets a chance to determine the state changes you want to make. Data flows into them in the form of an action. This reducer can be a single function or a combination of reducers. This becomes clearer as you look at some code examples.

The npm package that you need to install is called "redux." The Redux object that you can access has only a few methods on it, so it is fairly easy to understand. Additionally, there are a few nuances to understand about each of the calls. Here is code showing the creation of a store and then some code showing the Redux functions:

```
import { createStore } from 'redux'
const store = store = createStore(reducer);
store.dispatch(action)
store.subscribe(listener)
store.getState()
store.replaceReducer(nextReducer)
```

The `createStore()` call needs one or more reducers given to it. It can also be passed middleware to be used. You can download other npm packages to act as middleware or you can write your own. This simply means that the flow can be intercepted and altered in some way, which is similar to what middleware does in Express with Node.js. I have not ever needed to use this capability.

The `dispatch()` call is what you use to cause a change in state to happen and acts to call a reducer. The `getState()` function is a way to fetch the state at that moment in time and the `subscribe()` is for receiving asynchronous notifications as state is changed.

In the NewsWatcher application codebase, I make the initial call to `createStore()` with some provided reducer functions where `dispatch()` is called a lot in the code.

# The flow of a state change

What you typically do in your React code is to code up a component and use local state wuth the `useState()` hook for keeping data that changes local to that component. As you make progress and find that the state is something that needs to be shared, you can move it to the Redux state storage. Just remember that you don't put everything in Redux because that can become overwhelming to manage.

To update the Redux state, you make a call to `dispatch(action)`. For example, if a component wants to update some state and make that available back to itself and to other components, it makes that call. The component calling `dispatch()` does not need to know about who will listen for that state change. This is a single direction of flow with the state change. There are no bi-directional messages flowing through Redux in any kind of conversation.

The simplest usage is to preconfigure Redux with what are called reducers. These act to filter data as it flows through. Dispatch calls happen and flow through what is called an "action". This then causes the particular affected state to change and become available in a component where needed.

Deep down, code receives the changes in the state via a `store.subscribe()` call. However, React provides a wrapper on top of that to use. You also can get a snapshot of the state at a given instant with the call `store.getState()`. Here is a visual representation of flow for a state change as it passes via the Redux store:

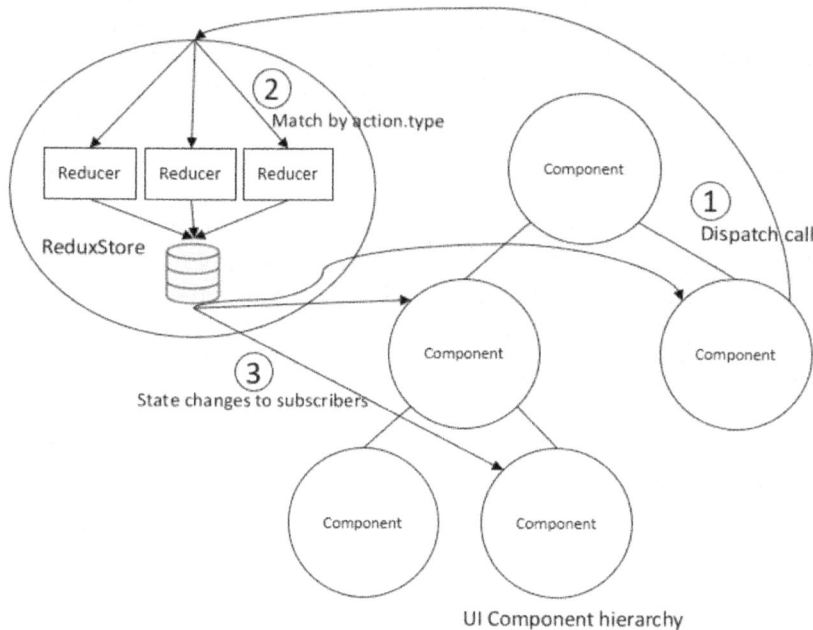

*Figure 97 - React state management with Redux*

Below is the simplest code that can be used to demonstrate the basic flow. Note that the action object must at least have a property named "type." This is used to find the correct router to use for processing. The other properties of the action object are up to you to pass and deal with. The following code creates the Redux store with the passed in reducer. The function usage has an initial object passed that sets the storage of the state property. You would also need to have done a npm install of redux for this to work.

```
import ReactDOM from 'react-dom/client';
import { createStore } from 'redux'
const initialState = {
  count: 0,
  currentMsg: "Hello Redux"
}

const startState = (state = initialState, action) => {
  switch (action.type) {
    case 'INCREMENT':
      return {...state, count: state.count + action.incAmount}

    default:
      return state;
  }
}
```

296

```
const store = createStore(startState);

const Counter = ({
  value,
  onIncrement
}) => (
  <div>
    <h1>How many bugs have you fixed today?</h1>
    <h1>{value}</h1>
    <button onClick={onIncrement}>+</button>
  </div>
);

const render = () => {
  const root = ReactDOM.createRoot(document.getElementById('root'));
  root.render(
    <Counter
      value={store.getState().count}
      onIncrement={    ()    =>    store.dispatch({type:    'INCREMENT',
incAmount: 1}) }
    />,
    document.getElementById('root')
  )
};

store.subscribe(render);
render();
```

The state for a given reducer can take any form you define for it. You can see in the previous code that it is defined with default values to be initialized. Redux will make an initial call to the reducer on its own and the defaults are set.

To alter the state, a dispatch call happens and passes the action object with the type and values to consume. In the Redux code, the state is immutable, so you cannot change it directly. You have to make a clone of it (e.g., see JavaScript `Object.assign()` usage, such as `Object.assign({}, state, { users: action.users });`), or use something like the immutability-helper.

You also need to return the complete state object back as it will all be replaced. The code above simply creates the new object from properties of the existing state using the object spread syntax of JavaScript. Then the count is updated.

If it is an object, for example, you replace the whole thing, not just one property. With what you saw with the React component state, you have one object that has independently updatable properties on it. Redux does not work the same way. You can, however, have different reducers that are independent, so updating one does not affect the other(s).

Most applications will want to split out what is stored in the Redux storage into separate concerns. You do this by calling `combineReducers()` to piece them all together. Each one receives all of the actions routed to them, but it is up to the reducers to match what `action.type` values they care about. You call createStore with this combined rootReducer instead of a single reducer function.

```
import { combineReducers } from 'redux'

const rootReducer = combineReducers({
  app,
  news,
  sharednews,
  profile
})
```

# Usage of Redux with react-redux

The Redux library is a reusable JavaScript library that can be used in your code. It is usable as-is with its methods. However, certain code design patterns of React are used over and over. One of these patterns involves state manipulations. In order to combine that pattern with the use of Redux in an easily consumable way, a library was created named react-redux.

To get started, you need to install the npm packages redux and react-redux. You then need to add some code to create the single Redux store. There is a special React component that is used to wrap your application component to make Redux available in all of the component hierarchy. Here is what the index.js file ends up looking like. The key changes are in bold.

```
import React from 'react';
import ReactDOM from 'react-dom/client';import { createStore } from 'redux'
import { Provider } from 'react-redux'
import reducer from './reducers'
import App from './App';
import 'bootstrap/dist/css/bootstrap.css';
import 'bootstrap/dist/css/bootstrap-theme.css';
import './index.css';

const store = createStore(reducer);
const root = ReactDOM.createRoot(document.getElementById('root'));
root.render(
  <Provider store={store}>
    <App />
  </Provider>, document.getElementById('root'));
```

298

# Chapter 16: Further Topics

The App component will be a child of the Provider component. The code from react-redux library will be able to use it and pass in additional props. It will pass in a prop that is a function named "dispatch." That means you don't need to pass the Redux store to the App component. You won't actually be coding up calls to `store.dispatch()`, just `dispatch()` as seen in previous code.

You will need to understand is how to take a functional component that you write and do the work to interact with redux to call the reducers and to also get asynchronous redux state updates when they happen and have the UI refreshed.

The state coming from the Redux store can be used in a component. Props are then changed via the dispatch to cause the behavior to change. You do *not* take the Redux state and transfer it to the `state` hook of a component because that is a waste since they both serve the same purpose, would be duplicated, and you would not have a single source of truth. Just to be clear, you can have the state from redux through props automatically passed down to children. There are two hooks you will learn about to make use of that make this a bit clearer. Here is what you need to do in a component to set some state in redux and to retrieve state:

```
import { useSelector, useDispatch } from 'react-redux'
function Message() {
  let message = useSelector((state) => state.app. currentMsg);
  const dispatch = useDispatch();

  dispatch({ type: 'MSG_DISPLAY', msg: "Registered" });

  return (
    <div>
      <h1>{message}</h1>
    </div>
  );
}
```

The `useDispatch` function is made available to get at the dispatch function that will then update the central redux state. The `useSelector` will set up to asynchronously update a local variable when the state is ever changed anywhere.

*Note: The usage of additional libraries like Redux is completely at your discretion. Only use a library if at any point you see that you can simplify your code. Sometimes a simple code base with no extra frills works just fine. You must decide when and how to refactor and introduce new libraries into your code, and what the cost versus benefit will be. In the case of Redux, I believe there is a clear benefit. You will see it used in the NewsWatcher application code.*

# Chapter 17: NewsWatcher App Development with React

Having covered the material in the previous chapter, you now have the concepts and tools to construct the NewsWatcher presentation layer code. If you followed along in that chapter's section on installation steps, you are all set to begin understanding the code.

In this chapter, you will learn about the files as they exist in the project posted on GitHub and how each one is pieced together. If you clone the project from GitHub, the top-level project folders will look as follows:

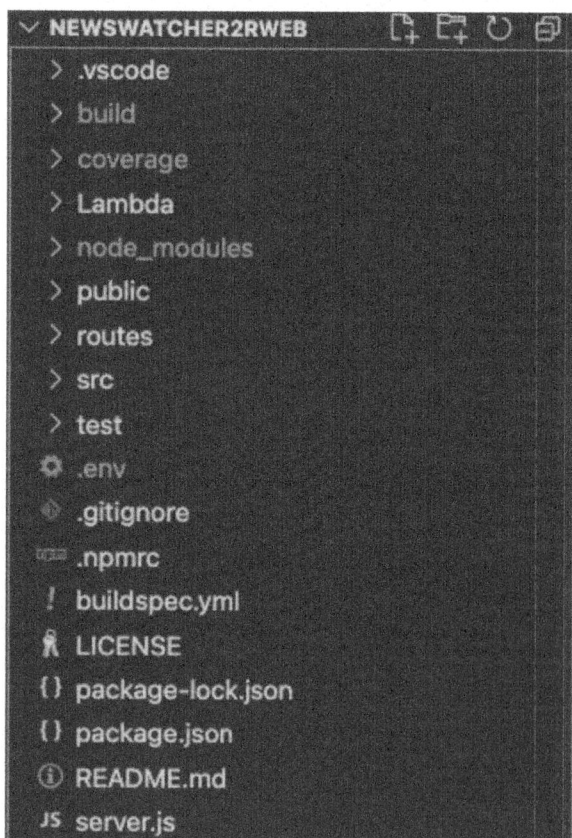

*Figure 98 - VS Code file tree*

**Note:** *Don't forget that you can access all of the code for the NewsWatcher sample application project at https://github.com/eljamaki01/NewsWatcher2RWeb.*

# 17.1 Where It All Starts (src/index.js)

You may have noticed that you have a placeholder index.html file that your Node.js service has served up in part two of this book. The existing Node.js project from Part Two of this book can now be altered to change it over to our actual React application needed.

Back in the sections on the Node.js web service development, you saw a few lines of code in the server.js file that served up a get request for the main HTML page. This allowed for the downloading and display of the NewsWatcher application site.

The Express routing for that file was separated from those providing access to the backend API. Here are the lines from server.js for the route handling that serves up the index.html file and its associated static resources:

```
app.get('/', function (req, res) {
  res.sendFile(path.join(__dirname, 'build', 'index.html'));
});
app.use(express.static(path.join(__dirname, 'build')));
```

The app.get() call gives a specific route for a get request to your root site and states that you will always serve up your index.html file for that. This is all you need to set up the use of React to get the React application sent to run on a client-side browser.

The app.use() call sets up middleware for a router path for all of the static files. You simply tell it that you have this directory named "build" for where they all are located.

You will note that this final index.hml file is not a file that you created, but it is generated for you in the build process from the index.js file that you provided and the template index.html file found in the public folder. Look in the build folder to find the generated index.html file that gets served up.

Here is the React index.js file under our control. The few additions you will find here are just the inclusion of CSS files that can be used across anything that is rendered. In particular, you see the ones for styling bootstrap elements. You also see the code to set up Redux to be used across the application. The Redux reducers are brought in, and the Provider component, from react-redux, provides Redux to the overall application.

```
import React from 'react';
import ReactDOM from 'react-dom/client';
import { createStore } from 'redux'
import { Provider } from 'react-redux'
import reducer from './reducers'
```

```
import './index.css';
import App from './App';
import 'bootstrap/dist/css/bootstrap.css';

const store = createStore(reducer);
const root = ReactDOM.createRoot(document.getElementById('root'));
root.render(
  <React.StrictMode>
    <Provider store={store}>
      <App />
    </Provider>
  </React.StrictMode>,
  document.getElementById('root')
);
```

The call to `root.render()` is the starting point call to React that renders to the root div element. All the UI gets its start from here and comes from the App component.

# 17.2 The Hub of Everything (src/App.js)

The App.js file is where we establish the UI that renders the views. This file provides things like the presentation of a navigation bar which allows for the navigating between the different page views - logging in, viewing news, and setting profile settings etc. This file sets up the App class as a functional component.

**Imports and hooks**

The imports up top bring in external libraries and the other components that are needed for views we navigate to.

Redux is used for the state settings that are central. You will find the use of the `dispatch()` function to store state data in Redux in a central way that is global to the application. The data that is kept in Redux is as follows:

- A Boolean to tell us if we are signed in.
- The session token that was retrieved from the server-side log in.
- The status message that is displayed at the top of the UI.

The `useEffect()` hook event method is where you do processing once the UI has been rendered and need to do things like fetch data. The code there checks the local browser storage to see if a token has been saved, and if so, gets it and sets a message that the user is logged in and sets the session token in Redux.

The user would need to have selected the option at login time to have the token saved. The dispatch places the token into Redux storage for global access from all other pages that need it, such as for profile retrieval.

The `loggedIn` flag is set to true in that same dispatch processing, which alters the menu rendering, as you will see. The value of `currentMsg` is set through the dispatch and shows up in the UI.

This React UI is rendered by retrieving the necessary files from the server side, with the initial HTML, CSS, JavaScript, and other resource files. The UI is completely self-sufficient in the client browser from that point on. There are no pages being rendered from the server-side after that unless you press the refresh icon and reload it. You can create a combination of server and client-side page rendering in a true isomorphic application, and this will be discussed later. Here is the code that has been discussed thus far.

```
// App.js
import React, { useEffect } from 'react';
import './App.css';
import {NavLink,HashRouter,Routes,Route,Navigate} from 'react-router-dom'
import { Navbar, Nav } from 'react-bootstrap';
import { useSelector, useDispatch } from 'react-redux'
import LoginView from './views/loginview';
import NewsView from './views/newsview';
import HomeNewsView from './views/homenewsview';
import SharedNewsView from './views/sharednewsview';
import ProfileView from './views/profileview';
import NotFound from './views/notfound';

function App(props) {
  let loggedIn = useSelector((state) => state.app.loggedIn);
  let session = useSelector((state) => state.app.session);
  let currentMsg = useSelector((state) => state.app.currentMsg);
  const dispatch = useDispatch();

  useEffect(() => {
    // Check for token in HTML5 client side local storage
    const storedToken = window.localStorage.getItem("userToken");
    if (storedToken) {
      const tokenObject = JSON.parse(storedToken);
      dispatch({ type: 'RECEIVE_TOKEN_SUCCESS', msg: `Signed in as
${tokenObject.displayName}`, session: tokenObject });
    } else {
    }
  }, []);

...code left out...
}
export default App
```

**Navigation bar**

One main purpose of this App component is to set up the navigation bar for users to get access to the other page views and see a status message. The DOM rendering makes use of some handy bootstrap styling to be usable on a mobile device. You can investigate the particulars of how it all works on your own through the usage of the react-bootstrap npm package. Here is an image of what would be rendered on a smartphone device footprint. It shows the UI state when the menu is opened.

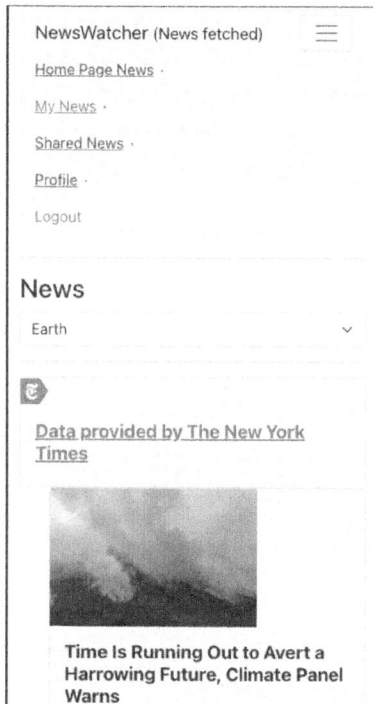

*Figure 99 - Sample image from a smartphone*

The react-boostrap usage can do what is called "responsive" web rendering and it creates a button that will be used to provide a drop-down menu when it is being run on a mobile smartphone. If you run it in a desktop browser, it does not appear this way, and the navigation bar is spread out. Bootstrap classes adapt to the size of the display.

***Note:*** *The great thing is that you don't need to use the raw bootstrap stylings with low-level HTML elements. React-bootstrap wraps all of that and presents it for your usage in React Components that are consumed.*

As part of the header of the navbar, a span element is used to display messages you want the user to see. For example, if a user's login fails, you want the user to know that. The state is set through Redux with the appropriate message string.

# Chapter 17: NewsWatcher App Development With React

Each `NavLink` represents the menu selections for navigating around the application. There is a clever way to show or hide each entry depending on the state of the app. The `loggedIn` Redux state is used to determine the display of navigation bar entries. When a user is logged in, they get more menu selections made visible. Until then, some will remain hidden.

The actual navigation click-handling is done using the react-router-dom npm package. At the very top of what is rendered is the usage of the `HashRouter` component. This is what wraps everything and gives us the ability to render in the DOM whatever we set up as a page view route. Then we can either let the navigation bar selections control the route to select or, in some cases, manage the navigation programmatically.

You can see the react-router-dom `Routes` component that is used in conjunction with the `Route` component to handle what page views get rendered. The interesting thing is that for each `Route`, you are specifying the `path` that it is for and the `element` (component) to render. There is both the use of passing in props for redux state, and in one case, with the LoginView (if you look at that code), you see the alternate way of using `useSelector` and `useDispatch` from `react-redux`. Here is the `return` for the App component that renders the navigation:

```
return (
  <HashRouter>
    <div>
      <Navbar bg="light" expand="lg">
        <Navbar.Brand href="/#">
          NewsWatcher {currentMsg &&
            <span>
              <small id="currentMsgId">({currentMsg})</small>
            </span>}
        </Navbar.Brand>
        <Navbar.Toggle aria-controls="basic-navbar-nav" />
        <Navbar.Collapse id="basic-navbar-nav">
          <Nav className="ml-auto" >
            {
              <Nav.Link>
                <NavLink exact to="/" style={({ isActive }) => ({ color:
isActive ? 'green' : 'blue' })} >Home Page News</NavLink>
                <span className="d-lg-none"> &sdot; </span>
              </Nav.Link>
            }
            {loggedIn &&
              <Nav.Link>
                <NavLink exact to="/news" style={({ isActive }) => ({
color: isActive ? 'green' : 'blue' })} >My News</NavLink>
                <span className="d-lg-none"> &sdot; </span>
              </Nav.Link>
            }
            {loggedIn &&
              <Nav.Link>
```

```
              <NavLink exact to="/sharednews" style={({ isActive }) => ({
color: isActive ? 'green' : 'blue' })} >Shared News</NavLink>
                <span className="d-lg-none"> &sdot; </span>
            </Nav.Link>
          }
          {loggedIn &&
            <Nav.Link>
              <NavLink exact to="/profile" style={({ isActive }) => ({
color: isActive ? 'green' : 'blue' })} >Profile</NavLink>
                <span className="d-lg-none"> &sdot; </span>
            </Nav.Link>
          }
          {loggedIn &&
            <Nav.Link>
              <Nav onClick={handleLogout}>Logout</Nav>
            </Nav.Link>
          }
          {!loggedIn &&
            <Nav.Link>
              <NavLink to="/login" style={({ isActive }) => ({ color:
isActive ? 'green' : 'blue' })} >Login</NavLink>
            </Nav.Link>
          }
        </Nav>
      </Navbar.Collapse>
    </Navbar>
    <hr />
    <Routes>
      <Route exact path="/" element={<HomeNewsView dispatch={dispatch}
/>} />
      <Route path="/login" element={<LoginView />} />
      <Route path="/news" element={<NewsView session={session}
dispatch={dispatch} />} />
      <Route path="/sharednews" element={<SharedNewsView
session={session} dispatch={dispatch} />} />
      <Route path="/profile" element={<ProfileView
appLogoutCB={handleLogout} session={session} dispatch={dispatch} />} />
      <Route path="/404" element={<NotFound />} />
      <Route path="*" element={<Navigate replace to="/404" />} />
    </Routes>
  </div>
</HashRouter>
);
```

**Note:** *Look at the top of the file App.js file to see the import statements. These will help you keep straight which components come from which npm packages. For example, those from react-bootstrap and those from react-router-dom.*

## Logging out

The navigation bar has a selection to log the user out. This simply places a call to the backend service, where it can do whatever it needs code-wise. The client-side just needs to set the state

306

to reflect that and redirect the view to the login. This is done through a browser capability with the use of setting `window.location.hash`.

```
const handleLogout = (event) => {
  event && event.preventDefault();
  fetch(`/api/sessions/${session.userId}`, {
    method: 'DELETE',
    headers: new Headers({
      'x-auth': session.token
    }),
    cache: 'default' // no-store or no-cache?
  })
    .then(r =>r.json().then(json => ({ok: r.ok, status: r.status, json})))
    .then(response => {
      if (!response.ok || response.status !== 200) {
        throw new Error(response.json.message);
      }
      dispatch({ type: 'DELETE_TOKEN_SUCCESS', msg: "Signed out" });
      window.localStorage.removeItem("userToken");
      window.localStorage.clear();
      window.location.hash = "";
    })
    .catch(error => {
      dispatch({ type: 'MSG_DISPLAY',
        msg: `Sign out failed: ${error.message}` });
    });
}
```

# 17.3 Redux Reducers (src/reducers/index.js)

You saw in the src/index.js file how the reducers were brought into the application when it started up. The line was as follows that sets up the Redux Store:

```
import reducer from './reducers'
const store = createStore(reducer);
```

*Note: As a reminder, if you have a require or import statement of a directory, it will by default look for a file named index.js and use that.*

Index.js in the reducers folder combines all the separate reducers into one and exports that. We only have one, but you could add more.

```
// src/reducers/index.js
import { combineReducers } from 'redux'
import app from './app'
const rootReducer = combineReducers({app})
export default rootReducer
```

**The App Reducer**

Let's now look at the App reducer bring used. The App reducer has three actions that it supports – MSG_DISPLAY, RECEIVE_TOKEN_SUCCESS, and DELETE_TOKEN_SUCCESS. We have seen these used in the src/app.js file where the redux state is used for each of these. In the case of the MSG_DISPLAY, it simply sets a new string for the text of the currentMsg property.

Notice the "...state" code that is a JavaScript way of taking an object and pulling in all of its properties. This way, you don't have to list them; you get whatever was set for session and loggedIn and then currentMsg is overridden.

```
// src/reducers/app.js
const initialState = {
  loggedIn: false,
  session: null,
  currentMsg: ""
}
const appLevel = (state = initialState, action) => {
  switch (action.type) {
    case 'MSG_DISPLAY':
      return {
        ...state,
        currentMsg: action.msg
      }
    case 'RECEIVE_TOKEN_SUCCESS':
      return {
        ...state,
        loggedIn: true,
        session: action.session,
        currentMsg: action.msg
      }
    case 'DELETE_TOKEN_SUCCESS':
      return {
        ...state,
        loggedIn: false,
        session: null,
        currentMsg: action.msg
      }
    default:
      return state
  }
}
export default appLevel
```

The state in the Redux store is immutable, so you must either replace the complete object or use a special module to help you manage that. That is why you see the spread operator being used to return all properties that exist and then provide an override for the ones being replaced. This is also true when you use the state hook in React, so you must learn how to handle that.

There is a npm module named immutability-helper that allows you to make a change to an immutable object. The immutability helper syntax can get a little tricky, so I prefer to stick to plain JavaScript means to make the updates. The main difficulties occur when you need to alter an array with a delete or insert. For example, in the sharednewsview.js file you will see the following code:

```
let newNewsArray = [...newsState.news];
newNewsArray[selectedStoryIdx].comments.push(
  {
    displayName: props.session.displayName,
    comment: comment
  });
setNewsState({ isLoading: newsState.isLoading, news: newNewsArray });
```

Another option is to clone the array and then alter it. Some array functions like `slice()` actually return a new array and don't mutate the old one. You can thus get the new array and then alter it as you wish.

You can also use the JavaScript `Object.assign()` that allows you to create a new object from the existing state object and then override the comments property of that. Here is how the code will look for an array using the `slice()` function if it were in a Redux reducer:

```
case 'ADD_COMMENT_SUCCESS':
  var newNews = state.news.slice(0);
  newNews[action.storyIdx].comments.push({ displayName: action.displayName,
comment: action.comment });
  return {
    ...state,
    news: newNews
  }
```

The rest of the sections in this chapter go over the code that exists in each of the views that are rendered – login, home news, news, shared news, and profile. We begin with the login page.

# 17.4 The Login Page (src/views/loginview.js)

To log a user in, the person's email and password are entered and verified by the backend service. The login page contains a checkbox for the user to specify that they want the local device to store the login token for later usage. If the user is not registered yet, the user can click to bring up a popup modal dialog with a form to register a new account. As part of the registration, the user must provide an email. Emails are unique in the system, so no duplicates are allowed in the data layer. Here is what the UI looks like for logging in.

*Figure 100 - NewsWatcher login form*

The HTML code for the form is set up to have a submit handler. The `handleLogin()` function of the component is the code that executes at that time. To get the data from the form, various Bootstrap components are used. These bind their data to the state of the component through `onChange` handlers that get set as each character is typed. Here is the render code for the login page as found in the loginview.js file:

```
if (session) {
  return (
    <h1 id="h1ExistID">Logged in...</h1>
  );
} else {
  return (
    <div>
      <h1 data-testid="login_heading_id">Log in Page</h1>
      <Form onSubmit={handleLogin}>
        <FieldGroup
          id="formControlsEmail2"
          type="email"
          icon={faUser}
          label="Email Address"
          placeholder="Enter email"
          onChange={handleEmailChange}
        />
        <FieldGroup
          id="formControlsPassword2"
          icon={faEye}
          label="Password"
          type="password"
          onChange={handlePasswordChange}
        />
        <FormCheck type="checkbox" label="Keep me logged in"
id="rememberMeChk" checked={remeberMe} onChange={handleCheckboxChange} />
        <Button id="btnLogin" bsstyle="success" bssize="lg" block="true"
type="submit">
          Login
```

310

```
        </Button>
      </Form>
      <p>Not a NewsWatcher user? <button style={{ cursor: 'pointer' }}
onClick={handleOpenRegModal}>Sign Up</button></p>
      {_renderRegisterModal()}
    </div>
  );
}
```

At the top of the code, we first see if we are logged in. When we are logged in, the session property from Redux is available and not null.

The other interesting thing going on is that you see the call to the function `renderRegisterModal()`. This pulls in the rendering of a modal dialog UI. This code is pulled out into a separate function to make it more manageable. I could have just placed all the HTML right there in the render function, but I did not do that as that function is already large enough. Breaking it out makes the code easier to read. This takes that out and makes it self-contained.

The other thing to note is that this UI for registering a user is in a modal dialog and uses the `<Modal>` component that is provided by Bootstrap for React. It is always rendered, but it is being shown or hidden by a state property and you can see that it starts out as being hidden. The modal registration dialog UI looks as follows:

*Figure 101 - NewsWatcher Registration Modal*

# PART III: The Presentation Layer (React/HTML)

Here is the code for the registration modal UI:

```
const _renderRegisterModal = () => {
  return (<Modal show={showModal} onHide={handleCloseRegModal}>
    <Modal.Header closeButton>
      <Modal.Title>Register</Modal.Title>
    </Modal.Header>
    <Modal.Body>
      <Form onSubmit={handleRegister}>
        <FieldGroup
          id="formControlsName"
          type="text"
          icon={faUser}
          label="Display Name"
          placeholder="Enter display name"
          onChange={handleNameChange}
        />
        <FieldGroup
          id="formControlsEmail"
          type="email"
          icon={faUser}
          label="Email Address"
          placeholder="Enter email"
          onChange={handleEmailChange}
        />
        <FieldGroup
          id="formControlsPassword"
          icon={faEye}
          label="Password"
          type="password"
          onChange={handlePasswordChange}
        />
        <Button bsstyle="success" bssize="lg" block="true" type="submit">
          <FontAwesomeIcon icon={faPowerOff} /> Register
        </Button>
      </Form>
    </Modal.Body>
    <Modal.Footer>
      <Button bsstyle="danger" bssize="lg"
onClick={handleCloseRegModal}><FontAwesomeIcon icon={faWindowClose} />
Cancel</Button>
    </Modal.Footer>
  </Modal>)
}
```

This UI uses a form and has a function that handles the submit.

**Component supporting code of LoginView**
The LoginView component has local state that holds the bound data from the form, such as email and password. The showModal state is used to show and hide the modal registration form.

312

At the top of the render code is a test to see if there is a session token and a null is returned, which React sees and ignores as previously explained. The login menu item is not shown if the user is already logged in. The session token comes through the usage of Redux.

Once the user types in their user name and password and then clicks "Login", the handleLogin() method makes the call to the backend to get a token. It sets the email and password in its HTTP POST request. If successful, the message RECEIVE_TOKEN_SUCCESS is sent through Redux with a dispatch() call. This sets the session token for the rest of the application to see. A change is made to the browser location to go to the news page view.

The handleRegistration() method makes the call to the backend to create a user account in the data layer. The other methods you see are for the binding of the data to the state and the opening and closing of the modal registration dialog via the state property for that. Here is the rest of the code, minus the render code that you have already seen:

```
import React, { useState } from 'react';
import { useNavigate } from "react-router-dom"
import { useSelector, useDispatch } from 'react-redux';
import { Form, FormCheck, Button, Modal } from 'react-bootstrap';
import { FontAwesomeIcon } from '@fortawesome/react-fontawesome'
import { faEye, faUser, faPowerOff, faWindowClose } from
'@fortawesome/free-solid-svg-icons'
import { FieldGroup } from '../utils/utils';
import '../App.css';

function LoginView() {
  let session = useSelector((state) => state.app.session);
  const dispatch = useDispatch();
  const [name, setName] = useState("");
  const [email, setEmail] = useState("");
  const [password, setPassword] = useState("");
  const [remeberMe, setRemeberMe] = useState(false);
  const [showModal, setShowModal] = useState(false);
  const navigate = useNavigate();

  const handleRegister = (event) => {
    event.preventDefault();
    fetch('/api/users', {
      method: 'POST',
      headers: new Headers({
        'Content-Type': 'application/json'
      }),
      cache: 'default', // no-store or no-cache ro default?
      body: JSON.stringify({
        displayName: name,
        email: email,
        password: password
      })
    })
  })
```

```
      .then(r => r.json().then(json => ({ok:r.ok, status:r.status, json}))))
      .then(response => {
        if (!response.ok || response.status !== 201) {
          throw new Error(response.json.message);
        }
        dispatch({ type: 'MSG_DISPLAY', msg: "Registered" });
        setShowModal(false);
      })
      .catch(error => {
        dispatch({ type: 'MSG_DISPLAY', msg: `Registration failure:
${error.message}` });
      });
  }

  const handleLogin = (event) => {
    event.preventDefault();
    fetch('/api/sessions', {
      method: 'POST',
      headers: new Headers({
        'Content-Type': 'application/json'
      }),
      cache: 'default', // no-store or no-cache ro default?
      body: JSON.stringify({
        email: email,
        password: password
      })
    })
      .then(r => r.json().then(json => ({ok:r.ok, status:r.status, json}))))
      .then(response => {
        if (!response.ok || response.status !== 201) {
          throw new Error(response.json.message);
        }
        // Set the token in client side storage if the user desires
        if (remeberMe) {
          var xfer = {
            token: response.json.token,
            displayName: response.json.displayName,
            userId: response.json.userId
          };
          window.localStorage.setItem("userToken", JSON.stringify(xfer));
        } else {
          window.localStorage.removeItem("userToken");
        }
        dispatch({ type: 'RECEIVE_TOKEN_SUCCESS', msg: `Signed in as
${response.json.displayName}`, session: response.json });
        navigate("/news")
      })
      .catch(error => {
        dispatch({ type: 'MSG_DISPLAY', msg: `Sign in failed:
${error.message}` });
      });
  }
```

314

```
const handleNameChange = (event) => {
  setName(event.target.value);
}

const handleEmailChange = (event) => {
  setEmail(event.target.value);
}

const handlePasswordChange = (event) => {
  setPassword(event.target.value);
}

const handleCheckboxChange = (event) => {
  setRemeberMe(event.target.checked);
}

const handleOpenRegModal = (event) => {
  setShowModal(true);
}

const handleCloseRegModal = (event) => {
  setShowModal(false);
}

const _renderRegisterModal = () => {
  return (<Modal show={showModal} onHide={handleCloseRegModal}>
    <Modal.Header closeButton>
      <Modal.Title>Register</Modal.Title>
    </Modal.Header>
    <Modal.Body>
      <Form onSubmit={handleRegister}>
        <FieldGroup
          id="formControlsName"
          type="text"
          icon={faUser}
          label="Display Name"
          placeholder="Enter display name"
          onChange={handleNameChange}
        />
        <FieldGroup
          id="formControlsEmail"
          type="email"
          icon={faUser}
          label="Email Address"
          placeholder="Enter email"
          onChange={handleEmailChange}
        />
        <FieldGroup
          id="formControlsPassword"
          icon={faEye}
          label="Password"
          type="password"
          onChange={handlePasswordChange}
```

```
      />
      <Button bsstyle="success" bssize="lg" block="true" type="submit">
        <FontAwesomeIcon icon={faPowerOff} /> Register
      </Button>
    </Form>
  </Modal.Body>
  <Modal.Footer>
      <Button bsstyle="danger" bssize="lg"
onClick={handleCloseRegModal}><FontAwesomeIcon icon={faWindowClose} />
Cancel</Button>
    </Modal.Footer>
  </Modal>)
 }
```

**...render() and _renderRegisterModal() left out...**

```
export default LoginView
```

# 17.5 Displaying the News
**(src/views/newsview.js and
src/views/homenewsview.js)**

The HomeNewsView and NewsView components displays the list of news stories. The only difference is that the NewsView has additional UI to select a profile filter for what to view. The UI looks as follows for the NewsView component:

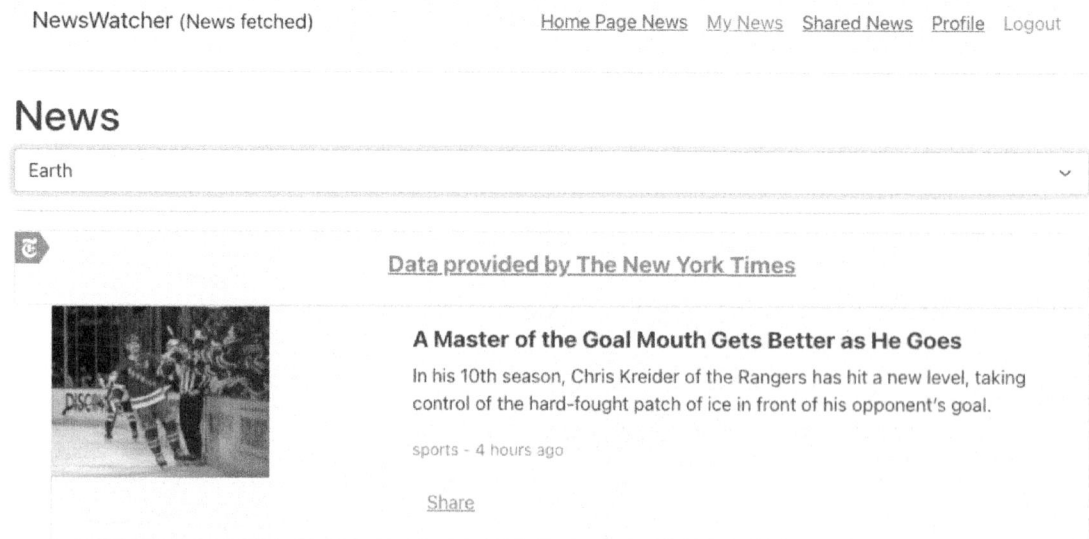

NewsWatcher (News fetched)                    Home Page News   My News   Shared News   Profile   Logout

## News

| Earth | ⌄ |
|---|---|

Data provided by The New York Times

**A Master of the Goal Mouth Gets Better as He Goes**

In his 10th season, Chris Kreider of the Rangers has hit a new level, taking control of the hard-fought patch of ice in front of his opponent's goal.

sports - 4 hours ago

Share

*Figure 102 - NewsWatcher application news page*

A dropdown list displays the list of news filters to select from. The JavaScript `map()` function is used to go through the array of filters and populate that dropdown. Each one is

316

given text from the `filter.name` property. Another use of the newsFilter array with the `map()` function is to render each of the news stories for the selected filter. A link is provided for each story, along with the image and URL to click and open. This code uses Bootstrap components. Here is the render code of NewsView:

```
if (newsState.isLoading) {
  return (
    <h1>Loading news...</h1>
  );
} else {
  return (
    <div>
      <h1>News</h1 >
      <FormGroup controlId="formControlsSelect">
        <FormSelect aria-label="News filter selection"
         onChange={handleChangeFilter} value={selectedIdx}>
          {newsState.newsFilters.map((filter, idx) =>
            <option key={idx} value={idx}>{filter.name}</option>
          )}
        </FormSelect>
      </FormGroup>
      <hr />
      <Card bg="light"
       key={newsState.newsFilters[selectedIdx].newsStories.length}>
        <div className="row g-0">
          <div className="col-md-4">
            <a href=http://developer.nytimes.com
             target="_blank" rel="noopener noreferrer">
              <img alt="" src="poweredby_nytimes_30b.png" />
            </a>
          </div>
          <div className="col-md-8">
            <Card.Body>
              <a href=http://developer.nytimes.com
               target="_blank" rel="noopener noreferrer">
                <h5 className="card-title">
                  <b>Data provided by The New York Times</b>
                </h5>
              </a>
            </Card.Body>
          </div>
        </div>
      </Card>
      <ul>
        {newsState.newsFilters[selectedIdx].newsStories.map((newsStory,
idx) =>
          <Card bg="light" key={idx}>
            <div className="row g-0">
              <div className="col-md-4">
                <a href={newsStory.link} target="_blank" rel="noopener
noreferrer">
                  <img alt="" style={{ height: 150 }}
```

```
          src={newsStory.imageUrl} crossOrigin="true" />
       </a>
     </div>
     <div className="col-md-8">
       <Card.Body>
         <h5 className="card-title"><b>{newsStory.title}</b></h5>
         <p className="card-text">{newsStory.contentSnippet}</p>
         <p className="card-text">
          <small className="text-muted">
            {newsStory.source} -
            <span>{newsStory.hoursString}</span>
          </small>
         </p>
         <p className="card-text">
          <small className="text-muted">
            <button type="button" className="btn btn-link"
            onClick={(event) => handleShareStory(idx,event)}>
               Share
            </button>
          </small></p>
       </Card.Body>
     </div>
   </div>
  </Card>
 )}
      </ul>
   </div>
 );
}
```

There is also the ability to click on a link to Share stories.

**Component supporting code**

When the component is opened, code is run to do the fetching of the news stories. This is done in the useEffect() method hook. You can see at the top of the code how the local state is set. The "newsState.isLoading" state is used for a UI indication that the news stories are being fetched. The message goes away once the data is available. The handleShareStory() method is provided so that a story can be sent to the backend to be put in a common location for all users to see and comment on.

Here is the code for the newsview.js with the render method left out since that already has been shown:

```
import React, { useState, useEffect } from 'react';
import { useNavigate } from "react-router-dom"
import PropTypes from 'prop-types';
import { FormSelect, FormGroup, Card } from 'react-bootstrap';
import '../App.css';
```

```
function NewsView(props) {
  const [selectedIdx, setSelectedIdx] = useState(0);
  const [newsState, setNewsState] = useState({ isLoading: true,
newsFilters: null });
  const navigate = useNavigate();

  useEffect(() => {
    if (!props.session) {
      return navigate("/")
    }

    const { dispatch } = props
    setNewsState({ isLoading: true, newsFilters: [] });
    fetch(`/api/users/${props.session.userId}`, {
      method: 'GET',
      headers: new Headers({
        'x-auth': props.session.token
      }),
      cache: 'default'
    })
      .then(r => r.json().then(json => ({ok:r.ok, status:r.status, json})))
      .then(response => {
        if (!response.ok || response.status !== 200) {
          throw new Error(response.json.message);
        }
        setNewsState({ isLoading: false, newsFilters:
response.json.newsFilters });
        dispatch({ type: 'MSG_DISPLAY', msg: "News fetched" });
      })
      .catch(error => {
        dispatch({ type: 'MSG_DISPLAY', msg: `News fetch failed:
${error.message}` });
      });
  }, []); // eslint-disable-line react-hooks/exhaustive-deps


  const handleChangeFilter = (event) => {
    setSelectedIdx(parseInt(event.target.value, 10));
  }

  const handleShareStory = (index, event) => {
    const { dispatch } = props
    event.preventDefault();
    fetch('/api/sharednews', {
      method: 'POST',
      headers: new Headers({
        'x-auth': props.session.token,
        'Content-Type': 'application/json'
      }),
      cache: 'default',
      body:
JSON.stringify(newsState.newsFilters[selectedIdx].newsStories[index])
    })
```

```
      .then(r => r.json().then(json => ({ ok: r.ok, status: r.status, json
}})))
      .then(response => {
        if (!response.ok || response.status !== 201) {
          throw new Error(response.json.message);
        }
        dispatch({ type: 'MSG_DISPLAY', msg: "Story shared" });
      })
      .catch(error => {
        dispatch({ type: 'MSG_DISPLAY', msg: `Share of story failed:
${error.message}` });
      });
  }
  ...RENDER CODE TAKEN OUT...ALREADY SHOWN...
}

NewsView.propTypes = {
  session: PropTypes.object.isRequired,
  dispatch: PropTypes.func.isRequired
};

export default NewsView
```

The home news story page is a stripped-down version of the code shown above, so no need to explain it further. That is found in the `HomeNewsView` component.

# 17.6 Shared News Page
**(src/views/sharednewsview.js)**

The shared news story view has the same type of news listing capability you have seen before. Here is the render code:

```
if (newsState.isLoading) {
  return (
    <h1>Loading shared news...</h1>
  );
} else {
  return (
    <div>
      <h1>Shared News</h1 >
      <Card bg="light" key={newsState.news.length}>
        <div className="row g-0">
          <div className="col-md-4">
            <a href=http://developer.nytimes.com
              target="_blank" rel="noopener noreferrer">
              <img alt="" src="poweredby_nytimes_30b.png" />
            </a>
          </div>
          <div className="col-md-8">
```

```
      <Card.Body>
        <a href=http://developer.nytimes.com
         target="_blank" rel="noopener noreferrer">
          <h5 className="card-title">
           <b>Data provided by The New York Times</b>
          </h5>
        </a>
      </Card.Body>
    </div>
  </div>
</Card>
<ul>
  {newsState.news.map((sharedStory, idx) =>
    <Card bg="light" key={idx}>
      <div className="row g-0">
        <div className="col-md-4">
          <a href={sharedStory.story.link}
           target="_blank" rel="noopener noreferrer">
            <img alt="" style={{ height: 150 }}
             src={sharedStory.story.imageUrl} crossOrigin="true" />
          </a>
        </div>
        <div className="col-md-8">
          <Card.Body>
            <h5 className="card-title">
             <b>{sharedStory.story.title}</b>
            </h5>
            <p className="card-text">
             {sharedStory.story.contentSnippet}
            </p>
            <p className="card-text">
             <small className="text-muted">
              {sharedStory.story.source} -
              <span>{sharedStory.story.hoursString}</span>
             </small></p>
            <p className="card-text">
             <small className="text-muted">
              <button type="button" className="btn btn-link"
               onClick={(event) => handleOpenModal(idx, event)}>
                 Comments
              </button>
             </small></p>
          </Card.Body>
        </div>
      </div>
    </Card>
  )}
</ul>
{newsState.news.length > 0 &&
  <Modal show={showModal} onHide={handleCloseModal}>
    <Modal.Header closeButton>
      <Modal.Title>Add Comment</Modal.Title>
    </Modal.Header>
```

```
        <Modal.Body>
          <Form onSubmit={handleAddComment}>
            <FormGroup controlId="commentList">
              <FormLabel><FontAwesomeIcon icon={faUser} />
               Comments
              </FormLabel>
              <ul style={{ height: '10em', overflow: 'auto', 'overflow-
x': 'hidden' }}>
                    {newsState.news[selectedStoryIdx].comments.map(comment =>
                      <li>
                        <div>
                          <p>'{comment.comment}' - {comment.displayName} </p>
                        </div>
                      </li>
                    )}
              </ul>
            </FormGroup>
            {newsState.news[selectedStoryIdx].comments.length < 30 &&
              <div>
                <FieldGroup
                  id="formControlsComment"
                  type="text"
                  icon={faUser}
                  label="Comment"
                  placeholder="Enter your comment"
                  onChange={handleCommentChange}
                />
                <Button disabled={comment.length === 0}
                 bsstyle="success" bssize="lg" block="true"
                 type="submit">
                  <FontAwesomeIcon icon={faPowerOff} /> Add
                </Button>
              </div>
            }
          </Form>
        </Modal.Body>
        <Modal.Footer>
          <Button bsstyle="danger" bssize="lg"
           onClick={handleCloseModal}>
            <FontAwesomeIcon icon={faWindowClose} />
            Close
          </Button>
        </Modal.Footer>
      </Modal>
    }
  </div>
  );
}
```

There is a capability to comment on each story. That is done through a modal dialog similar to the one used for user registration. Here is the image of the viewing and entering of comments:

322

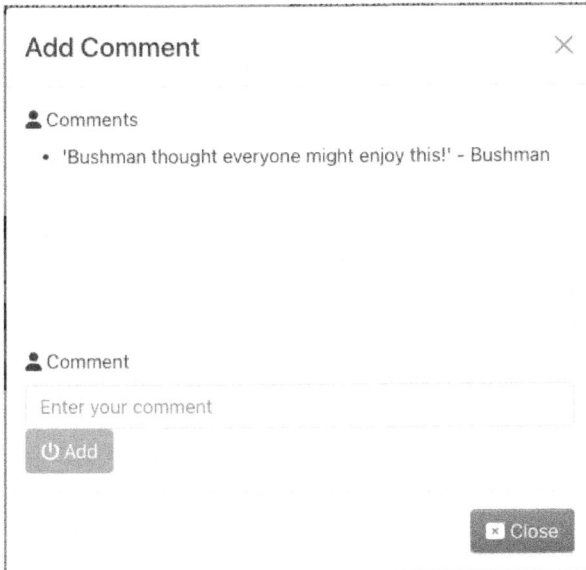

*Figure 103 - NewsWatcher add comment UI*

## Component supporting code

This code is very similar to the `NewsView` component. The difference is in the code for adding a new comment. Here is the code:

```
import React, { useState, useEffect } from 'react';
import { useNavigate } from "react-router-dom"
import PropTypes from 'prop-types';
import { Form, FormGroup, FormLabel, Button, Modal, Card } from 'react-
bootstrap';
import { FontAwesomeIcon } from '@fortawesome/react-fontawesome'
import { faPowerOff, faWindowClose, faUser } from '@fortawesome/free-solid-
svg-icons'
import { FieldGroup } from '../utils/utils';
import '../App.css';


function SharedNewsView(props) {
  const [comment, setComment] = useState("");
  const [selectedStoryIdx, setSelectedStoryIdx] = useState(0);
  const [showModal, setShowModal] = useState(false);
  const [newsState, setNewsState] = useState({ isLoading: true, news: null
});
  const navigate = useNavigate();

  useEffect(() => {
    if (!props.session) {
      return navigate("/")
    }
```

```
    const { dispatch } = props
    setNewsState({ isLoading: true, news: [] });
    fetch('/api/sharednews', {
      method: 'GET',
      headers: new Headers({
        'x-auth': props.session.token
      }),
      cache: 'default'
    })
      .then(r => r.json().then(json => ({ok:r.ok, status:r.status, json})))
      .then(response => {
        if (!response.ok || response.status !== 200) {
          throw new Error(response.json.message);
        }
        setNewsState({ isLoading: false, news: response.json });
        dispatch({ type: 'MSG_DISPLAY', msg: "Shared News fetched" });
      })
      .catch(error => {
        dispatch({ type: 'MSG_DISPLAY',
                   msg: `Shared News fetch failed: ${error.message}` });
      });
  }, []); // eslint-disable-line react-hooks/exhaustive-deps

  const handleOpenModal = (index, event) => {
    setSelectedStoryIdx(index);
    setShowModal(true);
  }

  const handleCloseModal = (event) => {
    setShowModal(false);
  }

  const handleAddComment = (event) => {
    const { dispatch } = props
    event.preventDefault();

fetch(`/api/sharednews/${newsState.news[selectedStoryIdx].story.storyID}/Co
mments`, {
      method: 'POST',
      headers: new Headers({
        'x-auth': props.session.token,
        'Content-Type': 'application/json'
      }),
      cache: 'default',
      body: JSON.stringify({ comment: comment })
    })
      .then(r=>r.json().then(json => ({ ok:r.ok, status:r.status, json})))
      .then(response => {
        if (!response.ok || response.status !== 201) {
          throw new Error(response.json.message);
        }
        let newNewsArray = [...newsState.news];
```

```
      newNewsArray[selectedStoryIdx].comments.push(
        { displayName: props.session.displayName, comment: comment }
      );
      setNewsState({isLoading: newsState.isLoading, news: newNewsArray});
      setComment("");
      setShowModal(false);
      dispatch({ type: 'MSG_DISPLAY', msg: "Comment added" });
    })
    .catch(error => {
      dispatch({ type: 'MSG_DISPLAY',
                 msg: `Comment add failed: ${error.message}` });
    });
  }

  const handleCommentChange = (event) => {
    setComment(event.target.value);
  }

  ...TAKEN OUT...ALREADY SHOWN...
}

SharedNewsView.propTypes = {
  session: PropTypes.object.isRequired,
  dispatch: PropTypes.func.isRequired
};

export default SharedNewsView
```

# 17.7 Profile Page (src/views/profileview.js)

The profile page allows the user to create one or more news filters. Each filter has a title and a list of keywords. You have the same type of dropdown button you have seen before. Then you have the form and the three buttons to save, delete, and create a new filter. The image and HTML are as follows:

*Figure 104 - NewsWatcher News Filter dialog*

There is a link to allow the user to delete their account. The HTML provides a modal dialog like you have used before. Here is the image:

*Figure 105 - NewsWatcher unregister dialog*

```
if (!user) {
  return (<h1>Loading profile...</h1>);
} else {
  return (
    <div>
      <h1>Profile: News Filters</h1>
      <Form>
        <FormGroup controlId="formControlsSelect">
          <FormSelect aria-label="News filter selection"
           onChange={handleChangeFilter} value={selectedIdx}>
            {user.newsFilters.map((filter, idx) =>
              <option key={idx} value={idx}>{filter.name}</option>
            )}
          </FormSelect>
        </FormGroup>
      </Form>
      <hr />
      <Form>
        <FieldGroup
          id="formControlsName"
          type="text"
          label="Name"
          placeholder="NewFilter"
          onChange={handleNameChange}
          value={user.newsFilters[selectedIdx].name}
        />
        <FieldGroup
          id="formControlsKeywords"
          type="text"
          label="Keywords"
          placeholder="Keywords"
          onChange={handleKeywordsChange}
          value={user.newsFilters[selectedIdx].keywordsStr}
        />
        <div className="btn-group btn-group-justified" role="group" aria-
label="...">
          <ButtonToolbar>
```

326

```jsx
          <Button disabled={user.newsFilters.length > 4}
           bsstyle="primary" bssize="sm"
           onClick={handleAdd}><FontAwesomeIcon icon={faPlus} />
             Add
          </Button>
          <Button disabled={user.newsFilters.length < 2}
           bsstyle="primary" bssize="sm"
           onClick={handleDelete}><FontAwesomeIcon icon={faTrashAlt} />
             Delete
          </Button>
          <Button bsstyle="primary" bssize="sm"
           onClick={handleSave}>
           <FontAwesomeIcon icon={faSave} />
             Save
          </Button>
        </ButtonToolbar>
      </div>
    </Form>
    <hr />
    <p>No longer have a need for NewsWatcher?
       <button id="deleteLink" style={{ cursor: 'pointer' }}
             onClick={handleOpenModal}>
         Delete your NewsWatcher Account
       </button>
    </p>
    <Modal show={showModal} onHide={handleCloseModal}>
      <Modal.Header closeButton>
        <Modal.Title>Un-Register</Modal.Title>
      </Modal.Header>
      <Modal.Body>
        <Form onSubmit={handleUnRegister}>
          <FormCheck type="checkbox"
            label="Check if you are sure you want to delete your
NewsWatcher account"
            checked={deleteOK}
            onChange={(event) => setDeleteOK(event.target.checked)} />
          <Button disabled={!deleteOK} bsstyle="success" bssize="lg"
           block="true" type="submit">
            <FontAwesomeIcon icon={faPowerOff} />
            Delete NewsWatcher Account
          </Button>
        </Form>
      </Modal.Body>
      <Modal.Footer>
        <Button bsstyle="danger" bssize="lg" onClick={handleCloseModal}>
         <FontAwesomeIcon icon={faWindowClose} />
         Cancel
        </Button>
      </Modal.Footer>
    </Modal>
  </div>
  );
}
```

# PART III: The Presentation Layer (React/HTML)

## Component supporting code
All the functions necessary to provide functionality behind the button clicks are made available and should look very similar to code you have seen before.

```
import React, { useState, useEffect } from 'react';
import { useNavigate } from "react-router-dom"
import PropTypes from 'prop-types';
import { Form, FormSelect, FormGroup, FormCheck, Button, Modal,
ButtonToolbar } from 'react-bootstrap';
import { FontAwesomeIcon } from '@fortawesome/react-fontawesome'
import { faPowerOff, faTrashAlt, faPlus, faSave, faWindowClose } from
'@fortawesome/free-solid-svg-icons'
import { FieldGroup } from '../utils/utils';
import '../App.css';

function ProfileView(props) {
  const [user, setUser] = useState(null);
  const [deleteOK, setDeleteOK] = useState(false);
  const [selectedIdx, setSelectedIdx] = useState(0);
  const [showModal, setShowModal] = useState(false);
  const navigate = useNavigate();

  useEffect(() => {
    if (!props.session) {
      return navigate("/")
    }

    const { dispatch } = props;
    dispatch({ type: 'REQUEST_PROFILE' });
    fetch(`/api/users/${props.session.userId}`, {
      method: 'GET',
      headers: new Headers({
        'x-auth': props.session.token,
        'Content-Type': 'application/json'
      }),
      cache: 'default'
    })
      .then(r => r.json().then(json => ({ok:r.ok, status:r.status, json})))
      .then(response => {
        if (!response.ok || response.status !== 200) {
          throw new Error(response.json.message);
        }
        for (var i = 0; i < response.json.newsFilters.length; i++) {
          response.json.newsFilters[i].keywordsStr =
            response.json.newsFilters[i].keyWords.join(',');
          response.json.newsFilters[i].newsStories = [];
        }
        response.json.savedStories = [];
        setUser(response.json);
        dispatch({ type: 'MSG_DISPLAY', msg: "Profile fetched" });
      })
      .catch(error => {
```

```
        dispatch({ type: 'MSG_DISPLAY', msg: `Profile fetch failed:
${error.message}` });
      });
  }, []); // eslint-disable-line react-hooks/exhaustive-deps

  const handleUnRegister = (event) => {
    const { dispatch } = props
    event.preventDefault();
    fetch(`/api/users/${props.session.userId}`, {
      method: 'DELETE',
      headers: new Headers({
        'x-auth': props.session.token
      }),
      cache: 'default'
    })
      .then(r => r.json().then(json => ({ok:r.ok, status:r.status, json})))
      .then(response => {
        if (!response.ok || response.status !== 200) {
          throw new Error(response.json.message);
        }
        props.appLogoutCB();
        dispatch({ type: 'MSG_DISPLAY', msg: "Account deleted" });
      })
      .catch(error => {
        dispatch({ type: 'MSG_DISPLAY',
                   msg: `Account delete failed: ${error.message}` });
      });
  }

  const handleNameChange = (event) => {
    setUser(prevUser => {
      let newNewFilters = [...prevUser.newsFilters];
      newNewFilters[selectedIdx].name = event.target.value;
      return { ...prevUser, newsFilters: newNewFilters };
    });
  }

  const handleKeywordsChange = (event) => {
    setUser(prevUser => {
      let newNewFilters = [...prevUser.newsFilters];
      newNewFilters[selectedIdx].keywordsStr = event.target.value;
      newNewFilters[selectedIdx].keyWords = event.target.value.split(',');
      return { ...prevUser, newsFilters: newNewFilters };
    });
  }

  const handleOpenModal = (event) => {
    setShowModal(true);
  }

  const handleCloseModal = (event) => {
    setShowModal(false);
  }
```

```
const handleChangeFilter = (event) => {
  setSelectedIdx(parseInt(event.target.value, 10));
}

const handleAdd = (event) => {
  const { dispatch } = props
  event.preventDefault();
  if (user.newsFilters.length === 5) {
    dispatch({type: 'MSG_DISPLAY', msg: "No more newsFilters allowed" });
  } else {
    var len = user.newsFilters.length;
    setUser(prevUser => {
      let newNewFilters = [...prevUser.newsFilters];
      newNewFilters.push({
        name: 'New Filter',
        keyWords: ["Keyword"],
        keywordsStr: "Keyword",
        enableAlert: false,
        alertFrequency: 0,
        enableAutoDelete: false,
        deleteTime: 0,
        timeOfLastScan: 0
      });
      return { ...prevUser, newsFilters: newNewFilters };
    });
    setSelectedIdx(len);
  }
}

const handleDelete = (event) => {
  event.preventDefault();
  setUser(prevUser => {
    let newNewFilters = [...prevUser.newsFilters];
    newNewFilters.splice(selectedIdx, 1);
    return { ...prevUser, newsFilters: newNewFilters };
  });
  setSelectedIdx(0);
}

const handleSave = (event) => {
  const { dispatch } = props
  event.preventDefault();
  fetch(`/api/users/${props.session.userId}`, {
    method: 'PUT',
    headers: new Headers({
      'x-auth': props.session.token,
      'Content-Type': 'application/json'
    }),
    cache: 'default',
    body: JSON.stringify(user)
  })
    .then(r => r.json().then(json => ({ok:r.ok, status:r.status, json})))
```

```
      .then(response => {
        if (!response.ok || response.status !== 200) {
          throw new Error(response.json.message);
        }
        dispatch({ type: 'MSG_DISPLAY', msg: "Profile saved" });
      })
      .catch(error => {
        dispatch({ type: 'MSG_DISPLAY',
                  msg: `Profile save failed: ${error.message}` });
      });
  }

  ...TAKEN OUT...ALREADY SHOWN...
}

ProfileView.propTypes = {
  session: PropTypes.object.isRequired,
  appLogoutCB: PropTypes.func.isRequired,
  dispatch: PropTypes.func.isRequired
};

export default ProfileView
```

You can see that the code limits the number of filters. What do you suppose will happen if the user employs the browser developer tools to mess with that code and then adds thousands of filters? It can create very large documents in MongoDB. Be mindful that you cannot ever rely on your UI side bounds-checking code to do the right thing on its own, as it is subject to tampering. Thus, place code in your service layer that will only take the first five filters.

It is also noteworthy that the function to handle logging out is not in this code. Instead, the App Component that uses the ProfileView component passes in a function as a property on the props and then the callback is used to get at code in App.js for that purpose. This seems to be more appropriate for the higher calling code to control that function to avoid duplicating code in this case. The logout is needed, as a user could unregister their account.

# 17.8 Not Found Page (src/views/notfound.js)

The Not Found page is something that will be displayed if the user tries to navigate by changing the URL in the browser to some page that does not exist. You can see this in action if you try and go to https://www.newswatcher2rweb.com/#/blah. Here is that Component code.

```
import React from 'react';
import '../App.css';

function NotFound(props) {
  return (
    <div>
      <h3>404 page not found</h3>
      <p>The page you are looking for does not exist.</p>
    </div>
  );
}

export default NotFound;
```

This completes the discussion of each of the files and their corresponding controller code. Everything can be zipped up and deployed to the AWS Elastic Beanstalk environment and used. On Windows, you will select the folders and files (as shown before) and then right-click and select **Send to->Compressed (zipped) folder**. Or, even better, run at the command line "npm run zipForEB" which makes sure all the right files are gathered.

# Chapter 18: UI Testing of NewsWatcher

With the NewsWatcher application code completed and running, you will want to thoroughly test it with every conceivable scenario. Manually doing this is great for your entertainment, but it will soon grow tedious if you have that as your only way of finding bugs. For example, what happens if you make any changes to your code base?

There are function and load testing, but you still need to run the UI through its paces to exercise the React JavaScript code. Not to fear, there are many great solutions to this problem, and I will present some as well as give general tips for debugging the code in Chrome.

Many different technologies are being used in the code and each has techniques that are unique. For example, the React Router usage and Redux code can be independently tested. Each individual component can be tested in ways that instantiate them and verify that correct DOM elements are present with the correct attributes and properties. Then you can also try testing frameworks that do an automated run through of the complete application in a browser. You use IDs of each UI element to identify what to click or inspect.

# 18.1 UI Web App Testing tools

There are tools you can use to record your interactions within a browser session and then play those back. Some tools record screen locations of your clicks and then rely on the UI elements to be in that exact same location later for the interaction to work. Other tools understand the DOM and can find elements you specify to click on.

One tool you can use to do automation testing of a UI is an older tool named Selenium built so long ago and supports many different languages. A new option would be Cypress, which is similar but only supports JavaScript.

The whole point of putting together a UI automation test is to mimic what you would normally do with human interactions in a repeatable test suite that you can run as part of a CI/CD script to save you a lot of time. The concept of a User Acceptance Test or UAT basically is the user script that you want to follow to prove that the UI can do everything it is supposed to do. Development teams can create a UAT for each code iteration they go through before a deployment can be approved. You can set up the test to go against your staging, production, or local hosted site.

# NewsWatcher Selenium tests

Here is some sample code using Selenium. I show it because you may end up working somewhere that requires you to use Selenium. This sample code I show launches the site, clicks on the login tab, and then attempts to log in with a user email that does not exist. The code tests that the correct error message appears. Note how you need to have delays put into the test code to wait for UI changes to happen.

```
describe('NewsWatcher UI exercising', function () {
  var driver;
  var storyID;
  var token;

  before(function (done) {
    driver = new webdriver.Builder().withCapabilities(
                webdriver.Capabilities.ie()).build();
    driver.get('http://localhost:3000');
    driver.wait(webdriver.until.elementLocated(
      webdriver.By.id('loginLink')), 10000).then(function (item) {
      done();
    });
  });

  after(function (done) {driver.quit().then(done);});

  it('should deny a login with a non-registered email', function (done) {
    driver.findElement(webdriver.By.id('loginLink')).click();
    driver.findElement(webdriver.By.id('formControlsEmail2')).
      sendKeys('T@b.com');
    driver.findElement(webdriver.By.id('formControlsPassword2')).
      sendKeys('abc123*');
    driver.findElement(webdriver.By.id('btnLogin')).click();
    driver.wait(webdriver.until.elementLocated(
                webdriver.By.id('currentMsgId')), 5000);
    setTimeout(function () {
      driver.findElement(webdriver.By.id('currentMsgId')).getText().then(
        function (value)
        {
        assert.equal(value,'(Sign in failed: Error: User was not found.)');
        done();
      });
    }, 5000);
  });
});
```

*Note: I originally wrote some tests with Selenium, however, when RTL (React Test Library) came along, I decided not to continue with the Selenium testing and instead go with RTL. Selenium still has its place and is useful for end-to-end user acceptance testing. You may want to investigate the use of Nightwatch, as a nice wrapper on top of Selenium.*

# 18.2 UI Testing with Jest and RTL

RTL (React Test Library) provides an API that you can use inside of Jest tests to test your React components at the code level. This component level testing is analogous to Unit testing of the service layer code. You isolate a given component and verify that it works as an individual piece of code, and then have a greater assurance when consumed, it will all work together.

You will want to start with tests for components at the lowest levels because it is sometimes best to test code where you can cause the most complete exercising of its functionality. This is the most efficient means of trying all the combinations of code paths (including error paths). As part of doing this, you should mock data that may be coming from HTTP calls or other sources. You should isolate a component as much as possible.

You can also test components that are at a higher level, such as the ones that consume other sub-components. This means you can even test the complete application from the highest App component and exercise all parts of it.

It is up to you to decide how rigorous you want the RTL tests to be. For example, you can test components and verify that when they are instantiated, they have the proper state, props, and styles. You can go further and code up tests for your Redux reducers. You can also do testing to mimic the full range of interactions with a component, such as mouse clicks, along with other UI interactions.

*Note: Jest is often used as the test runner for the RTL tests. I have chosen Jest (built on top of Jasmine) because that is what is installed by default with the React application that was created. It also has some great features such as built-in code coverage and parallel running of tests. You may have used Mocha and will not really see much difference between a Mocha test file and a Jest one. They both have 'describe' and 'it' or 'test' blocks of code and both have the ability to use the done() function to do asynchronous testing.*

RTL provides an easy-to-use API you use inside of each test case in code that Jest will run. With it, you instantiate a component with the `render()` function. This will render the UI element into the DOM and return.

If you are testing something like a component that is wrapped by Redux, you need to provide the created store to be used, just as you would in your regular code. Your test will then have a full virtual DOM, and a redux state store for its use. You will see how this is done.

As mentioned, RTL gives you the ability to render the UI in a test setup. You do that with the following line in your test code - `render(<App cartId={777} />)`.

By doing that, all your actual code in the functional component will be run and be using a virtual DOM. This means that the App gets instantiated, and thus all the components that App is using underneath are also used. As with the regular functionality of React, the `useEffect()` function of the rendered components will be called, along with any code such as network `fetch()` calls that are made to retrieve data.

You could have the components do their normal fetching of data, or you could mock the calls to fetch data. Mocking will allow you to control the data in a way to best exercise the rendering and test that everything works for all circumstances of data, including errors with returned data. This is where we make use of a capability of Jest to override the `fetch()` call and return results of our own. That will be set up in the `beforeAll()` call and then using the call window.fetch.mockResolvedValueOnce(). This will then intercept the low-level network call and return the results we want for the test.

We use the screen object to find UI elements to be tested and interacted with. You can see several examples of the testing of values. For example, there could be the clicking of the radio button or the entering of text. The bulk of the testing of results are done using the `expect()` call that is provided by the @testing-library/jest-dom/extend-expect import.

In certain key places in your tests, you will need an `await` statement because. At the point the code begins to run, the UI is being rendered and we need to wait for a certain UI element to appear. The alternative is to set up the use of timers, but that is certainly not as efficient.

# NewsWatcher RTL tests

Think back to the home page UI that had news stories served up. In that React code, you can find a `useEffect()` function hook that goes to the backend to fetch the list of news stories. The data returned from that fetch call gets placed into the local component state through the `setState()` call and then the render code updates the UI with the latest data. Here is some test code for verifying this functionality using some mocked data. You see one test case for when data is properly fetched and a second test case for when the fetch call fails.

```
import React from 'react';
import { render, screen } from '@testing-library/react'
import '@testing-library/jest-dom/extend-expect'
import { createStore } from 'redux';
import reducer from '../reducers';
import HomeNewsView from './homenewsview';

describe('<HomeNewsView /> (render with mocked data)', () => {
  beforeAll(() => jest.spyOn(window, 'fetch'))
  afterAll(() => window.fetch.mockClear())

  test('news stories are displayed', async () => {
    window.fetch.mockResolvedValueOnce({status: 200, ok: true,
```

```
      json: async () => ([{
        contentSnippet: "The launch of a new rocket by Elon Musk SpaceX.",
        date: 1514911829000,
        hours: "33 hours ago",
        imageUrl: "https://static01.nyt.com/images/blah.jpg",
        link: "https://www.nytimes.com/2018/01/01/science/blah.html",
        source: "Science",
        storyID: "5777",
        title: "Rocket Launches and Trips to the Moon",
      }]),
    })

    const store = createStore(reducer)
    render(<HomeNewsView dispatch={store.dispatch} />);
    expect(await screen.findByTestId('homepage_heading_id'))
        .toBeInTheDocument();
    expect(window.fetch).toHaveBeenCalledTimes(1);
    expect(window.fetch).toHaveBeenCalledWith('/api/homenews',
                        { "cache": "default", "method": "GET" });
    expect(store.getState().app.loggedIn).toEqual(false);
    expect(store.getState().app.currentMsg)
        .toEqual('Home Page news fetched');
    expect(screen.getByText('Home Page News')).toBeInTheDocument();
    expect(screen.getByTestId('homepage_heading_id'))
        .toHaveTextContent('Home Page News');
    expect(screen.getAllByTestId('story-name_id')[0].textContent)
        .toEqual('Rocket Launches and Trips to the Moon');
    const items = screen.getAllByTestId('story-name_id');
    expect(items).toHaveLength(1)
  });

  test('handle fetch error', async () => {
    window.fetch.mockResolvedValueOnce({
      status: 500,
      ok: true,
      json: async () => ({
        message: "Error: This is bad",
        error: {}
      }),
    })

    const store = createStore(reducer)
    render(<HomeNewsView dispatch={store.dispatch} />);
    expect(await screen.findByTestId('loading_id')).toBeInTheDocument();
    expect(screen.getByText('Loading home page news...'))
        .toBeInTheDocument();
    expect(store.getState().app.currentMsg)
        .toEqual('Home News fetch failed: Error: This is bad');
  });
});
```

Right after the render call there is an `await` statement for the `screen.findByTestId()` call. The code execution will not proceed until a certain element appears in the DOM. You can add specific ids in the HTML that you would be looking for. In the actual code in the file homenewsview.js the following header element was defined. Note the `data-testid` attribute that is there.

```
<h1 data-testid="homepage_heading_id">Home Page News</h1>
```

The `expect()` function of Jest is used for the validations. The component rendered can be searched in the virtual DOM, using the `screen` functions to find sub-elements and inspect them. You can see where we find an `h1` element by using the `findByTestId` function and `getByTestId` to verify it exists and to verify the text it contains. You can also inspect the state properties of Redux. This is done to verify what news stories were fetched. You can see the test that inspects the Redux store with a `getState()` call. You can look at anything in the store as you see it happening in the verification of what the `app.currentMsg` is set to. However, you would not be able to verify the actual `useState` state hook values. You can find elements and verify their settings and contents, so that the recommended approach.

*Note: You may be tempted to use a setTimeout() in test code and delay for a few seconds until the UI has rendered. This actually will not work in Jest, and is a bad idea anyway. If you have hundreds of tests and each has delays, you will increase the time it takes to run your tests and cannot be guaranteed any race conditions will work out.*

Be aware that the actual `useEffect()` code would normally get called here, and it would do things like make the fetch call and then set the state in Redux with a dispatch call. Since we want to control the actual returned response, we mock all of this up and override the fetch function. You see that with the use of `jest.spyOn`. Mocking can get complex, such as replacing calls to the Stipe payment or to override AWS service calls, such as SQS.

We also set up Redux here because that is normally done in the App component, and we are working at a level below that. You see that we create the Redux store and use that with reder of the App component.

# Login UI testing

The code to test the login also uses Redux and so you see the same code there. The one addition is the actual clicking of a UI control. The button to do the login is clicked and then the validation is done that the fetch happened with correct input and response and that the UI reflects that the user was logged in.

Here is the code that renders the `LoginView` component and then tests that a login was successful.

```
import React from 'react';
import { render, screen, fireEvent } from '@testing-library/react'
import '@testing-library/jest-dom/extend-expect'
import { createStore } from 'redux';
import { Provider } from 'react-redux'
import { HashRouter as Router } from 'react-router-dom';
import reducer from '../reducers';
import LoginView from './loginview';

describe('<LoginView /> (render with mocked data)', () => {
  beforeAll(() => jest.spyOn(window, 'fetch'))
  afterAll(() => window.fetch.mockClear())

  // Need to mock the local storage call as well
  global.localStorage = {
    setItem: () => { },
    removeItem: () => { }
  }

  test('User can log in', async () => {
    window.fetch.mockResolvedValueOnce({
      status: 201,
      ok: true,
      json: async () => ({
        displayName: "Buzz",
        userId: "1234",
        token: "zzz",
        msg: "Authorized"
      }),
    })
    const store = createStore(reducer);

    render(
      <Provider store={store}>
        <Router>
          <LoginView />
        </Router>
      </Provider>
    );
    expect(await screen.findByTestId('login_heading_id'))
    .toBeInTheDocument();
    expect(store.getState().app.session).toEqual(null);
    expect(screen.getByText('Log in Page')).toBeInTheDocument();

    fireEvent.click(screen.getByText('Login'));
    expect(await screen.findByText('Logged in...')).toBeInTheDocument();
    expect(window.fetch).toHaveBeenCalledTimes(1);
    expect(window.fetch).toHaveBeenCalledWith('/api/sessions', {
      "body": "{\"email\":\"\",\"password\":\"\"}",
      "cache": "default",
      "headers": {
        "map": {
          "content-type": "application/json",
```

```
        },
      },
      "method": "POST",
    });
    expect(store.getState().app.currentMsg).toEqual('Signed in as Buzz');
    expect(store.getState().app.session).not.toEqual(null);
  });
});
```

As before, we override fetch again. Notice how we also need to provide a stubbed out `localStorage` calls to not do anything, as `handleLogin()` is expecting that functionality to exist.

You will notice we don't even fill in the email and password, as our fetch mocking does not pay attention to anything passed in and returns what we want under our control. You could provide other tests that type in text to the fields and then actually use those value in the test code to determine what mock data is returned. For the email text input element, we can simulate the text entry. In this case, we can get an `onChange()` to happen automatically for us.

Here are some code examples that show how to interact with UI elements.

```
// HTML anchor link href testing
expect(getByText("Click Me").href).toBe("https://www.test.com/")
expect(screen.getByRole('link'))
      .toHaveAttribute('href','https://www.test.com');
const { getByTestId } = render(<a data-testid='link'
                                  href="https://test.com">Click Me</a>);
expect(getByTestId('link')).toHaveAttribute('href', 'https://test.com');

// Making sure a button was clicked
import {render, fireEvent, screen} from '@testing-library/react'
const onClick = jest.fn();
const { getByText } = render(<Button onClick={onClick} />);
fireEvent.click(getByText(/click me/i));
expect(onClick).toHaveBeenCalled();
```

# Getting set up to use RTL and running tests

When you use create-react-app, you get all the necessary npm packages installed to be up and running with Jest and RTL. For example, if you look in the package.json file, you will see the @testing-library family of packages needed.

To run the tests for NewsWatcher, you just run the following:

```
npm run test-react
```

This is what is in the package.json file. Part of what is specified is that we want the run it to include code coverage numbers.

```
"test-react": "react-scripts test --coverage --watchAll=false --env=jsdom",
```

Good test code can sometimes be as difficult or even more difficult than the actual product code. I have worked on several projects where twice as much test code was written as production code.

If you do some searching online, you will find all kinds of creative ways to test with Jest and RTL. For example, I did not mention the ability of RTL to compare DOM snapshots from one test run to another.

*Note: To see the code coverage report, go to the coverage/lcov-report directory and open the index.html file. On that screen, you can click on a file and drill into the source code to see the actual line coverage.*

Code coverage numbers would look as follows:

```
 :/c/Users/eljam/Documents/GitHub/NewsWatcher2RWeb

Ran all test suites.

File                 | % Stmts | % Branch | % Funcs | % Lines | Uncovered Lines
---------------------|---------|----------|---------|---------|----------------
All files            |   27.94 |    24.17 |   23.64 |   29.02 |
 src                 |    5.26 |        0 |       0 |    6.45 |
  App.js             |       5 |        0 |       0 |    5.26 | ... 51,74,78,88
  index.js           |       0 |        0 |       0 |       0 | ... 7,8,9,11,13
  setupTests.js      |     100 |      100 |     100 |     100 |
 src/reducers        |   62.16 |    46.15 |     100 |   62.16 |
  app.js             |   85.71 |       80 |     100 |   85.71 |              22
  homenews.js        |   66.67 |       50 |     100 |   66.67 |            9,14
  index.js           |     100 |      100 |     100 |     100 |
  news.js            |   66.67 |       50 |     100 |   66.67 |            9,14
  profile.js         |      40 |       25 |     100 |      40 | ... 22,40,49,58
  sharednews.js      |   57.14 |       40 |     100 |   57.14 |        11,16,22
 src/utils           |   85.71 |    66.67 |     100 |   85.71 |
  utils.js           |   85.71 |    66.67 |     100 |   85.71 |               9
 src/views           |   23.68 |    20.31 |   20.65 |   23.89 |
  homenewsview.js    |     100 |      100 |     100 |     100 |
  loginview.js       |   58.97 |    58.33 |      55 |   59.46 | ... ,96,104,108
  newsview.js        |    5.56 |        0 |       0 |    5.88 | ... ,96,107,133
  notfound.js        |       0 |      100 |       0 |       0 |               6
  profileview.js     |    3.45 |        0 |       0 |    3.64 | ... 143,149,210
  sharednewsview.js  |       5 |        0 |       0 |    5.26 | ... 108,134,174

> newswatcher@0.0.1 posttest C:\Users\eljam\Documents\GitHub\NewsWatcher2RWeb
> echo All tests have been run!

All tests have been run!
```

*Figure 106 – Code coverage details*

# 18.3 Debugging UI Code Issues

It can be straightforward and simple to debug your server-side node.js code locally on your computer. To do so, you open the VS Code editor and place some breakpoints and do your run. With React code running in the browser you need to use the Chrome JavaScript debugging capabilities. Browsers have their own debugging capabilities that you must learn and make use of. With Chrome, for example, there is a selection in its menu under **More tools** to launch **Developer tools**.

Once open, you can click on the **Console** tab and see errors that occur in your code. This image shows a code error for an undefined property:

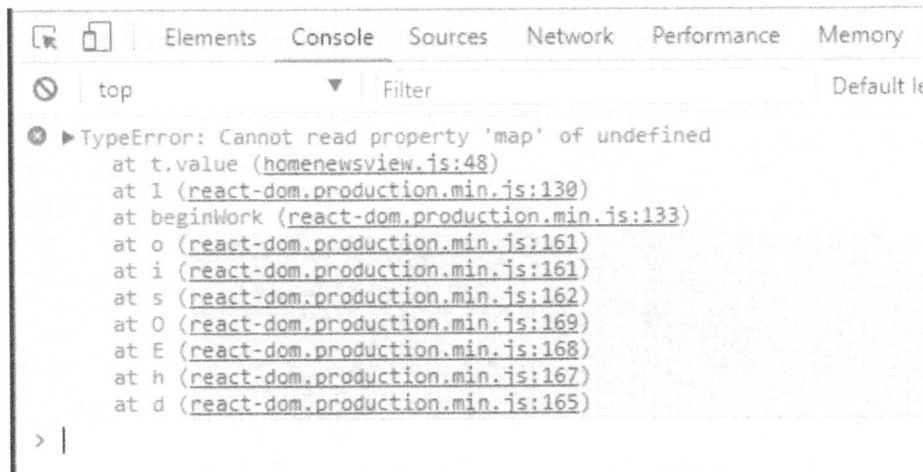

```
        Elements    Console    Sources    Network    Performance    Memory

   top                    ▼   Filter                                Default l

 ⊗ ▶ TypeError: Cannot read property 'map' of undefined
        at t.value (homenewsview.js:48)
        at 1 (react-dom.production.min.js:130)
        at beginWork (react-dom.production.min.js:133)
        at o (react-dom.production.min.js:161)
        at i (react-dom.production.min.js:161)
        at s (react-dom.production.min.js:162)
        at O (react-dom.production.min.js:169)
        at E (react-dom.production.min.js:168)
        at h (react-dom.production.min.js:167)
        at d (react-dom.production.min.js:165)
 > |
```

*Figure 107 - debugger console in Chrome*

This window is handy to keep open to see problems you may not otherwise notice. It is often quite easy to go and fix the bugs you find. Other times, you will need to step through the lines of code to tell you what is happening. Simply click the **Sources** tab and you can browse your code and set breakpoints in the browser. The experience is typical of any debugger in that you can inspect variables and set watches on them.

This next image shows how I used the debugger to set a breakpoint and then inspected the response data being returned from an HTTP request. The left pane has a list of files that you can look through. I opened the JavaScript code for the home news viewing. In the pane where the code is, you can click on a line number and set a breakpoint. You may need to hit the browser refresh button or click around in your application to make the code run to hit your breakpoint.

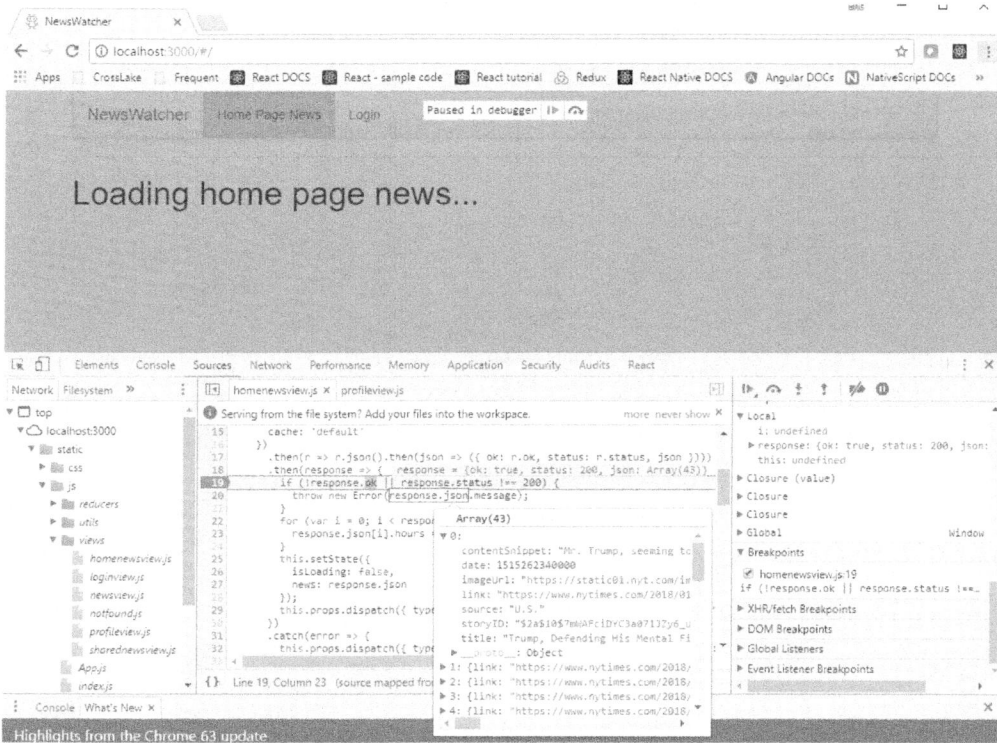

*Figure 108 - NewsWatcher HTTP request debugging*

You can have VS Code running the node server and place breakpoints in the server-side code and debug back and forth in both environments.

If you have written HTML markup, you know the capabilities of the browser developer tools and will know how to inspect the DOM elements and their associated CSS. Click the **DOM Explorer** tab to try this out. For example, you may find that an element is not styled or functioning as expected. A quick inspection can usually turn up the issue.

The **Network** tab shows you the HTTP traffic going back and forth like what you see if you are using Postman. The **Performance** tab can let you capture the calls going on and then view that in a call graph that shows the time of everything in details. This allows you to check for and fix performance issues. The **Memory** tab lets you set up a run that captures memory snapshots so you can investigate memory leaks in your code.

# PART III: The Presentation Layer (React/HTML)

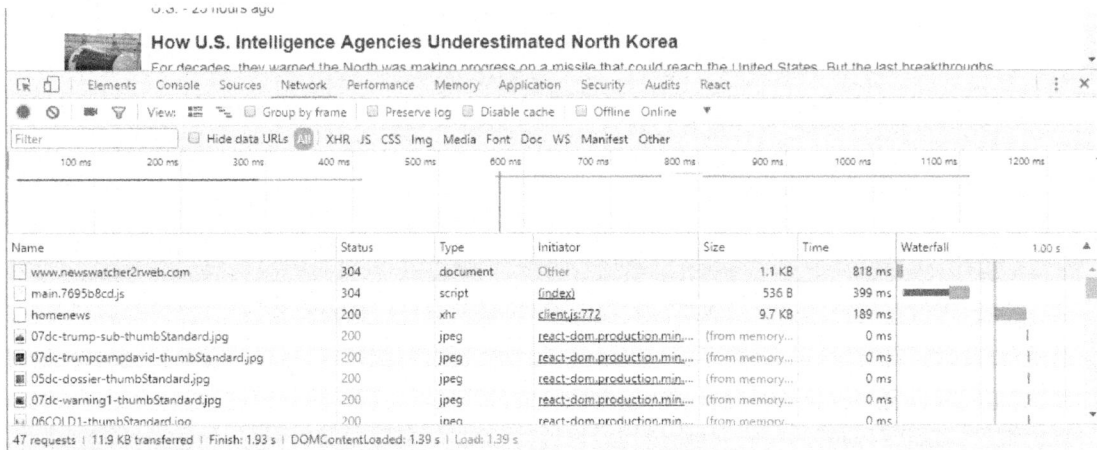

*Figure 109 - Chrome Network capturing*

# Chapter 19: Server-Side Rendering

The NewsWatcher application has been developed as a client-side SPA application that loads and renders on a client device. It is run in the context of a browser, whether that is on a mobile smartphone, tablet, laptop, or personal computer. This works quite well. And offloads some of the processing that would have happened with server-side rendering of a web site.

A SPA application has great performance and mostly minimizes traffic back and forth to the data fetching. All the UI is always rendered client-side and is never loaded from the server once it is all brought over. There is a bit of resistance in the development community to fully sanction SPAs as the way to go, and their objections are based on two points as follows:

1. Performance is better if HTML is rendered on the server side and sent in a response from a server. A server rendered app can collect all the data on the server-side and serve it up faster than a SPA, especially when it comes to the initial page loading if there is a lot that goes with it.
2. SEO (Search Engine Optimization) is more efficient for server rendered HTML pages because of how Google indexes them.

Some people are advocating to move to an SSR design to address these two issues. There is a hybrid solution to give you the best of a SPA and an SSR solution. Let's first take a look at the objections one by one.

As far as the first point, this can be mitigated by implementing code splitting so that not everything is sent to the client at the start. This is where you only send up UI as needed. The code is effectively split up into parts. The initial UI code is sent and renders very fast and then other parts are brought in as needed. This then allows just as fast of a page load for a SPA as for an SSR site. As you navigate through the UI, pages are brought to the client just in time.

The second point is also debatable in that Google will crawl SPA applications and can even look through JavaScript code while doing web indexing.

*Note: To do justice to an SSR, you also need to put in meta tags in the HTML as helpers for the web indexers to help with SEO.*

If you decide to, you can write the site to be completely rendered as an SSR site. You can even have Redux and React Router used on the server. As a compromise, you can make a kind of hybrid implementation, where some of the UI is rendered on the server-side and returned as HTML/JavaScript to the client browser. Once there, that code can also contain React code and have that code run on both the Browser side as well as on the SPA. This is

345

what a Universal application design pattern is all about – being able to run the same type of code (React Router, Redux, etc.) on the server and on the client.

The hybrid Universal JavaScript solution certainly addresses both issues and may be what you choose to go toward. This means that you have the initial page rendered with SSR, including its Redux store data along with the initial page so that it renders instantly. This gives you the speed for the initial page rendering. The SEO issue is taken care of for this initial page for Google to index. The page loads immediately and does not have to go back to fetch any additional data until further user interactions are happening.

A Universal hybrid approach is great if you have an initial route page with critical information to be indexed. From there, that page can even function as a SPA, with sub-navigation that goes along with it. You can decide what pages you want served up as SSR routes and then have each of those be SPA pages functioning on their own.

*Note: If you are thinking that a lot of your users will be on mobile devices that have cellphone or Wi-Fi connection anyway, you might as well develop native applications. This means you have a purely native application they download. This is in essence functioning like the SPA does, except that the application is always instantly there on the device. You get the benefit of writing the native application to work in a mode where it is offline, and you can also investigate what a Progressive Web application is for doing that in the browser.*

You will also want to implement some caching on the server-side if you are doing SSR. This is a huge performance advantage. For example, if the home news page is going to stay the same for everyone for a few hours, you render it and serve that up from a cached copy and then update it when a new set of news is ready to go. There are NPM modules you can use that make this seamless and are easy to put into your code.

# 19.1 NewsWatcher and SSR

I decided to implement the initial home news page as a server-side rendered page and then let the rest of the application being a client SPA. It turned out to work well and is an option for those who decide to go this route. It may even turn out that someday `create-react-app` generates code that allows you the option of doing code as SSR. I set this up to be configurable through an environment setting just so the code could be used either way.

I set up for the route to the home news page to be rendered with SSR. The rest of the application still works the same as a SPA. The idea is that an initial call comes in for the NewsWatcher application site and an HTML page is rendered on the server-side for it. This initial page is the home news page. That HTML can also contain a special section that contains all of the actual fetched data for the news stories, and this also goes up. You may wonder why in the world that is going on. The answer has to do with how React works. React will only render UI that has changed. Thus, an initial page of the home news is rendered on the server side and then React also works as if it is a SPA on the client side and takes the data sent up and tries to create a virtual DOM for the home news from that. It turns out to be identical with what was already being shown in the actual DOM and so it throws it away.

It turns out that the page instantly loads and there is no delay; it all just appears. Other views, such as the user news page will show the loading text while the useEffect() code chunks away, but the home news page never shows the loading text.

To get started, you need some new npm modules, because you will now have React JSX style code on the server side, and Node will not understand that to be able to run that as it is not regular JavaScript. This is where babel comes in. It is what is called transpiler. It takes the JSX files and converts them into regular javaScript to be used on the Node side V8 engine. The added installs are as follows:

```
npm install --save @babel/register @babel/preset-env @babel/preset-react
ignore-styles
```

One change that needs to happen is that the server.js file needs to serve up the route for the main '/' route that is the UI application itself. That code will end up being in a new file that you create named ssrrender.js. Here is the server.js file code to return the server-rendered main page:

```
require('ignore-styles')
require("@babel/register")({
  ignore: [/(node_module|build|Lambda)/],
  presets: ["@babel/preset-env", "@babel/preset-react"],
});

...skipping code...

const SSRRender = require('./ssrRender');
app.use("^/$", SSRRender);
app.use(express.static(path.join(__dirname, 'build')));
```

The code to render the static page is going to look familiar. You will even see the use of the Provider component and the Redux store and reducers. It is all there as before. This means that the exact process of rendering a page on the client is now happening on the server. There are two interesting lines that do string replacements on the HTML template. The first one does the actual replacement of the HTML that was rendered by React and the second one is the one that adds to the Redux state. It is transferred with the news story list up to the client to be used there. This is because the useEffect() is not used to retrieve it anymore. Here is the code:

# PART III: The Presentation Layer (React/HTML)

```
// ssrrender.js
const path = require('path')
const fs = require('fs')
const React = require('react')
const { renderToString } = require('react-dom/server')
const { StaticRouter } = require('react-router-dom/server')
const { createStore } = require('redux')
const { Provider } = require('react-redux')
const { default: App } = require('./src/App')
const { default: reducer } = require('./src/reducers')

module.exports = function handleSSR(req, res, next) {
  req.db.collection.findOne({ _id: process.env.GLOBAL_STORIES_ID },
                            { homeNewsStories: 1 },
                            function (err, doc)
  {
    if (err)
      return next(err);

    let preloadedState = { homenews: { isLoading: false,
                           news: doc.homeNewsStories } }
    // Create a new Redux store instance
    const store = createStore(reducer, preloadedState)
    const context = {}
    const html = renderToString(
      <Provider store={store}>
        <StaticRouter
          location={req.url}
          context={context}
        >
          <App />
        </StaticRouter>
      </Provider>
    )

    // Grab the initial state from our Redux store
    const finalState = store.getState()

    const filePath = path.resolve(__dirname, 'build', 'index.html')
    fs.readFile(filePath, 'utf8', (err, htmlData) => {
      if (err) {
        console.error('read err', err)
        return res.status(404).end()
      }

      // We're good, so send the response
      res.send(htmlData.replace('{{SSR}}',
html).replace(`console.log("REPLACE")`,
          `window.__PRELOADED_STATE__ =
${JSON.stringify(finalState).replace(/</g, '\\u003c')}`))
    })
  });
};
```

348

You see that on the server-side code in ssrrender.js that the store is also preloaded with that data. You can look there, and you will find the call to the database. There is no HTTP call since we have the database connection directly on the server-side. There is one minor tweak to the homenewsview.js file. You simply comment out the `useEffect()` function because it already has the news stories from the initial load in the Redux store that is used.

*Note: JSON.stringify can be subject to script injections. To counter this, we can remove HTML tags and other dangerous characters. This is done with a simple text replacement on the string, e.g. JSON.stringify(state).replace(/</g, '\\u003c').*

The file src/index.js has a minor tweak where the `window.__PRELOADED_STATE__` property thas the news stories and loads that into the Redux store on the client-side. We still do the call to have react render the page, but the call is change to be `root.hydrate()`. You see that we set up the redux store from what was set available in the browser window object.

```
if (window.__PRELOADED_STATE__) {
  let store = createStore(reducer, window.__PRELOADED_STATE__);
  delete window.__PRELOADED_STATE__
  let root = ReactDOM.hydrateRoot(document.getElementById('root'));
  root.hydrate(
    <React.StrictMode>
      <Provider store={store}>
        <HashRouter>
          <App />
        </HashRouter>
      </Provider>
    </React.StrictMode>,
    document.getElementById('root')
  );
} else {
  let store = createStore(reducer);
  let root = ReactDOM.createRoot(document.getElementById('root'));
  root.render(
    <React.StrictMode>
      <Provider store={store}>
        <HashRouter>
          <App />
        </HashRouter>
      </Provider>
    </React.StrictMode>,
    document.getElementById('root')
  );
}
```

*Note: SSR is not for everyone. You will have to weigh the costs versus the benefits to help you make the right decision for your project. It can be tricky to get right and can certainly make your code more complicated to understand going forward. There are complete frameworks you can use instead of Node that prefer this static client-side rendering.*

# Chapter 20: Mobile Device Usage

The existing NewsWatcher React SPA client code was built with the use of Bootstrap and could just have easily used material-ui. Thus, it is ready to go on any mobile device and sizes itself appropriately. The performance is nice, and you can pin the browser page onto your phone screen for quick launching. The React Web app was built to be responsive, meaning it can adapt to the screen size that it is running on. The use of Bootstrap gives it a collapsible menu when running on a mobile phone in the browser. This still does not compare to what performance and UI capabilities can be accomplished with a truly native application under certain use cases.

One benefit of Native applications is that they can be designed to run in an offline mode when no data or Wi-Fi connection is available. Sure, you have the possibility of doing a progressive web application, but why not just stick with the real thing and go completely native? Native applications have much better performance than a browser-based application running on a phone. The other big differentiator is that Native applications can access all the hardware features of the phone itself, such as the camera, contacts, gyroscope, and many other features.

To get the existing SPA code to be a native application, you would create what is called a React Native app. There is even a create-react-native-app npm package you can use.

A React Native application runs the low-level mobile platform code through a JavaScript "bridge" layer to interact with the phone for accessing all its functionality and presenting a UI. There is a JavaScript engine interface that exists on both iOS and Android. This is a low-level layer that then interacts with the OS for the phone to do the same things that other platform languages on the phones do. You may have heard of Objective-C, Java, Xcode, or Swift.

React web sites write to the HTML DOM as that is the layer for all browsers, but for phones, it is the JavaScript engine layer as the point of interaction. This means you get the most performant applications possible and access to all the phones OS and hardware features.

The great benefit with React Native as the core technology is that the code can be very similar to that of your React web application. This is because the React Native library is directly related to the web version. You have all the familiar concepts, like the usage of JSX, lifecycle events, Component class, render method, usage of modules like Redux, and many other React capabilities.

The one big difference you will run into is that you have no HTML elements available in React Native. No <div>, <span>, <ul>, etc. There is also a different approach to CSS for styling. All of this seems a pain because you will have gone through a lot of work to get all of that written up in your React Web application. Yet it does not turn out to be too much of a hassle, partly because you need to re-think your UI design when you are writing an application for a small mobile device screen. The website pages in many cases do not make sense to be rendered the same on a mobile device. Even when you are doing a responsive design with libraries like Bootstrap, the result often is still not good enough.

Then ideal is to write your mobile code once and be able to run the application on both iOS and Android. While it is not always possible to write the code and run it on both iOS and Android, you can come close if you work at it.

***Note:*** *You can integrate low-level native code into your application. This is a technique where you actually write code in Objective-C for iOS or Java for Android.*

# CONCLUSION

A lot of material has been covered and I hope you feel good about your newly acquired knowledge and skills. You began this journey by learning about what a three-tier architecture is and how a MERN JavaScript full-stack implementation is a great choice for an implementation.

The NewsWatcher sample application is featured throughout the book as a full end-to-end implementation of the architecture you have been learning about. You now have the code that you can refer to and base any of your own work on.

The three parts of the book covered the layers of a three-layer architecture and gave you foundational knowledge about each layer. You also learned about the SDKs to use and specific usage scenarios of everything for the NewsWatcher application.

One last thing to mention. If you have a React and Node application in production, you need to be aware the npm packages do age and get out of date. These are constantly being updated to fix bugs and even more importantly, fix security vulnerabilities. You can run the following command to see how out of date you are:

```
npm audit
```

Then you can run the following command to get all the latest updates for everything found in your package.json file. Just be aware that if there is a major update on something you use, it most likely means you will have breaking changes that you will need to research and fix.

```
npx npm-check-updates -u
```

You can now go forward and do your own learning for your own specific needs and be successful. You may choose to just work in one of the layers or you may be able to contribute across all layers. It is always good to be knowledgeable overall as you will know better how everything works internally and realize how what you do in one layer affects another layer.

# INDEX

## V

## X

Made in the USA
Las Vegas, NV
04 May 2022